WATER

# WATER
## METAPHYSICS, MYSTICISM, AND MECHANICS IN SEARCH OF REALITY

Ronald A. Case

MCP

Mill City Press, Inc.
2301 Lucien Way #415
Maitland, FL 32751
407.339.4217
www.millcitypress.net

© 2018 by Ronald A. Case

All rights reserved. No part of this publication may be reproduced, stored in a retrieval system, or transmitted, in any form or by any means, electronic, mechanical, photocopying, recording, or otherwise, without the prior written permission of the author.

Printed in the United States of America

ISBN-13: 978-1-54561-378-8

# CONTENTS

PHOTOS AND MAPS . . . . . . . . . . . . . . . . . . . . . . . . . . . .viii

PREFACE . . . . . . . . . . . . . . . . . . . . . . . . . . . . . . . . . . . . . .xxi

INTRODUCTION . . . . . . . . . . . . . . . . . . . . . . . . . . . . . xxvii
    Wading Into Water. . . . . . . . . . . . . . . . . . . . . . . . . xxvii
    The Finger Lakes . . . . . . . . . . . . . . . . . . . . . . . . . . xxx
    The Cottage . . . . . . . . . . . . . . . . . . . . . . . . . . . . . .xxxiii

1. FLUID REALITY . . . . . . . . . . . . . . . . . . . . . . . . . . . . . . .1
    Dancing Shadows . . . . . . . . . . . . . . . . . . . . . . . . . . . .1
    Other Philosophers Weigh In . . . . . . . . . . . . . . . . . . . 7
    The "Process" Thinkers. . . . . . . . . . . . . . . . . . . . . . . 12
    "Openness" Theology . . . . . . . . . . . . . . . . . . . . . . . . 15
    Cause, Effect, and Time . . . . . . . . . . . . . . . . . . . . . . 19
    More Openness . . . . . . . . . . . . . . . . . . . . . . . . . . . . 23
    Foreknowledge and Freedom . . . . . . . . . . . . . . . . . . 28

2. ESSENCE . . . . . . . . . . . . . . . . . . . . . . . . . . . . . . . . . . . 35
    The Idea of Self . . . . . . . . . . . . . . . . . . . . . . . . . . . . 35
    Eastern Thinking . . . . . . . . . . . . . . . . . . . . . . . . . . . 39
    Locke and Hume . . . . . . . . . . . . . . . . . . . . . . . . . . . 42
    Hegel and Selah . . . . . . . . . . . . . . . . . . . . . . . . . . . 49
    Determinism . . . . . . . . . . . . . . . . . . . . . . . . . . . . . . 52
    Pan(en)theism . . . . . . . . . . . . . . . . . . . . . . . . . . . . . 57
    Existentialism . . . . . . . . . . . . . . . . . . . . . . . . . . . . . 62
    Whence Cometh Life . . . . . . . . . . . . . . . . . . . . . . . . 65
    Dogs and Computers . . . . . . . . . . . . . . . . . . . . . . . . 72
    A Philosophical Recap . . . . . . . . . . . . . . . . . . . . . . 75
    Transcendentalism . . . . . . . . . . . . . . . . . . . . . . . . . . 77
    Back to Impermanence . . . . . . . . . . . . . . . . . . . . . . 82

3. THE REGION OF AWE . . . . . . . . . . . . . . . . . . . . . . . . 86
    Drawn to The Islands . . . . . . . . . . . . . . . . . . . . . . . 86
    Point Peninsula . . . . . . . . . . . . . . . . . . . . . . . . . . . .91
    Stony Point . . . . . . . . . . . . . . . . . . . . . . . . . . . . . . . 93

    Additional Perspectives. . . . . . . . . . . . . . . . . . . . . . . . . 97
    Mysticism and Metaphor . . . . . . . . . . . . . . . . . . . . . . 103
    Scripture and Spirit. . . . . . . . . . . . . . . . . . . . . . . . . . . 108
    Maps and Mariners . . . . . . . . . . . . . . . . . . . . . . . . . . . 112
    Intuition and Exposition . . . . . . . . . . . . . . . . . . . . . . .117
    Chemical Mysticism. . . . . . . . . . . . . . . . . . . . . . . . . . .121
    Summer Camp . . . . . . . . . . . . . . . . . . . . . . . . . . . . . . 125
    Romans 1: 20 . . . . . . . . . . . . . . . . . . . . . . . . . . . . . . . 128
    Breathing Water . . . . . . . . . . . . . . . . . . . . . . . . . . . . .131
    Approaching the Absolute Systematically . . . . . . . . . 134
    Soul Searching . . . . . . . . . . . . . . . . . . . . . . . . . . . . . 142

4. PARADIGM PARALYSIS . . . . . . . . . . . . . . . . . . . . . . . . 149
    The Canal and Other Wonders . . . . . . . . . . . . . . . . . 149
    Normal and Abnormal Science. . . . . . . . . . . . . . . . . . 153
    The Big Bomb. . . . . . . . . . . . . . . . . . . . . . . . . . . . . . . 160
    Science and Religion Generally . . . . . . . . . . . . . . . . . 180

5. PROVING GOD . . . . . . . . . . . . . . . . . . . . . . . . . . . . . . .191
    Induction and Deduction. . . . . . . . . . . . . . . . . . . . . . .191
    Where Thought Experiments Lead . . . . . . . . . . . . . . 203
    The Mother of All Thought Experiments . . . . . . . . . .211
    Where We Stand Now . . . . . . . . . . . . . . . . . . . . . . . . 216

6. RELATIVITY . . . . . . . . . . . . . . . . . . . . . . . . . . . . . . . . . 219
    The Initial Epiphany . . . . . . . . . . . . . . . . . . . . . . . . . 219
    Going Farther . . . . . . . . . . . . . . . . . . . . . . . . . . . . . . 226
    "Block Time" and Eternity. . . . . . . . . . . . . . . . . . . . . 232
    On the Trampoline . . . . . . . . . . . . . . . . . . . . . . . . . . 236
    Singularities . . . . . . . . . . . . . . . . . . . . . . . . . . . . . . . 245
    Black Holes and Beyond. . . . . . . . . . . . . . . . . . . . . . 250
    Reflection . . . . . . . . . . . . . . . . . . . . . . . . . . . . . . . . . 253

7. RADIATING REALITY . . . . . . . . . . . . . . . . . . . . . . . . . 258
    Gone Fishing. . . . . . . . . . . . . . . . . . . . . . . . . . . . . . . 258
    Schrodinger's Cat . . . . . . . . . . . . . . . . . . . . . . . . . . . 268
    Speaking to God . . . . . . . . . . . . . . . . . . . . . . . . . . . . 277
    Prophecy . . . . . . . . . . . . . . . . . . . . . . . . . . . . . . . . . . 285
    Quantum Indeterminism . . . . . . . . . . . . . . . . . . . . . . 288

## CONTENTS

    A Predestinarian Footnote . . . . . . . . . . . . . . . . . . . . . . 292

8. IDENTITY AND ORIENTATION . . . . . . . . . . . . . . . . . . . 302
    Smith Hollow . . . . . . . . . . . . . . . . . . . . . . . . . . . . . . . 302
    The Essence of Postmodernism . . . . . . . . . . . . . . . . 305
    The Effluvia of Postmodernism . . . . . . . . . . . . . . . . 319
    Out of the Mist . . . . . . . . . . . . . . . . . . . . . . . . . . . . . 330
    In the Book . . . . . . . . . . . . . . . . . . . . . . . . . . . . . . . .341

9. CONCLUSIONS. . . . . . . . . . . . . . . . . . . . . . . . . . . . . . . 343

EPILOGUE. . . . . . . . . . . . . . . . . . . . . . . . . . . . . . . . . . . . 363

ACKNOWLEDGMENTS. . . . . . . . . . . . . . . . . . . . . . . . . . 367

BIBLIOGRAPHY. . . . . . . . . . . . . . . . . . . . . . . . . . . . . . . .371

SUBJECT INDEX . . . . . . . . . . . . . . . . . . . . . . . . . . . . . 383

ILLUSTRATION CREDITS . . . . . . . . . . . . . . . . . . . . . . .411

*Fig. 1. Me and Dad and the Chevy*

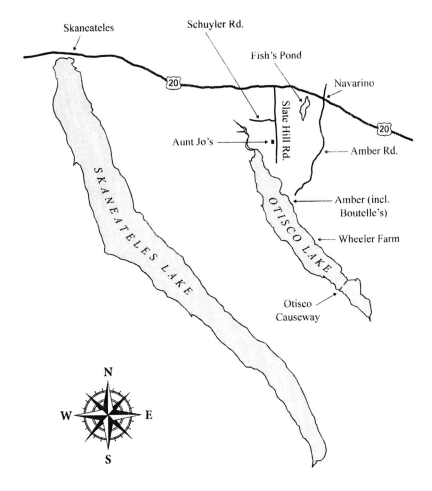

*Fig. 2 Otisco and Skaneateles lakes*

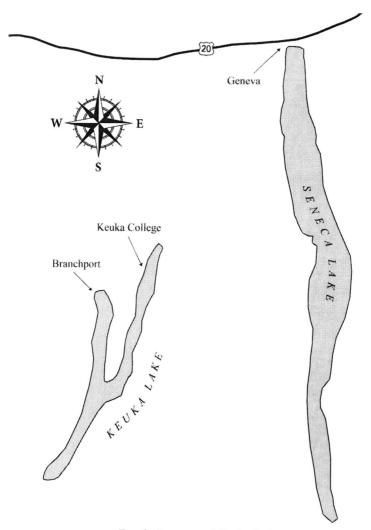

*Fig. 3. Seneca and Keuka Lakes*

*Fig. 4. The Cottage*

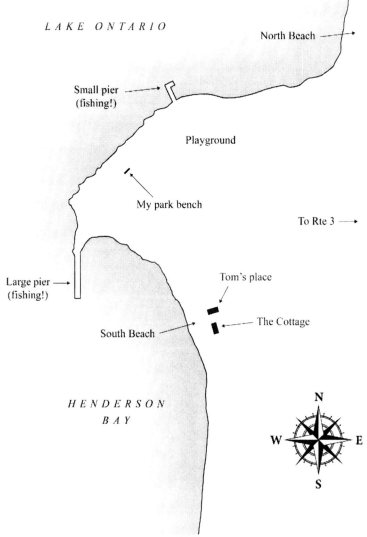

*Fig. 5. Campbell's Point*

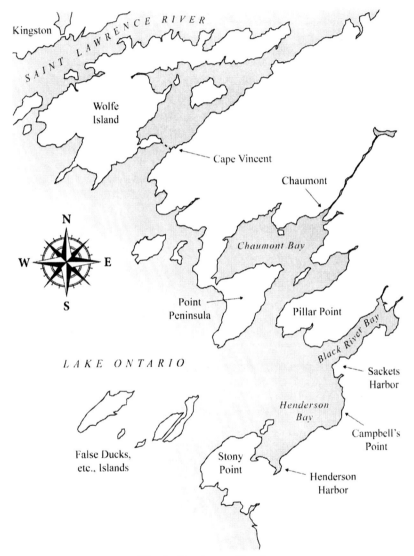

*Fig. 6. Henderson Bay Area*

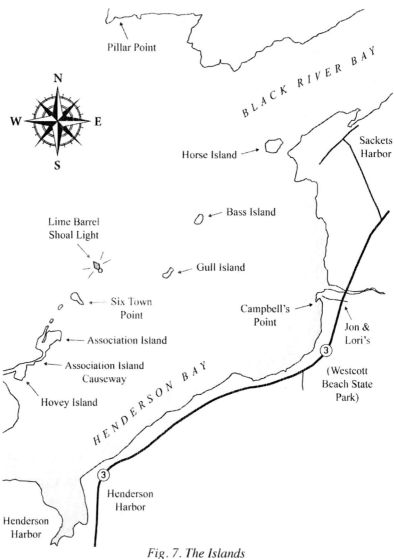

Fig. 7. The Islands

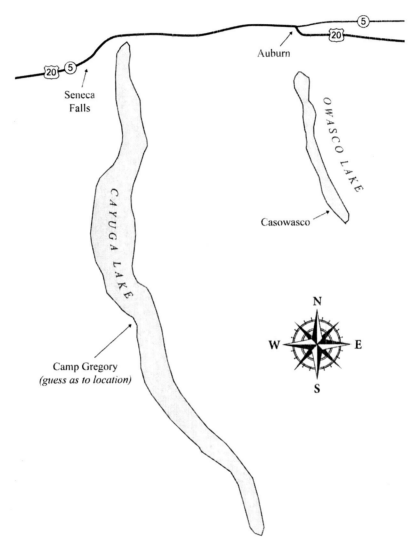

*Fig. 8. Cayuga and Owasco Lakes*

*Fig. 9. Campbell's Point in winter*

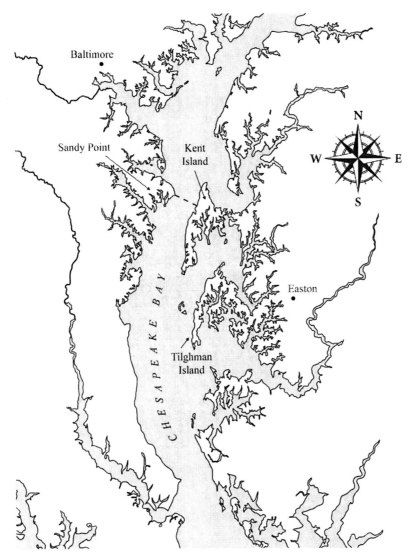

*Fig. 10. Chesapeake Bay*

*Fig. 11. Navarino*

*Fig. 11 (cont). Navarino*

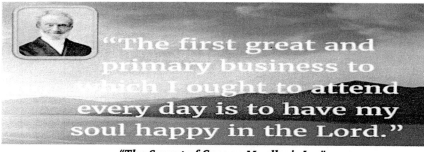

"The first great and primary business to which I ought to attend every day is to have my soul happy in the Lord."

### "The Secret of George Mueller's Joy"

While I was staying at Nailsworth, it pleased the Lord to teach me a truth, irrespective of human instrumentality, as far as I know, the benefit of which I have not lost, though now... more than forty years have since passed away.

The point is this: I saw more clearly than ever, that the first great and primary business to which I ought to attend every day was to have my soul happy in the Lord. The first thing to be concerned about was not, how much I might serve the Lord, how I might glorify the Lord; but how I might get my soul into a happy state, and how my inner man might be nourished. For I might seek to set the truth before the unconverted, I might seek to benefit believers, I might seek to relieve the distressed, I might in other ways seek to behave myself as it becomes a child of God in this world; and yet, not being happy in the Lord, and not being nourished and strengthened in my inner man day by day, all this might not be attended to in a right spirit.

Before this time my practice had been, at least for ten years previously, as an habitual thing, to give myself to prayer, after having dressed in the morning. Now I saw, that the most important thing I had to do was to give myself to the reading of the Word of God and to meditation on it, that thus my heart might be comforted, encouraged, warned, reproved, instructed; and that thus, whilst meditating, my heart might be brought into experimental communion with the Lord. I began therefore, to meditate on the New Testament, from the beginning, early in the morning.

The first thing I did, after having asked in a few words the Lord's blessing upon His precious Word, was to begin to meditate on the Word of God; searching, as it were, into every verse, to get blessing out of it; not for the sake of the public ministry of the Word; not for the sake or preaching on what I had meditated upon; but for the sake of obtaining food for my own soul. The result I have found to be almost invariably this, that after a very few minutes my soul has been led to confession, or to thanksgiving, or to intercession, or to supplication; so that though I did not, as it were, give myself to prayer, but to meditation, yet it turned almost immediately more or less into prayer.

When thus I have been for awhile making confession, or intercession, or supplication, or have given thanks, I go on to the next words or verse, turning all, as I go on, into prayer for myself or others, as the Word may lead to it; but still continually keeping before me, that food for my own soul is the object of my meditation. The result of this is, that there is always a good deal of confession, thanksgiving, supplication, or intercession mingled with my meditation, and that my inner man almost invariably is even sensibly nourished and strengthened and that by breakfast time, with rare exceptions, I am in a peaceful if not happy state of heart. Thus also the Lord is pleased to communicate unto me that which, very soon after, I have found to become food for other believers, though it was not for the sake of the public ministry of the Word that I gave myself to meditation, but for the profit of my own inner man.

The difference between my former practice and my present one is this. Formerly, when I rose, I began to pray as soon as possible, and generally spent all my time till breakfast in prayer, or almost all the time. At all events I almost invariably began with prayer.... But what was the result? I often spent a quarter of an hour, or half an hour, or even an hour on my knees, before being conscious to myself of having derived comfort, encouragement, humbling of soul, etc.; and often after having suffered much from wandering of mind for the first ten minutes, or a quarter of an hour, or even half an hour, I only then began really to pray. I scarcely ever suffer now in this way. For my heart being nourished by the truth, being brought into experimental fellowship with God, I speak to my Father, and to my Friend (vile though I am, and unworthy of it!) about the things that He has brought before me in His precious Word.

It often now astonished me that I did not sooner see this. In no book did I ever read about it. No public ministry ever brought the matter before me. No private intercourse with a brother stirred me up to this matter. And yet now, since God has taught me this point, it is as plain to me as anything, that the first thing the child of God has to do morning by morning is to obtain food for his inner man.

As the outward man is not fit for work for any length of time, except we take food, and as this is one of the first things we do in the morning, so it should be with the inner man. We should take food for that, as every one must allow. Now what is the food for the inner man: not prayer, but the Word of God and here again not the simple reading of the Word of God, so that it only passes through our minds, just as water runs through a pipe, but considering what we read, pondering over it, and applying it to our hearts....

I dwell so particularly on this point because of the immense spiritual profit and refreshment I am conscious of having derived from it myself and I affectionately and solemnly beseech all my fellow-believers to ponder this matter. By the blessing of God I ascribe to this mode the help and strength which I have had from God to pass in peace through deeper trials in various ways than I had ever had before; and after having now above forty years tried this way, I can most fully, in the fear of God, commend it. How different when the soul is refreshed and made happy early in the morning, from what is when, without spiritual preparation, the service, the trials and the temptations of the day come upon one!

**Taken from the Autobiography of George Müller**

# PREFACE

This book is written to help roust the resisters, inspire the indifferent, and educate the elite. It arose out of a realization that there are a number of us who call ourselves Christians but haven't the slightest notion of how we would go about explaining or defending our faith. Lacking knowledge, we appear to lack motivation. Purely by reason of ignorance, we're unable to rise in defense of the Savior who died a cruel death to "[take] away the sin of the world" (John 1:29). I accuse such persons, myself included to the extent that the tag may fairly be applied, of being CINOs — Christians in name only. CINOs read only non-religious books and magazines inasmuch as they persist in the belief that there's no need to seriously consider the question of God and His place in our lives. They consider themselves learned but regarding spiritual matters are abysmally ignorant.

I was a CINO for better than half my life to date. As may be inferred from the Introduction that follows, religion for me was mainline, denominational Christianity. I was married to an institution rather than to the reality behind it. I tried to be a responsible churchperson, and yet avoided the issues of sin, sacrifice, and sovereignty (my, Christ's, and Father God's, respectively). As someone brought up in the institutional church (Methodist), I suffered from a kind of myopia that seems to be common in mainliners: I couldn't see past all the "good" things the church does, could never get the atoning-sacrifice Jesus in focus. I fought hard against the idea of putting my life, here or in the hereafter, in the nail-scarred hands of Jesus, the crucified and risen One. Eventually, thanks be to God, I lost that fight.

Please don't understand me, because of the way I spoke of "learned" men and women above, as saying that you can't be a formally educated person and still be saved. Salvation may take a bit more openness to the working of the Holy Spirit under such circumstances, that's all. A bit more receptivity to the Book, instead of preoccupation with the Look. Image is a principal focus of attention under postmodernism, which I define as the present unwillingness of Western society to acknowledge that there is any such thing as truth. For postmoderns, what you appear to be rather than what you are is the important thing.

But skin-deepness is not the worst of it. Sadly, evangelical Christianity, traditionally the wing of the church most oriented toward salvation by grace through faith (Ephesians 2: 8-9), is presently in the process of doing away with its central document, the Bible, by adopting the mindless mysticism of a loosely-configured Emergent Church and re-writing the Gospel to allow for what many of us see as aberrant behavior. So it's not entirely clear where the terra firma of orthodoxy is, if your eye is on the church rather than the written Word of God.

Regardless of the cause of his or her ignorance—church, no church, or what-have-you—the poor lost soul who is trying to avoid drowning in postmodern spiritual and moral ambivalence can't stay out in the deep treading water forever; he or she must at some point reach the beach. Our floundering non-swimmer will almost certainly need someone's help to make it to shore, which is where you and I come in, gentle reader. We're training as spiritual lifeguards, as I see it, and the individuals we'll be seeking to deliver from danger will be in water over their heads, at locations to which they've been carried by some strong currents. Getting to them will not be as easy as pulling a clumsy preschooler out of the shallows (see, in the latter regard, the Wading Into Water section of my Introduction, infra).

The "mindless mysticism" that I mention above is something that has arisen in reaction to what passes for worship and congregant education in modern "seeker-friendly" churches, which trivialize the Gospel by packaging it as entertainment and relying on a "market-driven" approach as a means of getting people into the church building.[1] By way of contrast, in the Emergent Church and related movements associated with Christianity's so-called "New Age," the idea is to embrace the past rather than make the church appealing by reason of its modernity. The past is embosomed by returning to the mystical practice of the early church, particularly of the so-called "desert fathers," who in the early Middle Ages fled to Egypt to escape the corruption of the Roman church. "Spiritual formation," the inculcation of a Christ-like character in the believer, is held up as the goal, and this objective is supposedly achieved through various mental and physical exertions.[2]

The "spiritual disciplines" recommended by Quaker pastor Richard Foster and other instructors in aid of "formation"—contemplative (non-interactive, non-mental) prayer, repetition of words and phrases, holy breathing, visualization, fasting, confession, saying the rosary, prayer wheel use, and various other expedients—are really nothing more than exercises that are mechanically performed, however. What the so-called spiritual disciplines amount to, in my view, is a program that excuses you from trying to communicate with God in a way (the "conventional" way that is deemed too challenging). You don't have to follow the time-honored protocol of asking God for something and then waiting for or trying to discern His answer, since no real dialogue is anticipated. (Indeed, how could there be dialogue of any kind if, as the New Agers suggest, you and God are one?)

---

[1] See generally, in this regard, Gary E. Gilley, *This Little Church Went to Market: The Church in the Age of Entertainment* (Darlington, U.K.: Evangelical Press, 2006); also Gary E. Gilley with Jay Wegter, *This Little Church Had None: A Church in Search of the Truth* (Darlington, U.K.: EP Books, 2009), 23-35.

[2] See generally Gary E. Gilley, *Out of Formation: Spiritual Disciplines of God and Men* (Darlington, U.K.: Evangelical Press, 2014); also, Richard J. Foster, *Celebration of Discipline: The Path to Spiritual Growth* (New York: HarperCollins, 1998); and Ray Yungen, *A Time of Departing: How Ancient Mystical Practices Are Uniting Christians with the World's Religions*, 2nd Ed. (Eureka, Mont.: Lighthouse Trails Publishing, 2006).

Before we proceed farther, I should mention something that I consider a matter of form rather than substance: gender neutrality. In the interest of maintaining smooth sentence flow and in keeping with a long tradition in discourse concerning the Bible, I've chosen in a number of instances to refer to members of the human race, both male and female, in the masculine. I do so because the Bible does so; no disrespect or slight to the world's women is intended. My focus in a work such as this one, I feel, has to be on biblical correctness first, political correctness second.[3] The two are not opposed in the present case; it's just that a profusion of pronouns included for PC reasons would be a potential distraction, in my opinion.

This is not to say that women did not play an important role in the early church; indeed, there are significant examples to the contrary—viz., Lydia (Acts 16:14-15), Priscilla (Acts 18:18, 26; Romans 16:3), and Dorcas [Tabitha] (Acts 9:36-42). I offer, for your assurance, the apostle Paul's declaration, in his letter to the Christians in Galatia, that "there is neither bond nor free, there is neither male nor female: for ye are all one in Christ Jesus" (Galatians 3:28).

Distinctions of which you *should* be aware involve the use of the term "nature." When "nature" is capitalized, it is in my opinion given a God-like status, which constitutes pantheism. This is particularly so when reference is made to the work of Dutch philosopher Baruch Spinoza and "Nature" is separated from "God" by the conjunction "or," but the problem is found in other contexts as well. To speak of "Mother Nature," on the other hand, is to personify nature but not necessarily make her a deity. Uncapitalized "nature" is similarly used as a means of allowing God's creation only a subordinate role with respect to happenings in His universe. Finally, there is the use of "nature" to refer to the set of qualities that makes a person, group of persons, or thing different from others. The use

---

[3] Political correctness, as I see it, is often nothing more than a means of pleasing a small but vocal group that succeeds at winning by intimidation. Policy makers cave in order to avoid being seen as hostile to these special interests. However, the fear engendered in the supposed bad guy (the caver, if you will) may be like a crocodile cowering when an otter appears on the scene.

# PREFACE

of the term in this last case (as, for example, in "in the nature of") is purely descriptive, no metaphysical status implied.

Another variety of the false deification problem is indicated by New England transcendentalism's capitalization of "Oversoul," an appellation attached to an entity in which God and man are supposedly united. Here again, God-like status for man would be implied from his being joined with the Almighty through Nature. We'll discuss this in further depth later.

The above should give you some idea, gentle reader, of my religious and philosophical inclinations. You also need to know a bit about my life before I embraced the true life. I was not always as committed to finding out and propounding God's truth as I am now.

# INTRODUCTION

*Wading Into Water*

I was four years old at the time. We were having a family outing at Boutelle's, a boat livery at Amber, on the east side of Otisco Lake (Fig. 2), which is 15 miles or so southwest of Syracuse. My parents were there, as were some of my aunts, uncles, and older cousins. I was wading in the water along the shoreline, clad only in my underpants and thus enjoying one of the prerogatives that go with being young and supposedly innocent. On that part of Otisco at that time, there was no wide sandy beach to soften the transition between land and lake. You thus had the grassy picnic area, then a modest dirt-and-gravel embankment, then the water, beneath the surface of which there was a stony bottom that dropped off quickly. I must have gotten my feet tangled up in the stones. Down I went, plunging fanny-first into the drink, which was deep enough that I was totally submerged.

I remember at that moment seeing air bubbles coursing upward through the water, which was brightly illuminated from above even though the shore where we were was lined with willow trees. This tells me that it must have been afternoon, when sunshine from the west would have been able to reach the part of the water's surface that was overhung by the trees. I also remember being more surprised than frightened as I looked up and realized that the still-somewhat-mysterious substance that had been under me was now over me. Then a giant air bubble, many times bigger than the tiny globules I had seen a moment earlier, shot upward, as I witlessly opened my mouth and gulped lake water. It was at that point that a hand, I know not whose, reached down, grabbed my skinny little arm, and plucked me from the deep.

As soon as my head broke the surface of the water, I must have begun coughing or crying or both, since the voices that had been muffled during my submersion were now heard asking me if I was all right. I don't know exactly what if anything I said right after I got the water out of my windpipe, but I do recall later going on about how I "fell in" and Mother and Dad nodding sympathetically during the ride home in our 1941 Chevrolet sedan (Fig. 1). What would otherwise have been a mere dunking now became a near drowning. I had been enjoying the water when I was first mucking around in it in my underpants, then was traumatized by it when I fell in and had some of it "go down the wrong way." Finally, I found that the lake, by swallowing me, had given me something that I could milk for all it was worth with my parents. Such was my early education regarding the polarity of water, as a source of both threat and benefit.

When it came time for me to learn to swim, several years after my "near drowning," Mother and Dad began taking me to a waist-deep (on adults) stream, not far from the north end of Otisco Lake, that was known to locals as Cow Flop. So named because the water that flowed there, under gracious willows such as those at Boutelle's, was pasture drainage. We were simple country folk who were not intimidated by cow manure. This was before there was public awareness of the e-coli bacterium.

What *was* recognizable as a negative, even then, was that we would sometimes after swimming have to remove from our bodies the leeches—"bloodsuckers," we called them—that burrowed in the silty creek bed. Bloodsuckers were more an issue for me than for Dad, who was protected from them by the chest-to-thighs knit bathing suit that he always wore. Mother, although she was ten years Dad's junior and owned a more modern (one-piece) suit, usually chose to sit on the stream bank in her house dress until called on to get up and assist with the removal of leeches. Here again,

# INTRODUCTION

there were both positives and negatives in my experience with water. I enjoyed thrashing around in Cow Flop Creek while Dad held me by the waist to supposedly keep me from sinking, until such time as I would become confident that I could stay afloat on my own. I liked the idea of our doing things like this as a family. I just didn't care much for the leeches.

We continued to go to leech-infested Cow Flop partly because it was easy to reach from our home in the hamlet of Navarino (Figs. 2, 11). Simply drive a mile or so west on U.S. Route 20, turn left onto Slate Hill Road, then right onto Schuyler Road, follow Schuyler to the bottom of the hill, and you were there. A third or quarter of the distance that it would be from Navarino to Boutelle's. In addition, Cow Flop was private, in the sense of being secluded. Although there was a grassy pull-off out by the road and a trail leading in to the swimming hole, we generally had the place to ourselves, and were protected from the view of passing motorists by the thickets of bushes that struggled to survive under the overhanging willows. Also, you didn't have to worry about property rights, since it was unlikely that the owner of the place would appear and throw you out. Otisco Lake, on the other hand, was almost completely enclosed by farmland and cottages (camps, in the lexicon of Central New York), which meant that there was a lack of places where ordinary people such as ourselves could get to the lake.

One of the exceptions to the no-access rule, besides Boutelle's, was at the causeway. Otisco is some six miles long and a mile across at its widest. The causeway is a little more than a mile from the south end of the lake, where the Otisco valley begins. The Otisco valley may easily be crossed by an east-west road just beyond the end of the lake; thus, it's unclear to me why this land bridge across the lake was ever built, unless the lake was at one time longer to the south. At any rate, the causeway had become washed out some time before my growing-up years, and by then was merely a half-mile long row of irregularly spaced wooded islands. To reach the nearest of the causeway's ersatz islands from the east side of the lake, you would have needed to get across a hundred yards of water. The easternmost portion of the causeway, accessible by Churchill Hill Road, projected into the lake as a generous, partially shaded peninsula that was a popular location for picnics (and, as I discovered on occasions when I was home from college in the summer, for making out).

The scenario, seen with the Otisco causeway, of water being conquered and later re-asserting its superiority, seemingly repeated for my benefit as a kind of mystical moral teaching, was not a problem for me. What I did find vaguely troubling about the Otisco causeway was that it laid bare my weakness. I would have liked to have been able to get to each of those treed land masses that lined up across the lake but couldn't. I had no boat and was afraid that I would hurt myself on submerged roots or rocks if I tried to swim across the gaps in the land bridge. Thus, I could never stand in the middle of the lake (as close to walking on water as I could hope to come), nor could I cross the lake. I would always have to go around.

Property rights are waived when an owner needs work done. Accordingly, another avenue by which I was able to get to Otisco Lake was through Dad's role as a handyman. As such, he was hired to close and open cottages, to winterize their plumbing in time to keep it from freezing and de-winterize it in the spring. I would go with him, sit on the seawall or dock in front of the cottage (camp) while he was working, and dangle a fishing line into the lake. At this point in my life, my preteen years, being part of a family that owned a cottage was a fantasy I dared not indulge. You had to have money to have a camp, and to have money you needed a college education. The college education part of it in itself seemed unattainable.

*The Finger Lakes*

Otisco is the easternmost of the Finger Lakes, which are spread across central and western New York and were probably so named because they all have elongated shapes, are roughly parallel, and yet appear to "fan out" as they run from north to south, much as the fingers of an open hand would. Science posits the southward movement of glaciers, acting like plows, as the explanation for the ridge-and-ditch topography seen in the Finger Lakes region, and biblical creationists see such glaciation as consistent with conditions that

# INTRODUCTION

would have existed following the worldwide flood described in Genesis.[4]

The village of Skaneateles, located at the head of the lake of the same name (Fig. 2), which is a few miles to the west of Otisco, bills itself as "The Eastern Gateway to the Finger Lakes;" however, this claim holds no water if glacier movement is accepted as the reason for the Finger Lakes, since the pattern of north-south lakes in north-south valleys is repeated with Otisco. Why should a body of water almost certainly formed in the same way as others nearby not be considered part of the same system? Some of us feel that Skaneateles's boast reflects its elitist tendencies. The Roosevelts once had a home there, and Hillary Clinton visited this upscale New York town when she was on the campaign trail.

Eventually I did go to college. The yokel venturing into a world that was way bigger than his experience to date, where lurked anxious prep and downstate public school graduates who were bent on besting you in class ranking so as to assure their admission to medical or law school. Although the intellectual and social terrain changed, the physical landscape didn't, which I found reassuring. My undergraduate institution, Hobart, was and is located in Geneva, New York, at the north end of Seneca Lake (Fig. 3), the largest of the Finger Lakes and two lakes west of Skaneateles. Seneca is some 35 miles long and three miles across at its widest; thus, it was as though I had graduated to the big time as far as water was concerned but nothing else about my surroundings had been altered.

My room during my sophomore and junior years was in a crumbling Victorian mansion, the Phi Phi Delta fraternity house, which sat on high ground overlooking the lake. The window by my desk faced southeast, so that in the evening I could often look at moonlight reflecting off the water below. Like the other fraternities on the lake side of South Main Street, Phi Phi was separated from the water's edge by a wooded slope and a railroad roadbed, which meant that the lake even at its closest was distant enough from

---

[4] See, in this regard, John C. Whitcomb and Henry M. Morris, *The Genesis Flood: The Biblical Record and Its Scientific Implications*, 50th anniv. ed. (Phillipsburg, N.J.: Presbyterian and Reformed Publishing Co., 1961), 292-303.

my window that light reflecting off the water appeared as dancing punctuation marks rather than wriggling blobs.

Either kind of radiant energy quantum would have been all right as far as I was concerned. Properly understood and spoken of (as by a poet; I was an English major, after all), shimmering Seneca, whether viewed from its shoreline or from afar, was a setting for romance. Real romance, I thought, something other than the tawdry advances that typically occurred during fraternity parties.

Another lake in the area did actually become for me the setting for the fantasized real romance. In my junior year, I met a girl who was a sophomore at Keuka, a women's college situated on the west side of the east branch of Keuka Lake (Fig. 3), one of the more diminutive of the Finger Lakes. Three of us brothers of Phi had piled into A. J. Davis's 1952 Pontiac and headed south to Keuka, twenty-five miles away, in search of dates (resist, please, the conclusion that we were trolling for trollops; at least yours truly wasn't). We went to the front desk in one of the residence halls and told the student who was manning the desk that we were looking for girls who wanted to go out. Her assessment of us was evidently that we looked reasonably respectable. We were not turned away. Campus security was not called.

My Real Romance, one of the three young women that eventually appeared, was tired of reading *Don Quixote*, as she later told me. As I look back on the above sequence of events, I marvel at the innocence in which our meeting proceeded. There were no date rape contracts or Title IX investigations in those days. It was assumed that girls of college age could extricate themselves from any potential sex situation that might be presented, and would want to do so unless they were slatterns or the relationship had progressed to the point where marriage was a real possibility.

It's not my intention to further reflect on how water washed over me during my younger years, however. Yes, I did eventually marry my Real Romance, after a courtship[5] that involved many hours spent on the shores of Keuka Lake. I proposed to her in a

---

[5] "Courtship," as I use the term above, means wooing with a view to marriage. Fundamentalist Christians regard courtship as a concept that excludes casual dating.

# INTRODUCTION

drinking establishment that overlooked the south branch of the same lake, while in a state of relative sobriety for that period of my life. I later gave the love of my life an engagement ring on a warm April evening as we sat on a grass-and-gravel bank on The Point, a spit of land that extends the Keuka College campus fifty yards or so out into the lake. That these things could ever have happened may have had something to do with the properties of water, but that's not the direction I'm headed in with this narrative. The important fact is that my marriage to Susan (Sue), now of some fifty-one years' duration, is what eventually brought us, not through any design on our part, to the point where the real story herein, or at least part of it, could begin.

## *The Cottage*

The real story herein involves our having a cottage, on a lake, a very large lake, where communion with wind and wave and sky could be had and I could try to play the mystic or do whatever else intuitional thinkers do. It was this circumstance that positioned me to start trying to address the kinds of philosophical and theological questions that are the central inquiry of this book, which we'll get into shortly. But first you need to know how we happened to have this cottage. That Sue and I and our two sons and their wives could ever have owned this place was the result of what in retrospect I see as an improbable chain of events.

After college and the United States Army, I became an attorney, but then proceeded to spend most of my professional life as a public agency child welfare advocate, hardly a situation in which one amasses great personal wealth. Sue's father, in the meantime, was an American Baptist minister who, as he approached his sixties, seemed destined to retire in genteel poverty.

Then Sue's maternal aunt, who had married well twice, passed away and left a sizable estate to Sue's mother. This changed everything. Sue's dad now took early retirement from the pastorate of his struggling urban church, and he and my mother-in-law bought a thirty-foot travel trailer, along with a burly Lincoln to haul it, and set out to see the United States and Canada. An early stop in their travels was a visit with some friends who had a place just north of Oswego, on Lake Ontario, the easternmost of the Great Lakes. Sue's parents were immediately enamored of the eastern Ontario area and were taken with the possibility of its providing them with a summer home where family members could come and visit. They now had the money to indulge such a fantasy.

Here, in the story, improbability is multiplied upon improbability. As a general rule, lakefront cottages pass from generation to generation; there is usually some sort of family connection between former and present owners. If a cottage should happen to be offered for sale to outsiders, the time it's on the market is generally miniscule. The dwelling that was eventually purchased by Sue's parents (shown in Figure 4), which I will from here on be calling "The Cottage," did not become available to my in-laws through the open real estate market or through any family connection, however.

What was involved, instead, was a familial disconnect: the sellers wished to avoid having a hated son-in-law as a neighbor. The cottage next door to the sellers' happened to be on the market—indeed, contrary to the normal pattern, had been listed with a realtor. It had been listed for a long enough time that the sellers' daughter and her husband had learned of the offering and had made the mistake of expressing to her parents that they were interested in the listed cottage, whereupon the sellers quickly put in an offer on that cottage in order to prevent their daughter and her detested husband from buying it.

By the time Sue's parents had been shown the listed cottage by their realtor and had decided that they would make an offer, the dysfunctional-family scenario described above had already played itself out: the owners of The Cottage had made an accepted offer on the camp next door. Assuming that the sellers' unlisted cottage would now be available, Sue's parents put in an offer on this property (The Cottage) without having ever been inside.

The purchase of The Cottage from strangers was in 1978. My father- and mother-in-law passed away in 2002 and 2003 respectively. My wife and I and our sons were able to obtain financing, and by 2005 had bought out Sue's sister and her two brothers, who had no wish to continue with the responsibility of a second home.

# INTRODUCTION

Surely a providential chain of events, I thought in 2005, as I looked back over the preceding thirty or so years.

The providence that I'm referring to would have been of divine rather than natural origin, or so I've chosen to view it. I'm assuming, from the fact that you've bought, borrowed, or been perusing this book, that you're either a believer or open to becoming one; thus, you probably have some idea of what I'm talking about when I speak of divine providence. It's the prescient (i.e., forward-looking) care exercised by God toward His creatures. It was God's plan, I believe, to have me enter adulthood with a combination of love and reverential fear of water, the same sort of attitude as we should each of us have toward Him—God, the Absolute. It was the working of His providence, exercised through a special combination of circumstances, that in my retirement I could spend a large part of every year on the shores of one of the Great Lakes.

To what end was the privileged life of a cottager conferred on me? So that I could simply relax and enjoy my last few years?[6] Hardly. The real God, not the God we want but rather the God who is (see Exodus 3: 14), expects service from His children, even though that's not what admits us to His kingdom (see Ephesians 2:8-9, "saved through faith;[n]ot of works, lest any man should boast."). You don't just go around saying that you know who Jesus is and what He did for you on the cross and then sit back and wait for the blessings to roll in. Faith without works is dead, epistle writer James tells us (2:17), even though we're saved through faith rather than because of our good deeds.

But I need to back up with my narrative. Yes, there did come a moment in my life when I suddenly realized that I was a believer, regenerated—made new by God's grace—through reliance on Christ's atoning sacrifice. An unreasonable peace swept over me at that point in time. Actually, it was a point out of time, since salvation lifts you out of the temporal and deposits you in th eternal, as we'll see later. My consciousness, I then knew, had expanded to include a dimension the existence of which I had not been aware

---

[6] You can do the math. I was four years old when my parents were driving a 1941 Chevrolet, which they purchased new the year before I was born. I'm not exactly a youngster.

of previously. I stopped being unnerved by the clock ticking away the minutes and seconds of my existence. I was no longer peering over the edge of a precipice into a dark and bottomless abyss. I was now—if you'll pardon my purple prose—looking into a beautiful sunlit valley.

In describing what happened to me when I became "in Christ, a new creature" (2 Corinthians 5:17), I hesitate to tell people that I was "born again." The term "born again" comes with a lot of baggage, after all. The compass of the problem includes more than just discourteous drivers with Jesus bumper stickers and street-corner preachers unceremoniously telling total strangers that they and their children are going to hell. One recalls published stories of televangelists helping themselves not only to money donated for specified ministries, but also to women not their wives.

It's also public knowledge that "born-again" people sometimes attend meetings at which curious things happen. There may be "signs and wonders" such as hip pain being cured when the "healer" simply yanks on a leg that looks like it's shorter than the other (unlicensed chiropractic?), believers being prostrated when "slain in the Spirit" (i.e., falling down as a result of hypnotic technique applied by the "slayer"), and believers laughing hysterically as a result of another new "move of the Spirit."[7] I finally decided not to let such associations deter me. Our Lord himself used the term "born again" in referring to salvation (John 3:3), and this is still the best description of what happened to me.

What does it mean, by the way, to say that someone who rejects Christ's gracious offer of salvation is going to hell? As used in both the Old and New Testaments, "hell" (Hebrew *sheol*), refers to a place where persons go after death and suffer eternal fiery torment and punishment for unforgiven sin. Jesus spoke repeatedly of such a place (Matthew 5:29-30; 10:28; 13:41-42; 18:8-9; 23:33; 25:41, 46; Mark 9:42-48; Luke 12:5; 16:19-25). This would be a destination quite different from the house of God the Father to which

---

[7] A poor imitation, obviously, of the healings and other "signs and wonders" that the apostles were given power to perform as a witness to the truth of the Gospel and their apostleship (Mark 16:15-18; Acts 2:43; 4:30; 14:3; Romans 15:18-19; 2 Corinthians 12:12; Hebrews 2:1-4).

# INTRODUCTION

Jesus was going to prepare a place for His followers (John 14:2, 3). Our Lord's testimony regarding this place is conclusive, since both Christ and the Father's word are truth (John 14:6; 17:17). As Christian believers, we can't reject the idea of hell any more than we can deny the reality of heaven.

My place in heaven was secured in January of 1981, a little over two years after Sue's parents bought The Cottage. By this time I was in my late thirties. As noted earlier, I grew up in a denominational church (Methodist). There I heard snippets from the Bible preached on; however, I never acknowledged the identity of Jesus as God the Son, sent to Earth to pay for the sins of all who would claim the benefit of His atoning sacrifice. I learned, from the preaching that preceded my trip to the altar on the Sunday of my conversion, that accepting Christ for who He is and what He has done is the way, the only way, in which you can satisfy God the Father's justice principle: punishment for wrongdoing. Someone has to pay for your sin, and that someone is the Lord Jesus Christ if you let Him accept that role for you.

I stepped forward, on the fateful Sunday, and knelt at the altar in a pentecostal church, whose way of worship was foreign to me at the time and to some extent still is (forget the tongues-speaking; I never could get the hang of it). The apparent full-bore commitment of this church's congregation to the Lord was what I needed, I realized.

Sue and I were there on the day I walked down the aisle because some friends had invited us. These friends, whom we had known for several years by reason of our sons being on the Rochester Recreation Swim Team together, "just didn't feel" that the time was right for their invitation until then. They were right. The frustration that I was experiencing as Sue and I trekked in and out of dead mainline churches, searching for something that could finally make me feel like I had religious faith, would not until then have been at a level high enough to support such a radical step as responding to an altar call. Only suggestible simpletons did that.

After my trip to the altar, an extraordinary thing happened: the words seemed to jump off the pages of the New Testament and Psalms that Mother had bought for me some thirty years earlier.

She had had the cover of it imprinted with my name in gold leaf, as if to create a special bond between me and the book. I had carried this pocket-size Bible to Vietnam and back without ever opening it; now I took the giant step of actually turning its pages. That I would have felt motivated to read Scripture was remarkable in itself.

Also, post-conversion, I felt a warmth toward the world—a generosity of spirit— that was simply not normal for me. Anyone who knew me growing up will tell you that I did not usually display overarching goodness. I was cruel toward younger children and animals, mouthy with my parents (I once made Dad cry), and lazy in carrying out chores unless there was something in it for me. The deeper traits that these behaviors represented did not simply go away when I reached adulthood; they went sub-surface and squeezed back out as other negative behaviors. Now, however, after I came forward for salvation and accepted God's gracious gift, I *wanted* to feel kindness toward people. I no longer had to pretend that I cared about them so that they in turn would treat me well. I wanted to display exemplary behavior, even if I didn't always do so.

There was nonetheless an aspect of the story that troubled me: although I believed with all my heart that Christ had redeemed me by His work on the cross, I felt that there were still things that I needed to understand more fully ("Lord, I believe; help thou my unbelief." Mark 9:24). I needed to get down to the core of things theologically and metaphysically, I thought, to achieve a "deeper knowledge" (my phraseology). Gaining this in-depth knowledge would require more than just faithful daily Bible reading and prayer, as I saw it. But I was in the middle of a demanding legal career (sixty-hour weeks) and helping my wife raise our two sons; thus, the study and meditation that were needed to make my faith complete would have to wait, I told myself. A poor excuse, probably, for deferring my spiritual development, but it was the best I could manage at the time.

There came a day, however, when the boys were out of the house and married, I was retired, and Sue and I had our recently-purchased cottage to ourselves for weeks at a time. As far as being free to do whatever I wanted with the rest of my life was concerned, the stars were all in alignment (just an expression; I don't

## INTRODUCTION

believe in astrology). God's idea in setting things up this way was simply to provide a peaceful environment in which I could read and write on the questions that had been gnawing at me, I thought.

The questions I'm referring to are not exactly five-centers. Exactly what is involved when we say that God sits astride time and eternity? How can we have free will if God already knows what's going to happen? Or does He *not* see the whole parade at once? Does He only know the probabilities of particular things happening in the future (making Him a quantum physics kind of God)? Does He experience genuine disappointment when we disobey Him, or was He expecting it all along? How can He be angry with us when our sin was for Him a foregone conclusion? If God inhabits eternity and at death we go to be with Him, do we then partake fully of His knowledge? Do we see all eternity at a glance, as a giant composite snapshot of all that is and was and ever will be, or do we instead enjoy an endless succession of pleasures or interactions with God? Framing the issue another way, is our experience in heaven linear (one event following another) or radial (all eternity at once)?

The above and other questions had been troubling me ever since I trusted Jesus for my salvation. Like my handyman father, I needed to know how things worked. The lapping waves and the starry sky at Campbell's Point were there to clear the cobwebs from my brain so that I could begin to answer some of these questions, or so I thought. Relaxation and reflection, two commodities hard to come by while working a job in the city, these were the ticket; indeed, they were my entire strategy. No epiphanies came, however. My thought content, day after day at the Point, consisted of what I was reading, or rather the thinking of the authors I was reading, nothing more.

One of the many nonfiction books that I read as research during time spent at Campbell's Point was Robert Pirsig's *Zen and the Art of Motorcycle Maintenance,* written some forty years ago but still in print as a modern literary classic.[8] Pirsig's account of a motorcycle trip he had taken across the western United States with his

---

[8] Robert M. Pirsig, *Zen and the Art of Motorcycle Maintenance: An Inquiry into Values* (New York: HarperCollins, 1974, 1999).

eleven-year-old son convinced me that the particular sights, sounds, and physical sensations being experienced by an essayist or expositor at any given moment are critical to his or her accomplishment of the task at hand, as important for the writer as a serene work setting. Even if the writer is trying to puzzle through a philosophical or spiritual question that seems totally unrelated to where he or she is and what he or she is doing at the moment, the writer can still profit directly from the right contemporary sensory exposure. Fresh sights, sounds, and smells bring fresh insights.

Since we think thoughts against the background of what we're perceiving with our senses at the moment, a particular thought may be recalled when a previously-experienced background picture or other sense impression becomes available to the perceiver again. I'm reminded of an account given by my lifelong friend John regarding his transporting of his aging mother to and from his home for visits on various occasions. Normally he drove the same route between his house at Onondaga Hill and her house in Navarino (which were separated by the "wide and spacious valley" referred to in the Onondaga Central School alma mater), and John's mother during the car ride would repeat the same stories over and over. On one occasion, John relates, he took a different route and got an entirely different set of stories.

More is called to mind by sensory experience than simply memories of day-to-day life or previously-thought thoughts. The senses can be a door to the divine, to something outside the realm of personal memory, as in the case of C. S. Lewis, a former atheist who eventually became Christianity's foremost man of letters. For Lewis the Christian, everything in the cosmos mirrored God's character, even though the universe does not embody God. The hugeness of the creation symbolized the greatness of God and the energy in matter His power, while plant and animal life were reminders of God's nature as living and infinitely creative and human beings were a reflection of His personhood in their ability to reason and love.[9]

---

[9] See C. S. Lewis, *Mere Christianity* (San Francisco: HarperSanFrancisco, 1952, 1980), 158-59; also David C. Downing, *Into the Region of Awe: Mysticism in C.S. Lewis* (Downers Grove, Ill.: InterVarsity Press, 2005), 45-46.

## INTRODUCTION

How is man able to recognize things in nature as emblematic of God's personality? Answer: through first having received direct revelation of God's attributes through the living Word of God, Christ Jesus (John 1:1, 14; 14:9).[10] This enables the believer or prospective believer to then look at the natural world and see its resemblance to its Creator. But more than recognition of correspondences is involved in our gaining knowledge of the Lord. The witness of nature is not just something that man finds lying around and attaches to the scriptural God like an ornament hung on a Christmas tree. The Creator *affirmatively* speaks to us through nature, in a way that is binding. In his letter to the Roman church, the apostle Paul states:

> For the wrath of God is revealed from heaven against all ungodliness and unrighteousness of men, who hold the truth in unrighteousness; because that which may be known of God is manifest in them; for *God hath shewed it unto them* (emphasis supplied). For the invisible things of him from the creation of the world are clearly seen, being understood by the things that are made, even his eternal power and Godhead; so that they are without excuse:
>
> (Romans 1:18-20)

So, Scripture tells us, God would be justified in condemning us in our unrepentant state even if we were never to hear a Gospel sermon, be handed a tract, or sleep in a hotel room in which a Gideon Bible had been placed. We have the witness of God's creation, even when it's not attached to His written or spoken Word, to show us the truth.

I'd like to think that I was following both C.S. (Clive Staples) Lewis and Romans 1 in what I was proposing to do some ten years ago when I began my quest, my pursuit of the holy grail of

---

[10] There is the living Word of God (John 1:1; Hebrews 11:13; 1 Peter 1:23; 1 John 1:2), and also the written or repeated Word (Matthew 4:4; Acts 6:4; 2 Timothy 4:2; James 1:22).

knowledge. It was my plan, in seeking the truth about God and the universe, to spend extensive time each day that I was at The Cottage exposing myself to the witness of God's creation. How much better an opportunity to hear God's voice could I have imagined!

When I initially formulated the above-described strategy, it was it was late September, our boys and their spouses and the grandchildren were gone, and no other guests were expected. I was therefore free to sit by the hour gazing at Lake Ontario, nicely framed for my viewing by the arch of trees and cottages and sandy beach on the south side of Campbell's Point, to spend time taking in the endless variations of surf and sky. I was also free to explore by car the various capes, points, inlets, and other geographical features of our part of Ontario. To my reading, which had increasingly come to include science books, I would now be adding research time spent roaming the east end of Ontario.

What, exactly, was I hoping to accomplish as I proceeded with the described pursuits? At a minimum, I would be looking for divinely-provided intuition (let's not call it mystical encounter at this point, I told myself), out of which hopefully would come some conclusions regarding spiritual or metaphysical reality. Would these conclusions come in the form of direct understanding or would they be pointed to metaphorically? It was hard to say. The apostle Paul clearly was approaching the truth by metaphor when he likened a believer's death and resurrection to the sowing of a seed followed by the bursting forth of a new and incorruptible plant, one that will not rot (1 Corinthians 15:35-55). But how are we to understand Galatians 2:20 ("I am crucified with Christ: nevertheless I live; yet not I, but Christ liveth in me")? The latter verse, read as referring to the death of the old (not-yet-born-again) man, is a literal statement of a spiritual reality even though Paul is not physically hanging on a cross when he makes this statement.

There was something in the above scheme besides the mysticism-metaphor ambivalence that made me wary: On the far side of the narrow spiritual strait that I was navigating, the shore opposite the one I was launching from, was pantheism. That could be a problem, I acknowledged at the beginning of my Quest. Start looking for God in nature and you may start seeing God *as* nature.

## INTRODUCTION

He's no longer a transcendent God, but rather is a god who can't be separated from the natural world, a god who's in the rocks and trees and everything else material, including you and me, but maybe nowhere else. A god who's stuck in time along with the rest of us, since things that have physical dimension are located with reference to time in modern physics (by "world line" and "space-time interval," specifically).

"Stuck in time" is where the "openness" theologians want God for some reason. They want Him to be limited in His knowledge, to be learning about events as they happen, just as we mortals are. My mind was telling me that pantheism was an untenable alternative—it tied God's life to the life of the universe—but where was my gut? Would instinct take me, without my even realizing it, to a place that the Bible does not? This was a bit scary. I really was sailing between Scylla and Charybdis on this voyage, I realized.

Whatever I was doing—whether it was hearkening to God's "still small voice" (1 Kings 19:12), reading what was "writ large" in my Bible, or seeking information from a source yet to be identified—I was involved in a *Chautauqua* of sorts, in the same way as was the author of the "modern literary classic" mentioned above, who was searching for Quality in the motorcycle trip that he took with his son.[11] A Chautauqua, for those unfamiliar with the term, was a course of lectures or other types of educational events given with the intent of broadening the cultural horizons of its attendees.

The Chautauqua system of adult education, which was begun in the late nineteenth century and flourished until well into the twentieth, took its name from the circumstance that it was founded at Chautauqua, New York, and was carried on in a complex of buildings located on the shores of a lake of the same name. The Methodist Episcopal Church, precursor in Navarino of the United Methodist Church in which I was raised, was one of the early players in the Chautauqua movement; thus, there were persons in our rural community, including my mother, who had attended one or more Chautauquas. The movement quickly evolved from being a mechanism for training Sunday school teachers to having a menu that included science, art, literature, drama, and even athletics.

---

[11] See generally Pirsig, *Zen*.

Equate the trip that you and I have embarked on, therefore, to a Chautauqua, a voyage of discovery in which science, philosophy, personal experience, and the Bible will occupy the quarterdeck together, although only one of these will ultimately be allowed to steer the ship. It's time now for us to begin this journey.

Before we proceed, however, a disclaimer would seem to be in order. I do not hold myself out as a professor of philosophy or physics, although both of these disciplines are represented in the present work. This is a book about Christianity, which I try to illuminate in the perspective not only of my contact with the natural world, but also in the light of my reading in subject areas outside of theology. Such reading is admittedly less extensive than I would like it to be but is still more than sufficient, I think, to allow me to draw the conclusions and comparisons that are the stuff of this book. In my intellectual wanderings I have been the proverbial common man stumbling his way through matters of complexity ("a very ordinary layman," in the words of C. S. Lewis[12]); yet I am still, I confidently assert, someone who ought to be heard.

Oh, and one other thing: we're changing tenses. This requires a bit of explanation: A work of nonfiction generally approaches its content in one of three ways: (1) as history, (2) as a narrative of contemporaneous events, or (3) as a statement of the theoretical in the present tense. Something that is said to be *history* is stated in the past tense ("The annual trip to Maryland always took place in hot weather"). A description of an action or condition that is represented as taking place at the time of the writer's recording it ("It's June and Sue and I are back at the Point") is said to be in the *narrative present* tense. Theory that is represented as present reality or perceived reality ("Empiricism holds that all knowledge derives ultimately from experience") is said to be stated in the *theoretical present* tense. What I'm suggesting (truthfully, ordaining) for the remainder of this book is that we make what otherwise would be expressions of history into narrative present tense statements. Thus, "I was out walking on a partly cloudy late October day" would become, as it already has, "I'm out walking on this partly cloudy

---

[12] Lewis, *Mere Christianity*, viii.

## INTRODUCTION

late October day," and so on. Theoretical present tense statements would be unaffected by what I propose to do with respect to history.

Why am I making this change in our way of doing business? Answer: to help you understand some things about my journey. I've personally traveled the waters charted in the remainder of *Water* and arrived at the ports that are the conclusions herein stated. You haven't. You nonetheless need to know exactly how I got to where I am today in my belief system. You'll have a better sense of how my inquiry played out, I think, if the rest of this book reads as though neither of us knows where all the stated observations and theories are going to lead. This approach, it seems to me, will give you a more complete comprehension of the intellectual and emotional struggles—of the "blood, toil, tears, and sweat"[13]—that are involved whenever a believer undertakes to think in depth about his or her faith.

---

[13] /Winston Churchill, concerning the prospect of war with Nazi Germany, May 13, 1940, quoted in John Bartlett, Justin Kaplan gen. ed., *Bartlett's Familiar Quotations: A collection of passages, phrases, and proverbs traced to their sources in ancient and modern literature*, 17th Ed. (New York: Little, Brown, and Company, 2002), 665. 19.

# 1
# FLUID REALITY

*Dancing Shadows*

The sun is down and the moon is up now. A customary way, even in modern times, of describing what regularly happens when night replaces day; yet the Bible is sometimes criticized as unscientific for speaking in this manner. Genesis 15 records that a deep sleep fell on Abram "when the sun was going down (v.12)," and Psalm 113 says that the Lord's name is to be praised "from the rising of the sun until the going down of the same" (v.3).[14] Such a choice of words by a present-day scientist would not be questioned even though no mention was made by the scientist of the rotation of the earth. In 1 Samuel 20:24 the new moon "come[s]," and at Amos 8:5 it "go[es]." Again we have action words that are valueless as explanations of the phenomena referred to but are nonetheless not viewed as indicators of scientific ignorance when mouthed by an "educated" person. It's all a matter of perspective.

---

[14] Quotes from sources and specific verse or title and page references are generally added to my thoughts after the observations to which they attach, as you may already have guessed. Thus, for example, if you find me describing the moon as I see it on a particular night at Campbell's Point and then summarizing or quoting what an issue of *Scientific American* had to say regarding the Apollo 11 crew's view of the earth, you can bet that the Apollo 11 material was added hours or days, maybe even weeks, later.

And consider the fact that Bible-believing Christians are sometimes called flat-earthers even though there's nothing in Holy Writ that could be seen as supporting a flat-earth view. Is it fair to put in the mouths of Christians words that are never said in the central document of their faith? What we find instead in the Bible are passages such as the prophet Isaiah's declaration that God "sitteth upon the circle of the earth" (40:22) and the ancient book of Job's observation that God "hangeth the earth upon nothing" (26:7). The above statements, which are easily reconciled with the idea of a globe floating in space, exemplify the many scientifically accurate writings found in the Bible.[15] Such explanations of the natural order, it must be admitted, are more sophisticated (and correct) than the description of Earth as a table land resting on "turtles all the way down," attributed to an anonymous little old lady quoted in the first chapter of Stephen Hawking's *A Brief History of Time*.[16] Persons who consider themselves well-informed are going out of their way to characterize Bible literalists as ignoramuses, it seems.

As I think the above thoughts, I'm sitting on a park bench at a spot in a Campbell's Point Association common area where grass and trees have not yet given way to the shale shelf that forms the tip of the Point (Fig. 5). The bench faces roughly northwest. Across the expanse of water before me, which is more audible than visible given the moon's position behind me, are the lights of Pillar Point. Beyond Pillar Point, many more miles distant, is Point Peninsula (see Fig. 6), silhouetted against the deep azure of the post-sunset western sky. Neither of these land masses is imposing in height; they dominate, instead, by reason of their horizontal reaches. Beyond Point Peninsula, hidden from view, is the mouth of the Saint Lawrence. Also still visible against the western sky is Bass Island, which is more to my left and closer to where I am than either Point Peninsula or Pillar Point is (see Fig. 7).

To my right, beginning a few hundred feet from where I'm sitting, is a sweep of shoreline that I describe to visitors as being

---

[15] See, in this regard, Henry M. Morris, *The Henry Morris Study Bible* (Green Forest, Ark.: Master Books, 2012), appx. 8.

[16] Stephen Hawking, *A Brief History of Time*, Updated and Expanded 10th Anniversary Ed. (New York: Bantam Books, 1988, 1996), 1.

as far east as you can go on Lake Ontario. One of the many facts or factoids, such as the pigs on Bass Island, which you can regale guests with. As the story goes, pigs were at one point placed on Bass Island by the island's owner in order to get rid of the cormorants that had taken over the place (pigs are fond of bird eggs of all kinds, it's said). One summer, the older of my two sons, a competitive swimmer, swam out to Bass Island, a distance of approximately a mile and a half, after which I brought him back to Campbell's Point in the rowboat. While we were out there, we observed no pigs on the island.

The evening air is warm tonight, even though it's late September (year unspecified; this writing is as much a juxtaposition of learning and experience as it is a journal). September is an enjoyable time to be at the Point not only because the weather is usually moderate, but also because there are not many other people around. Gone, for the most part, are the families with school-age children and almost everyone else but the retired and semi-retired. I can therefore, as I take my evening stroll on the sidewalk that leads around the Point, walk past unlit cottages and offer thanks for the tranquility that now prevails. In the daytime, except on weekends, there are no jet skis (so-called personal watercraft, which for years have been recognized as a problem on the bays and inlets of the Great Lakes for reasons that include noise, air and water pollution, and interference with the rights of swimmers and other boaters). At night there are no noisy partyers, no folks setting off firecrackers. I'm not going to say that the quiet will hopefully allow me to "get in touch with my muse," since that would be trite talk coming from someone who holds himself out as an author. To hear from my Creator is more what I have in mind.

The months of July and August at Campbell's Point were never all that quiet, it appears. The history of Campbell's Point has been masterfully summarized by a long-time resident of the Point, in a book brimming with flavorful information, old photographs, and the author's Irish wit. Describing what it was like at the Point in the early part of the twentieth century, he tells of there being a hotel with an adjacent waterslide, church picnics, band concerts, steam launches from Watertown and Oswego docking at the large pier (the one projecting southward from the end of the Point, which acts as a breakwater), et cetera. This was obviously once a more public place than it is now. The style of some of the cottages on Campbell's Point, those that have steeply slanted gable roofs and two-over-two windows, evokes the above-referenced era, although porches have been enclosed and other "modernizing" improvements have been made.

Only our historian's cottage still has gingerbread on it. I define "gingerbread" as lavish architectural detail, especially around eaves. The builder went crazy with a jigsaw or a lathe, it seems. There's still much gingerbread in evidence on the turn-of-the-century buildings at Chautauqua. Closer to here, there's the village of Thousand Island Park, located on the southern end of Wellesley Island, which has summer homes of vintage and appearance similar to those at Chautauqua.

Even the bench that I'm sitting on, which consists of cast cement uprights connected by heavy planking and decorated with pebbles pressed into the cement, is evocative of the era referred to above. It's the kind of bench that's still found in older city parks, and was no doubt placed where it is before the advent of television, at a time when (I'm guessing) people needed little more for their entertainment than what somebody's musical skill or Mother Nature could provide. In those days, I surmise, you needed to be able to play an instrument so that there would be something other than adult conversation to listen to when the family sat in the parlor in the evening. A change of what your senses were being offered could also be gotten by taking a leisurely stroll in the night air (rather than changing the channel or putting in a DVD), or by sitting outside on a bench like this one.

The seat of this particular bench (there are two others like it in the expansive grassy area on this part of the Point) is barely a foot above the ground. This probably indicates that the uprights have begun to sink into the earth beneath them rather than that the bench was built too low to begin with. More relativistically, it might be said that the earth has begun to swallow the bench, just as it swallows us all ("for dust thou art, and unto dust shalt thou return" [Genesis 3: 19]). Change, evident in the situation of our sinking park bench, is one of the things we'll be talking about in our Chautauqua. The basic problem, as I see it, is that nothing stays the same from one moment to the next. I'm not the same person I was when I started writing this sentence. I've acquired new memories, and more of my cells have died or deteriorated. The earth, also, has changed. The oceans and other bodies of water have added or taken away shoreline. Earth inhabitants have died or been born. And out in space, stars have burned out, our telescopes inform us.

In the fifth century BC, the Greek philosopher Parmenides (c. 510-?440 BC) tried to deal with the issue of change by saying that things or persons that have passed away continue to exist because we can still think of them.[17] Likewise, he said, if you can think of something that you suppose will exist in the future, then that thing, also, has existence by reason of its being in your mind. Coming into being and passing away are therefore illusions; everything—past, present, and future—exists now and eternally, as an unchanging indivisible whole. A view not widely accepted by modern thinkers, but it had considerable influence for a while, and in a sense has been revived in modern times by relativity theory, which joins past, present, and future by viewing time and space as a continuum.

The twin realizations that my future was as a resident of the ground and that the day of my interment was upon me—even in the days of my young adulthood—may have been pivotal in my conversion to Christianity, I'll admit. At the same time, based on happenings in my life that seemed more than serendipitous, I had begun to think that I was on Earth for a reason, which implied the existence of a Reason Giver. In any event, I did, at the age of thirty-nine, finally accept God's gracious offer of salvation, and the first stop on my journey as a new convert was pentecostalism.

As I sit here on my old-style park bench in the moonlight, something happens that makes me think of my "holy roller" days, of Sue's and my time in pentecostalism. The moon is at my back, so that I see on the ground before me the shadows of the trees in the stately grove behind me. There's little wind to move their leaves tonight, but there is one area of vigorous movement in the shadows

---

[17] See Philip Stokes, *Philosophy: 100 Essential Thinkers* (New York: Enchanted Lion Books, 2002), 17.

# WATER

that fall on the ground in front of me. It looks like an angel dancing over my head. In my mind's ear I can hear a raspy-voiced speaker, his speech rising to a crescendo, describing this as an encounter with a messenger from God ("and an angel of the Lord came and hovered over me, and I said, 'Thank you, Jesus'"). Hallelujahs and amens ring out from all over the hall.

Such is the hangover (or afterglow, depending on your point of view) that you get from pentecostalism: you never get over expecting miracles in particular situations. I certainly haven't. I turn around and discover that the "angel," the dancing shadow, is a tree branch that's too broken or flimsy to be able to resist the light air.

The dancing shadow makes me think of the cave imagined by the Greek philosopher Plato (c. 427-347 BC),[18] and, feeling the need for research in this regard, I now retreat inside, to continue my reverie surrounded by my books and index cards. The cards, which have topic lines I've written across the top of them, allowing them to be arranged alphabetically and retrieved easily, are my way of assuring that significant things that I've turned up in my reading are locatable later, as are written-down thoughts concerning things in the natural environment that surrounds me. My notes are handwritten on the cards. No laptop or dictating machine here. If pencil and paper were good enough for Hemingway, then they work for me also.

I've already told you, in a footnote, that my research and ruminations don't necessarily get attached to my raw observations at the time of the observation. My methodology in researching and writing might be called "observe and swerve": view nature or mankind in the present, then let associations from your reading or mere blind intuition take you wherever. In the meantime, until I return to the narrative flow, my writing looks like a treatise rather than a diary.

---

[18] See, in this regard, Allan Bloom, ed., *The Republic of Plato*, 2d ed. (New York: Basic Books, 1968, 1991), 193-96, 403-05; also Stokes, *Essential Thinkers*, 23; and Mark Stephens, *The Philosopher's Notebook: A Creative Journal for Thinkers and Philosophers* (New York: Sterling Publishing, 2015), 16-17.

Explaining his theory of how we can know what we believe we know, Plato asked readers of *The Republic* to imagine a group of persons chained from birth inside a cave in such a way as to allow them to look only in the direction of an interior wall of the cave. What little light there was in the cave would be coming from behind the hapless chainees. Objects of which the chainees might wish to have knowledge would likewise be behind them. Thus, the chainees—"ordinary" people as opposed to philosopher-kings in Plato's scenario—would be able to know these objects only as flickering shadows cast on the wall before them, imperfect copies of the real objects.

Plato used the allegory of the cave to illustrate his theory of Forms. The beautiful rose that you see is not the real rose, he said; it's only a flawed approximation of the real thing, of the *ideal*. A similar duality applies with respect to everything else, according to Plato, including human love, the earthy variety of which is a poor substitute for the highest kind of love, in which you contemplate in purity the object of your love. The view that Platonic love is the highest and best kind of love is not shared by everyone; in fact, I eventually lost the interest of a girl I dated in high school because I was afraid I would spoil things by trying to kiss her.

*Other Philosophers Weigh In*

Like his predecessor Parmenides (c. 510–?440 BC), Plato seems to have been trying to show that things have a reality even though they constantly change. Even though, to put it another way, they exist in time, which is defined by change. For Plato, temporal things are real only to the extent that they partake of an unchanging ideal Form of the thing, which is not only imaginable but also exists on its own, somewhere "out there." For Parmenides, on the other hand, things are validated—at least, they're given a certain shadowy kind of existence—just by being imagined. What if something exists in the mind of God? All things do, don't they? This gets us to a higher level of existence than the imagination of Parmenides, does it not?

I'm tempted to say that there's a logical connection between Parmenides with his all- or semi-powerful imagination and English theologian Anselm (1033-1109) with his "ontological" argument for the existence of God.[19] Anselm, a scholar of the late Middle Ages, reasoned that if when we use the term "God" we mean something than which nothing greater can be thought of, then it necessarily follows that God exists, since it would be contrary to our definition of "God" to say that He does not exist. A God that is the greatest possible thing or person can't have the potential to not exist because if that were the case it would be possible to think of a God or entity that doesn't have the potential to not exist, which would necessarily be greater than the god who has the potential to not exist. Potential nothing can't prevail against guaranteed something.

The reasoning of both Parmenides and Anselm is flawed, it seems to me. For one thing, both of these thinkers put man in the driver's seat, to use his imagination in making the call as to ultimate ontological reality, as to what's real—man, mind you, who sees "through a glass darkly" during his stay here on Earth (1 Corinthians 13:12). Do we want God to be limited in this way?

Similar to Parmenides and Anselm in his thinking was Protagoras (c. 480-421 BC), the foremost of the Sophists, a group of itinerant Greek philosophers who appeared to be more interested in money than metaphysics. The wealthy of Athens could become the wise, it was supposed, by buying the services of these purveyors of wisdom. Protagoras tried to make man "the measure of all things," the determiner of the qualities of things by his perceptions. He assumed, apparently, that man's perceptions will never be faulty (say that in my good ear?).

Of the four philosophers mentioned above, Plato comes the closest to Christian truth in thinking about reality, it seems to me. The Bible says that man was "[made] in the image of God" (Genesis 1: 27), and who could be more real, as a Platonic archetype, than God! The rub, already hinted at herein, is that the ideal realm described by Plato, his "heaven," is comprehended intellectually and therefore does not have to change as time passes, whereas the physical world that we perceive here and now with our senses is in a constant state of flux and therefore is but an illusion, as Plato would say. For this reason, even Plato has been unable to

---

[19] See, in the latter regard, Stokes, *Essential Thinkers*, 48-49.

give me any help with the foundational question of how a God who is outside of time and immutable can relate to subjects who at least for now are neither extratemporal nor changeless.

Nor do I get any enlightenment from Aristotle (384-322 BC), a pupil of Plato who, it might be said, failed to show his mentor proper respect. Aristotle began by attempting to classify everything in the known universe into *categories*, a major category being "substance" (along with "quantity", "quality", and "relation"). Whereas Plato had seen matter as an imperfect rendering of his ideal Forms, Aristotle said that the operative principle as far as substance was concerned was *hylomorphism*—form *uniting with* matter so as to produce creatures and other things that have varying positions in the hierarchy of being.[20]

The form for the human physical body was the soul, according to Aristotle, and body and soul together comprised a living person. Animals and plants also had a soul, which was the basis of all biological life in Aristotle's scheme, but their souls differed from human souls. The plant soul—found, for example, in the trees that surround me and the grass under my feet here at Campbell's Point—helps its hosts grow, Aristotle said. Animals and humans have a soul that additionally allows them to experience sensations, according to Aristotle, and the human soul also allows its host to reason.

A corollary of Aristotle's hylomorphism was his view that individual things exist but "Forms" do not—the opposite of Plato's idea that the only thing that's real is the ideal.[21] For Aristotle, everything is made up of *unique* substances, which have *essential* properties (you can't be Aristotle unless you're both a man and a philosopher) and *accidental* properties (Aristotle has a beard, but he'll still be Aristotle even if he shaves it off).

The above brings us back to the question of change. A potential problem with Plato's ideal (and thus presumably unchanging) Forms was that natural things do grow and change. Aristotle sought

---

[20] Chris Rohmann, *A World of Ideas: A Dictionary of Important Theories, Concepts, Beliefs, and Thinkers* (New York: Random House, 1999), 27, 174-75.

[21] See, for a summary of Aristotle's philosophy, Stokes, *Essential Thinkers*, 24-25; also Rohmann, *Ideas*, 26-28; and Stephens, *Philosopher's Notebook*, 28-29.

to overcome this difficulty with his concept of *potentiality*.[22] An acorn is not an oak, he noted, but it has the potential to become one. Potential in this instance is of the "native" variety, he said, meaning that the help of another being is not required for the potential to be realized (I guess Aristotle didn't count as help the fact that God would be assisting by providing rain, sunshine, soil, and air). It's a different story, Aristotle reasoned, if we're talking about an oak tree becoming a plank: the oak can't become a plank on its own.

The difference between Aristotle and Plato is as much a matter of methodology as of metaphysics (the ontology branch thereof).[23] Aristotle was empirical in his approach, relying on the five senses to guide him to the truth about things. For Aristotle, ideas were something that you extracted from your or someone else's experience; they were not just some archetype that existed "out there." Nothing could be in the intellect, he said, that was not first in the senses.

Plato's method, a radically different one, was to attempt to discover ultimate truth through the use of reason. Implicit in Plato's way of doing business was the concept of *innate ideas*: in thinking our thoughts we're recognizing pre-existing, eternal truths. People like to say that ever since the time of Plato and Aristotle (fourth century BC) philosophers have had either Platonic or Aristotelian tendencies. You'll have to draw your own conclusions, as we go along, regarding which group, as an amateur philosopher, I fall into.

As I'm sure you can sense, I'm still feeling uneasy on the issue of change, which is really the question of where the ultimate reality of things lies. The conflicting positions of Plato and Aristotle have served merely to muddy the waters, and I haven't been helped much by the creative thinking of Parmenides, Anselm, and Protagoras. Establishing the reality of things by imagining them doesn't give them ontological stability sufficient to allow us to gauge whether they change, as far as I can see.

Another Greek, Heraclitus (?600-?540 BC), famously said that you can never step into the same river twice, because everything

---

[22] See Rohmann, *Ideas*, 27.

[23] "Metaphysics," as I understand the term, involves study or thought regarding the nature and origin of reality. If done in the context of disciplined Bible study, no grander enterprise can be undertaken.

is continually in a state of flux.[24] With a river, one knows this as a matter of simple physics: water obeys gravity by invariably flowing from higher to lower locations. The Great Lakes, which together hold nearly a fifth of the world's fresh water, are essentially a river of Brobdingnagian proportions. This river's headwaters are found at Canada's Lake Nipigon, which feeds Lake Superior, the highest, coldest, and northernmost of the Great Lakes. From Superior the waters of the system descend through the Saint Mary's River and the Soo canal locks to fill Lakes Michigan and Huron, then flow through the Saint Clair River, Lake Saint Clair, and the Detroit River into Lake Erie, then further descend through the Niagara River and Niagara Falls into Lake Ontario, from whence they enter the Saint Lawrence River and, a thousand miles later, the Atlantic Ocean.

Surprisingly, water was not the main player in Heraclitus's scheme.[25] Theorizing that there were three main elements of nature—Fire, Earth, and Water—he taught that it was Fire that was responsible for the continual transformations seen in the universe. Fire could do this, he said, because it was the principal of the three elements and thus able to control or modify the other two.[26] "All things are an exchange for Fire, and Fire for all things," was his mantra.[27] (Interesting. Sounds a bit like the first law of thermodynamics [equivalence of mass and energy; each can be converted to the other], the basis of Einstein's formula that the energy in a particular reaction equals mass times the speed of light squared.)

The fire that supposedly pervaded the cosmos was said by Heraclitus to have a counterpart in the human soul. A virtuous soul could survive the death of the physical body and be united with the cosmic fire, according to Heraclitus; however, the separation of the soul from the body and its unification with the cosmic fire was not something that was accomplished once for all time; instead, it was a process that happened continually.[28] Christians will recognize that

---

[24] See, in this regard, Stokes, *Essential Thinkers*, 15.

[25] Ibid.

[26] Ibid.

[27] Ibid.

[28] Ibid.

this is very different from how it works with Christ's redemptive work on the cross, which not only freed us from the need to gain salvation through good works, but also was done "once for all,"[29] meaning that one substitutionary death of our Lord was sufficient to redeem all believers for all time.[30] In Heraclitus's scheme, we see life being meted out moment by moment by a pantheistic cosmos, as I would describe his system.

*The "Process" Thinkers*

Heraclitus's idea that reality is process rather than something firmly established anticipates twentieth-century British philosopher Alfred North Whitehead (1861-1947).[31] According to Whitehead, becoming rather than being is the fundamental principle of the world. Whitehead's system is basically panentheistic (supreme deity present in but exceeding all things), and God is important in this system only as the source of the "pure potentials" without which nothing would be possible.[32] Whitehead's God doesn't foreordain; instead, he experiences events as they happen, just as man does. Thus, Whitehead's saying that "the World transcends God"[33] appears to mean that the becoming of any creature is not divinely determined in the sense that it is God-controlled, and that until the creature's becoming is accomplished, by the status of such creature's existence being changed from becoming to being, the creature is not prehended by (i.e., taken into) the divine consciousness.[34] If you continue to wheel, you are not the deal.

---

[29] Hebrews 10:10.

[30] See also Romans 6:10; Hebrews 7:27; 9:28; and 1 Peter 3:18.

[31] For concise and understandable summaries of Whitehead's thinking, see the Charles Hartshorne article on pantheism and panentheism in *The Encyclopedia of Religion* (New York: Macmillan, 1987), also Prof. John W. Cooper's article on pantheism, *Encyclopedia Americana*, vol. 21 (Danbury, Ct.: Scholastic Library Publishing Inc., 2006), 363.

[32] Hartshorne, Pantheism and Panentheism, *Encyclopedia of Religion*, 169.

[33] Alfred North Whitehead, *Process and Reality*, corr. ed., David Ray Griffin and Donald W. Sherburne, eds. (New York: The Free Press, 1978), 348.

[34] See Hartshorne, Pantheism and Panentheism, *Encyclopedia of Religion*, 169.

God therefore has a bipolar nature, in Whitehead's view.[35] His "primordial" nature is the infinity of possibilities of what the world could become.[36] His "consequent" nature is what the world actually is.[37] The process of world evolution actualizes God's primordial nature, making it his consequent nature.[38] In this process, God changes. This is contrary to Malachi 3:6 ("I am the Lord, I change not;"); however, that doesn't seem to matter. The controlling principle for Whitehead and others of his persuasion is that if man is to have free will, God *must* be changeable. God can't be immutable as Scripture says He is, according to "process" thinking, since any addition to God's knowledge—as, for example, by His learning of something of which He was not previously aware—would change Him.

Just as you and I are changed by seeing what the future brings. Every day here at the Point gives me a different Ontario—waves more translucent on some days than on others because the light isn't the same, waves coming in from different directions because of wind shifts, varying shapes of the clouds day to day, each day's sunset (if there is one) unique in its arrangement of orange, red, and violet—and I'm enlarged as a person by the flux, just as I'm enlarged by visiting places in the eastern Ontario area that I haven't seen before. Whitehead has thus called into question the permanence of both God's and my beings, it seems.

American theologian Charles Hartshorne, another process thinker of the twentieth century, likewise sees God as bipolar and panentheistic. Hartshorne begins with the idea that every *event* has both an eternal or abstract element and a temporal or concrete element. Thus, there is both permanence and change in every thing as well as every person. Since God is part of the world, He's just as bipolar as everything and everybody else. There is, then, both an absolute pole and a relative pole in His nature. Hartshorne speaks of God's "A-"(absolute) perfection and His "R-"(relatedness) perfection. The former means "that which in no respect could conceivably

---

[35] Cooper, Pantheism, *Encyclopedia Americana*, 363.
[36] Ibid.
[37] Ibid.
[38] Ibid.

be greater, and hence is incapable of increase."[39] The latter means "that individual being than which no *other individual* being could conceivably be greater, but which *itself*, in another 'state,' could become greater"[40] Another way of putting it is that God's perfection (which no rational person would dare deny) is "an excellence such that rivalry or superiority on the part of other individuals is impossible, but self-superiority is not impossible."[41] God can always better Himself!

Hartshorne additionally acknowledges the Augustinian idea that because God is timelessly eternal, He sees what we call the future as part of one eternal present, but Hartshorne then refuses to give what to us is the future the same degree of reality as that which to us is present or past. We experience the future as unsettled, indeterminate, and nebulous, and so does God, Hartshorne suggests, because that is indeed the real character of the future.[42] Bottom line: we're once again left without anything of permanence, since the content of God's mind is never totally established in Hartshorne's model.

The "process" God is easily confused with Teilhard de Chardin's evolving Christ. Teilhard (1881-1955), a French Jesuit priest, paleontologist, and philosopher who has enjoyed great popularity in the decades since his death and has become an icon of the "New Age" movement, advanced the idea that Jesus needed to be re-made in view of so-called scientific developments over the past hundred years. Specifically, Christ's persona needed to be brought into line with the theory of evolution, the truth of which Teilhard appears to have believed without qualification.[43] The cross, Teilhard proclaimed, could continue to stand

---

[39] Charles Hartshorne, *The Divine Relativity: A Social Conception of God* (New Haven: Yale University Press, 1941), 19-20.

[40] Ibid., 20.

[41] Ibid.

[42] See Charles Hartshorne, *Man's Vision of God and the Logic of Theism* (New York: Harper & Brothers, 1941), 98-99.

[43] See Pierre Teilhard de Chardin, *Christianity and Evolution*, transl. by Rene Hague (New York: Harcourt Brace Yovanovich, 1971), 77, 139.

on one condition, and one only: that it expand itself to the dimensions of a new age, and cease to present itself to us as primarily (or even exclusively) the sign of a victory over sin—and so finally attain its fullness, which is to become the dynamic and complete symbol of a universe in a state of personalizing evolution.[44]

In this Christology, Jesus became as much a moral leader as a redeemer (as well as a poster boy for the theory of evolution, it would seem).

## *"Openness" Theology*

A religious cousin of "process" philosophy is "openness" theology, although "openness" people deny that they've been influenced in any way by the "process" thinkers.[45] Openness theists contend that while God has complete knowledge of the past and similar exhaustive knowledge of what's happening in the present, he doesn't know the part of the future that involves human will as opposed to the playing out of natural processes such as hurricanes, earthquakes, and the like. In most cases, He has no idea what any human being is going to do until that person actually decides and acts, though He may be able to make an educated guess as to what the person is going to do. God knows for sure that the tornado is coming, but He doesn't know whether John Jones will be smart enough to go hide in his basement, although John's performance in the past suggests that he may instead jump in his car and try to outrun the twister.

Under the openness theism view, God's not having complete knowledge of the future doesn't mean that He's less than all-knowing. After all, God still knows everything that can be known, goes the openness argument.[46] The future is not something that has

---

[44] Ibid., 219-20.

[45] See Millard J. Erickson, *What Does God Know and When Does He Know It?: The Current Controversy over Divine Foreknowledge* (Grand Rapids, Mich.: Zondervan, 2003), 151-56.

[46] See ibid., 176.

any reality; thus, there's nothing to be known. God's foreknowledge or His lack thereof is relevant on the question of human free will, which is arguably lacking if God knows what is going to happen and yet does nothing to prevent an adverse outcome.

Free will isn't the only thing at stake, however. Openness theists maintain that the core issue is God's love. God does not just love; He *is* love (1 John 4:16). He loves us so much that He sacrificed the Son part of Himself on a cruel cross so that we might have life eternal with Him (John 3:16). He wants us to love Him in return, and therefore created us as free beings, capable not only of loving and therefore obeying Him but also of doing the opposite. This makes our love for Him, as C. S. Lewis put it, a "love worth having."[47] If God knew in advance that we were going to love and obey Him, ours would not be for Him a "love worth having," since we would effectively have no choice but to love and obey Him. Why would God want the love of robots! the openness theists argue.

Like "process" thought, the "openness" position is not without problems. It has been Christian orthodoxy, at least since Augustine (354-430), that God is outside of time.[48] The orthodox position is supported by modern cosmology to the extent that science declares that time had a beginning (indeed, it had to have had, since space had a beginning and Einstein's theory of relativity tells us that time and space are inseparable). Openness theology puts God squarely within time even though traditional theism says that He created it. By this stroke, openness theology takes away God's omniscience, as does process philosophy, since "within time" implies lack of knowledge of that which does not yet exist—i.e., the future.

The openness theists' answer that God still knows "everything that can be known" does not confer complete all-knowingness on God, it should be obvious. "All-knowing" is not "*all*-knowing" under the openness theology definition. God is learning about things as they happen, just as we mortals are, the same difficulty as you encounter under process philosophy. Further, openness theism seems to allow that God changes as He experiences new

---

[47] Lewis, *Mere Christianity*, 47-48, 183.

[48] See, in this regard, Saint Augustine, *Confessions*, transl. by Albert C. Outler (New York: Barnes & Noble Books, 2007), xxxviii.

developments in His universe. This would make Him the embodiment of existentialism ("existence precedes essence") and, once again, not the immutable God of the Bible (see Malachi 3:6; James 1:17).

Returning to the question posed when we first began talking about openness theology, what *is* the difference between it and process philosophy? The best answer this retired attorney can articulate, one that I give you because it's been given by open theist Gregory Boyd, is that the process view of God is that He needs the world in order to exist; He could not have existed without it.[49] It's not a matter of His needing love from His creatures because His essence is love (1 John 4:8, 16). The openness God, who is omnipotent, doesn't have the "needs love" problem; He can exist even if the world doesn't, theologian and author Millard Erickson indicates.[50]

But why would God want to exist without the world? Augustine effectively takes up this question in his response to the query, "What was God doing before He made heaven and earth?" Augustine resists answering facetiously, as he reports someone else did, that God was "preparing hell for those who pry too deep."[51] Augustine's answer, instead, is essentially that we were always the center of God's attention because there was no "before" when the universe didn't exist.[52] God, as "the creator of all times," had not yet established the temporal medium in which He could be said to have been doing either something or nothing.[53] Therefore, there was no foundation for the question of what God was doing. He was always focused on us.

Amazingly, Augustine's conclusion that space and time are intertwined was arrived at more than 1500 years before Einstein appeared on the scene with his theory of relativity. Augustine seems

---

[49] See Gregory A. Boyd, *God of the Possible: A Biblical Introduction to the Open View of God* (Grand Rapids: Baker, 2000), 170; also, Erickson, *What Does God Know*, 152-53.

[50] See Erickson, *What Does God Know*, 153.

[51] Augustine, *Confessions*, 11.12.14.

[52] See ibid., 11.12.14, 11.13.15.

[53] Ibid., 11.13.15.

to say, in effect, that there was never a possibility of the world's not existing, and that in fact it and mankind have always existed in God's eternal present. And if God sees everything that has ever happened and ever will happen from the vantage point of His lofty throne room outside of time, then there's no way His knowledge can be added to, no possibility of His changing in any way.

If God's knowledge can't be added to, how can there be such a thing as effectual prayer, which is recognized as a spiritual principle throughout the Bible (see, e.g., Joshua 10:12-14; 1 Samuel 1:27; Nehemiah 9:27; Matthew 7:7; John 15;7; James 4:3; 5:16)? Prayer that is interactive, as we would say in the computer age. How can God answer prayer if He already knows what the outcome will be in the situation being prayed about? And how can one better his individual spiritual situation—as, for example, by obtaining salvation through praying the Sinner's Prayer—if God already knows who eternity's winners and losers will be? I'm praying for insight in these matters, but so far I'm not getting it. Does this mean there's no effectual prayer? Does prayer bring about change or does it not? (It's getting warm in here, gentle reader, isn't it?)

Fifth-century BC Greek philosopher Zeno of Elea (490 BC - ?) is known for a series of paradoxes (apparently self-contradictory propositions) having to do with the question of change. When he formulated his famous paradoxes, Zeno was defending his mentor Parmenides against the followers of Pythagoras, so truth may not have been his agenda entirely; embarrassing the Pythagoreans by presenting them with unsolvable conundrums may also have been part of the program. Parmenides had taken the position that the common-sense notions of change and plurality are illusory, and Zeno set out with his paradoxes to show that natively intelligent ideas regarding change and plurality can often lead to nonsensical conclusions.

One of Zeno's better-known paradoxes had to do with the idea that in order to go from point A to point B, one first has to cover half the distance between the two points, and that in order to reach a new halfway point after that, one must once again cover half the intervening space, and so on ad infinitum. Movement from A to

B (adopted as the definition of change) is never possible because there's always some remaining distance to be traversed.

The "no A to B" contradiction is appealingly acted out in the movie *I.Q.*,[54] in which Albert Einstein (Walter Matthau) attempts to persuade his mathematician niece (Meg Ryan) to dance with a young auto mechanic (Tim Robbins) whom he has singled out as a fitting candidate for the affections of his niece. Describing the above paradox as she moves ever closer to Tim Robbins' character, Meg Ryan's character attempts to make the point that because of Zeno's paradox, she can never get from where she is (as a bright young graduate student at Princeton) to where the mechanic is (in a lower socio-economic stratum, some would say, if society's normal social criteria are applied). Eventually the guy does get the girl, nonetheless, which would seem ipso facto to overthrow the paradox.

*Cause, Effect, and Time*

But wait a minute, you say. In talking about Parmenides and Zeno and Meg Ryan and Tim Robbins, you're just addressing change in general; you're not getting to the specific question of whether God changes. Aren't we going somewhat astray in worrying about change or the lack thereof as a general principle? Why is the position of Parmenides and Zeno (the Eleatic school of philosophy), right or wrong, something that has to be noted in our discussion? Because, I reply, saying that there's no change makes it harder to say that time exists. Change defines time. That's why we think of the second law of thermodynamics (which states that the universe is "running down" as far as the amount of energy available to do useful work is concerned) as an "arrow of time," an indicator that there is such a thing as a flow of time. It's why another "arrow of time" is seen in the supposed continuing expansion of the universe, the galaxies getting farther apart even though this is not observable with the naked eye when one is sitting on the front stoop of The Cottage on a starlit night.

---

[54] Fred Schepisi, dir., *I.Q.* (DVD) (Hollywood: Paramount Pictures, 1994); and see Stokes, *Essential Thinkers*, 19; also Rohmann, *Ideas*, 435-36.

If you nonetheless choose to believe that there's no change, as the Eleatics did, then time itself becomes a fantasy. But God's very God-ness depends on time's existence, does it not? If there's no such thing as time, then we mortals even now inhabit the same milieu as God. The Bible says that God inhabits eternity (Isaiah 57:15). Wouldn't the nonexistence of time, which we've thought of as our dwelling place for so long, mean that we've never really been separated from God, that our human existence is really only a matter in God's eternal mind? That being the case, what would have been the point of Christ's enduring the crucifixion? Christian teaching says that God the Son came to Earth, to pay for our sin by being nailed to a cross and dying for us, in order to overcome a separation that existed between us and God the Father.[55] Why would what Christ did on the cross have been necessary if we and God were already in the same place metaphysically or spiritually, if we human folk were already where God is? Was the crucifixion pointless?

Saying that we exist solely in the eternal mind of God and not in time would not only render meaningless the plan of salvation offered by Scripture; it would also make it impossible to carry that scheme out, since it would deny us mortals the ability to affect our outcomes and the outcomes of others for eternity through a cause (confession of faith) producing an effect (salvation). Cause and effect, physicist/writer Paul Davies notes, "are concepts that are firmly embedded in the notion of time;"[56] however, God—Aristotle's "uncaused cause"—stands outside the law of cause and effect.

Davies finds it curious that God should be exempted from the requirement of having a cause. If God can be uncaused, he queries, why can't the universe? Eighteenth-century Scottish philosopher David Hume similarly found the so-called cosmological argument—that everything but God needs a cause—to be self-contradictory.

---

[55] Based on John 3:16; 1Corinthians 15:3; 2 Corinthians 5:18; and many other such verses.

[56] Paul Davies, *God and the New Physics* (New York: Simon & Schuster, 1983), 38.

Davies quotes Hume as saying, with regard to the granting of exemptions from the requirement of causation:

> If we stop, and go no farther (than God), why go so far? Why not stop at the material world? By supposing it to contain the principle of its order within itself, we really assert it to be God.[57]

There's no need to be concerned, regardless, since God's plan of salvation rather than man's logic controls. This God that we worship, remember, is three Persons: Father, Son, and Holy Spirit (2 Corinthians 13:14; 1 John 5:7-8). The Spirit, the member of the Trinity who works within us to bring us to the saving knowledge of the other two persons of the Trinity (Ephesians 2:18), doesn't care about antecedents and consequences in the normal sense. If He did, our salvation would merely be a matter of our performing prescribed acts of supplication, earning rather than learning our way into God's kingdom. The truth of the matter, as expressed by our Lord, is otherwise:

> The wind bloweth where it listeth, and thou hear the sound thereof, but canst not tell whence it cometh, and whither it goeth: so is every one that is born of the Spirit. (John 3:8)

We can't tell exactly what causes the results that we see in the spiritual regeneration process, in other words. We therefore can't knobtwiddle our way into the kingdom of God, the Master is in effect telling us. Righteous acts by themselves are of no significance as far as salvation is concerned (Isaiah 64:6; Titus 3:5). There is, however, one antecedent that's absolutely necessary for our redemption: faith.

> For God so loved the world, that he gave his only begotten Son, that whosoever *believeth* in him should

---

[57] Ibid.

not perish, but have everlasting life (emphasis supplied). (John 3:16)

Because there's a cause-and-effect sequence going on here as far as faith and salvation are concerned, it's necessary that the prospective proselyte be a resident of time. Only because the subject is operating in the context of time is he or she able to affect the outcome of his or her existence for eternity. Swaying one's future assumes a "then" and a "now," essential features of time. Both a change and a change agent are involved, but neither can be effective unless there is a "then" and also a "now."

With respect to salvation, the *change agent* is faith, which comes courtesy of the Holy Spirit (Ephesians 2:8-9; 1 Corinthians 15:53). The *change* in a person who is saved involves a transformation, effected through the renewal of his or her mind, which brings him into "that good, and acceptable, and perfect, will of God" (Romans 12:2). One's reformation, like his or her redemption, happens in time. We go from what we were then (preredemption) to what we are now (postredemption).

We mortals will, of course, undergo the ultimate transformation when we meet Christ, at which time we "shall be like him, for we shall see him as he is" (1 John 3:2). "(A)s he is" implies a no-change property in God's character, which is evidenced at various places in Scripture, but I still need to work on being able to express with clarity how it is that God can be eternal and changeless and yet miraculously be able to interact with a temporal and temporary world and call its inhabitants to communion with Himself. Some would say such mental gymnastics are an unnecessary exercise for a believer. I disagree, obviously, at least as that statement is applied to me.

It's now many nights after I observed the dancing tree branch, and I'm working in The Cottage, at the dining room table, rather than sitting out on the beach by a campfire. This is so that I can be checking references using the library and notes I brought with me when we came up from Rochester. Every day's efforts on this book lately involve my first living life by the water, trying to see if God is offering me any lessons in what I experience, and then doing my best to combine experience and scholarship in such a way as to be able to state intelligible things regarding reality. I'm forming sentences that sound like book writing, but unfortunately none of this qualifies as a metaphysical or theological bottom line that I can teach to anyone. I need to be more patient in trying to determine what God wants me to say, I guess.

My concentration is being disturbed, much of the time tonight, by the bugs—I think they're gnats of some kind—that keep throwing themselves against the bulbs in the oversize chandelier that illuminates our big oak table. The bugs get in because the sliding screen door to the back patio doesn't quite meet up with its frame. This screen door was installed some thirty years ago and The Cottage has settled since then. Install a new screen door? Too expensive, and besides, this is a cottage. You tolerate some things here that you wouldn't put up with at home. The sound of the waves washing onto the beach every night compensates for the annoying insects. In the daytime, the beach and the azure of the sky and water give us recompense for the infiltrating chipmunks (the darned things can come in through the smallest of openings). Where am I going with all of this? I'm not sure, except maybe to conclude that the working conditions here aren't as good as they could be, but they're tolerable nonetheless. The apostles didn't always have it easy, either.

*More Openness*

Another day, morning. I'm sitting in a lounge chair on the patio that faces away from the Lake (we have two patios, the other being on the lake side and referred to as the front stoop), my lukewarm coffee cup cradled in my hands and lap. Just beyond the edge of this back patio, there begins a grouping of shade trees that the early morning sunlight has up until now penetrated only in patches. The trees behind our cottage aren't all that close together or densely leaved; it's just that in screening out the sun they're getting help from the woods, a couple of hundred yards away. Since we have trees, we also have birds. At the moment the actors on the stage before me are hummingbirds and goldfinches (we've put up feeders for each).

Even in the setting described above, I'm not totally relaxed. The questions, metaphysical and theological, are still there. I wish I could simply conduct Einstein-style thought experiments—work out each issue in my mind, no verification other than a logic test being necessary, and have that solution in each case be the one the world cleaves to, as happened with Einstein's famous "twins" paradox. In that paradox, that seemingly self-contradictory proposition, a twin traveling through space at close to light speed ages more slowly than his earthbound sibling, because of velocity-based time dilation. This really is essentially the kind of thing I'm trying to do (conduct a thought experiment, not hurtle through space).

I'm still concerned with the question of change, but feel that I'm not getting anywhere with this issue. Can we at this point, therefore, just jettison the suggestion that things in the world don't change, and focus instead on whether God does? It seems to me that Parmenides and Zeno were playing pointless head games, with no prospect of arriving at anything defensible. The apostle Paul warned against such pursuits in his letter to the church at Colossae, urging the Christians there to "(b)eware lest any man spoil you through philosophy and vain deceit"(Colossians 2: 8). Arriving at the truth of things is not just a matter of resolving paradoxes. At some point common sense must come into play. The idea that the universe and its inhabitants don't change is counter intuitive, whereas the idea that God is changeless is not so easily dismissed.

For one thing, if God changes there's no bedrock on which principles of conscience or criminality can rest, since what's right or wrong will change as God changes. There is no moral law argument for the existence of God in that case inasmuch as it is the existence of law, not anarchy, that implies the existence of a lawgiver. So we have countervailing considerations. On the one hand, we have the need for us to have free will and thus a choice regarding such matters as salvation. A God who changes is presumably a necessary feature of that scheme of things, as is a God with whom we have assurance that prayer will be answered in some way. On the

# FLUID REALITY

other hand, God's existence as the lawgiver seems to require that He be immutable.

The openness theists, of whom I spoke earlier, cite various Bible passages in support of their position that God changes. The scripture relied on by them falls into a number of categories, as analyzed by Millard Erickson, whose book on open theism is one of the volumes I've brought with me to the Point.[58] There are the "repentance" passages, for example Genesis 6:6 ("[I]t repented the Lord that he had made man on the earth, and it grieved him at his heart"); and 1 Samuel 15:11 ("It repenteth me [the Lord] that I have set up Saul to be king"). Contrastingly, we find verses such as Numbers 23:19, where Scripture has the prophet Balaam saying, "God is not a man, that he should repent; neither the son of man, that he should repent." Prof. Erickson observes that regardless of what meaning you give to "repent" or which passage you give weight to, the simple fact is that knowing in advance that a bad thing will happen does not necessarily make the event less painful.[59] You don't have to be surprised to be hurt, even if you're God. Indeed, the likelihood of God's suffering hurt as a result of an outcome that's bad for man would seem to be increased by the fact of His great love toward man (Jeremiah 31:3; Hosea 11:1, 4; John 3:16; Romans 5:8; Ephesians 2:4-5; Titus 3:4-5; 1 John 3:1).

Then there are "change of mind" passages such as Genesis 18:20-33 (Abraham seemingly successful in negotiating God down to His finding ten righteous persons in the cities of Sodom and Gomorrah as His price for not destroying these cities); and 2 Kings 20:1-6 (the gravely ill King Hezekiah asking God for prolonged life and being told by God through the prophet Isaiah that he will be granted fifteen more years). We could say, in this category of situations, that God knew all along what He would wind up "settling" for but wanted to leave man with the impression that there is value in asking Him for things, a conclusion that would certainly fit if we hold to a "simple foreknowledge" premise.

We also see God "testing" individuals, as, e.g., in Genesis 22:1-18 (Abraham told to sacrifice his son Isaac, which Abraham

---

[58] See Erickson, *What Does God Know*, 17-40.

[59] Ibid., 22.

demonstrates he's willing to do, after which the angel of the Lord tells him that he doesn't need to) and 2 Chronicles 32:31 (God leaving King Hezekiah to his own devices re protection of his treasure "to try him, that he [the Lord] might know all that was in his [Hezekiah's] heart"). It may be theorized that God knew the way in which each of the tested men would respond, and that the testing in each case was more in the nature of providing His servant an opportunity for self-discovery—leading, in the case of Hezekiah, to the foolishness of the servant (in showing his treasure to the Babylonians) being made manifest to him through the mouth of Isaiah the prophet.[60]

We likewise see God "testing" Israel as a whole, as in Deuteronomy 8:2 (considering the Jews' conduct in the adversity of the wilderness); Deuteronomy 13:3 (the Jews' reaction to false prophets); and Judges 3:4 (the Jews' dealing with the nations already inhabiting Canaan). As with the testing of individuals, it could be said that since God knew what test grades would be received, the testing was more for the benefit of the mortals involved (to humble them, perhaps) than for God's information.[61] This interpretation requires us to go beyond what Scripture says, but it is also not contrary to Scripture.

What about prophesies that are characterized (though not by the Bible) as "failed" ones? Erickson states that if some prophecies are never fulfilled, this would be a telling consideration against the traditional view of divine foreknowledge. He then examines some specific claims of prophecy unfulfillment and essentially concludes that for the most part these interpretations reflect bad hermeneutics (i.e., erroneous Bible interpretations). Erickson rejects out of hand the contention of Gregory Boyd that the "failed" prophecies were mere probability statements.[62]

We also sometimes see God asking questions that might be seen as betraying just plain ignorance about the future, as in Numbers 14:11 ("How long will this people?"); Hosea 8:5 ("[H]ow long will it be?"); and 1 Kings 22:20 ("Who shall persuade Ahab?"). Such

---

[60] See ibid., 25; also, 2 Kings 20:12-19; and Isaiah 39:1-8.

[61] Erickson, *What Does God Know*, 26.

[62] Ibid., 27-30.

questions are easily seen as rhetorical questions, it seems to me. Even God's statement at Jeremiah 7:31 ("[N]either came it into my heart [that they would sacrifice their children]") can be disposed of as rhetoric: I knew this would happen, but it's still unthinkable.[63]

Erickson also examines "frustration" statements, such as at Ezekiel 22:30-31 (God says, "I sought for a man among them, that should make up the hedge,...but I found none"); and 2 Peter 3:9 (God is "not willing that any should perish,"). God does indeed seem committed to preventing from happening that which does in fact happen. Why? Because He *doesn't know* what's going to happen, say the open theists. Erickson concedes that the classical (simple foreknowledge) view has not offered much response to the "frustration" interpretation, but hints that the problem may be that "frustration" is a misnomer as applied to such situations.[64] But does God, as a person—indeed, the Supreme Person—have to be stoic when His wishes are not fulfilled? Why can't He be frustrated? Would it be more in keeping with His character if He simply looked down the road, saw futility, and accepted it?

Less problematic than most of the verses referred to above, it seems to me, are the Bible's conditional statements, such as at Exodus 13:17; Ezekiel 12:3; Jeremiah 26:3; and Jeremiah 26:19, statements that are framed in an "if, then" form. A possible response regarding the "if, then" passages is that they simply show God saying how it will be under specified circumstances, although He knows which of the alternatives will occur. But what are we to make of Jesus' prayer in the garden of Gethsemane, at Matthew 26:39 ("O my Father, if it be possible, let this cup pass from me")? Jesus, God the Son, didn't know that His execution was certain? This, even though He was present at the creation (John 1:1-3) and was "slain from the foundation of the world" (Revelation 13:8)? My answer, one not stated by Erickson, is that Jesus' Gethsemane prayer probably more reflects His anxiety and fear regarding His coming crucifixion than any thought that His suffering could be avoided. He was, after all, fully man (and therefore emotionally vulnerable) as well as fully God (John 1:14; Hebrews 4:15).

---

[63] Ibid., 31-32.

[64] Ibid., 33-34.

With all due respect to Professor Erickson and the other authorities to whom he refers, I find the responses outlined above, including those that I personally formulated, to be less than wholly satisfying. It seems to me that there must be some other principle(s) operating here. I'm reaching a crescendo of confusion in my thinking, I must admit. I need to walk down the beach or do whatever else I can to clear my head. When I was first sitting out on the back patio this morning, I heard Sue talking on the phone with our Rochester daughter-in-law, inquiring if she and our Rochester son and the kids would be coming up to The Cottage this weekend. I need for that to happen if I'm going to keep my sanity while thinking about God and change. Grandchildren running around is a good brain cleanser. Please, Lord, let them come

*Foreknowledge and Freedom*

Monday morning, after a fun visit from the Rochester grandchildren (and their parents). Our time at The Cottage with the girls, ages two and four as of this writing, consists more than anything of playing in the sand and going out in the rowboat when the weather is favorable, which it was this past weekend. Also, going to the playground that the Campbell's Point Association maintains. The girls aren't quite ready for fishing yet. When they are, I'll have the role of baiting hooks and freeing fish for return to the lake, I expect.

Anyway, where were we? Oh, yes. The question of whether God's knowledge changes. As I indicated in the course of earlier reflections, the divine knowledge conundrum—is God's knowledge truly exhaustive or is it limited to the present and past?—brings into the discussion the issue of whether we creatures have free will. It's counterintuitive to say that we don't, but maybe that doesn't settle the matter. It certainly doesn't if you want to be able to say that God's all-knowingness includes exhaustive foreknowledge—i.e., knowledge of the future that's not limited by the vagaries of human behavior.

But if God has exhaustive foreknowledge, doesn't what He knows about our individual futures foreclose our being able to change those futures? Haven't our futures already in effect been decided for us by what's in God's mind, what He sees from His position outside of time? How could God, I ask again, if His knowledge of the future is exhaustive, ever *respond* to any prayer, including the prayer for salvation? His response, if it changed anything—i.e., if it was genuinely a response—would render His foreknowledge faulty, making it not really foreknowledge at all! In that case, He's no longer omniscient, even though the Bible says He is (Job 24:1; 36:4; Psalm 147:5; Isaiah 42:9; 46:9-10; Matthew 24:26).[65]

The most pressing "free will" discussion for most people, I would guess, has to do with where one's home for eternity will be. The question of whether we're free to choose heaven or hell, or are instead predestined to one or the other, has been debated, on and off, for over nineteen centuries, since Paul authored his letter to the church at Rome, in which epistle he declared, regarding God and the new Christian believers,

> For whom he did foreknow, he also did predestinate to be conformed to the image of his Son, that he might be the firstborn among many brethren. Moreover, whom he did predestinate, them he also called: and whom he called, them he also justified: and whom he justified, them he also glorified. (Romans 8:29-30)

Some five years later, the apostle Peter, apparently following Paul's lead, addressed his fellow believers as those "(e)lect [i.e., chosen] according to the foreknowledge of God" (1 Peter 1:2).

Pelagius, an English monk living in Rome in the third and fourth centuries (?360-?420 AD), was among the first to challenge Paul's doctrine of predestination. His teaching occasioned an attempt by Augustine, bishop of Hippo (in North Africa), to reconcile Paul's talk about predestination with the seemingly more intuitive idea that man has free will. Only trouble was, Pelagius was teaching that the human race was not rendered sinful and lost

---

[65] You'll have to excuse my habit of citing authority. That's what we lawyers do. Later on, if we get more knee-deep in citations of authority, you may see me putting Bible cites in my footnotes along with all the other material.

as a result of the sin of Adam in disobeying God's command to not eat the fruit of the tree of the knowledge of good and evil (see Genesis 2:16-18; 3:1-24).[66] Adam simply set a bad example, that was all, Pelagius argued. Man on his own, rejecting what Adam did and following instead the exemplar of Jesus, can live a sinless life and by that route get to heaven, Pelagius said.

To Pelagius's assertion, Augustine replied that our redemption is based completely on God's grace (unmerited favor, for any reader unfamiliar with the term), not on anything man can do. Repenting (turning away from sin) and accepting God's gracious offer are still necessary for individual man's salvation, Augustine maintained; however, he said, God already knows who will get into heaven by doing these things and who will not. We, the elect, do what we need to do to be saved because God's foreknowledge allows us no leeway to do otherwise.

Does the above mean that man can't change anything by exercising free will? we ask. Not exactly, Augustine replies. Don't get hung up on the "pre-" in predestination, he says in effect. God is outside of time, which He created when He created the world. Thus, He doesn't look ahead and foreordain things. Rather, He sees events as happening in His eternal present. The alternatives in question, heaven or hell, can therefore each happen because of a choice on man's part. To be saved, man must choose to repent and accept God's gracious offer of salvation through Christ (Luke 13:3, 5; Acts 3:19; 26:20). Thus, man does have freedom in the matter, as Augustine sees it, although God has present knowledge of the choice man will make regarding his soul's residence.

I'm not sure the above formulation solves the problem, however. Augustine has preserved God's omniscience and timelessness, certainly, but mankind's freedom? How can individual man be genuinely free if where he will spend eternity is a foregone (although not foreordained) conclusion as far as God is concerned? It just doesn't sound right. Man's repentance and acceptance of God's grace seems more like his playing a scripted role than his exercising free choice, the deftness of Augustine's argument notwithstanding.

---

[66] From the cited verses, it will be seen that Adam's crime, abetted by Eve, was essentially that of trying to acquire knowledge on a par with God's.

Frenchman John Calvin's predestination has an even sharper edge on it than Augustine's, or at least so it feels to me. Calvin was a leading figure in the second generation of the "the Reformation," so called because it was the original aim of this movement to reform, rather than cause an exodus from, the Roman Catholic church. Martin Luther in October of 1517 had posted on the door of the university chapel at Wittenberg, Germany, his ninety-five theses, giving reasons why the Church's sale of indulgences (tickets out of purgatory for deceased relatives, supposedly) should be stopped. That was the beginning of the Reformation.[67] Within twenty years, Calvin had published the first edition of his *Institutes of the Christian Religion*,[68] which set forth the basics of what was to become five-point Calvinism.[69]

The first of Calvinism's five points has to do with what Calvinists label "total depravity." Because of Adam's fall, man is unable on his own to savingly believe the Gospel, it is said. The sinner is dead, blind, and deaf to the things of God, because his heart is desperately wicked (Jeremiah 17:9). His will, rather than being free, is in bondage to his evil nature; therefore, he will not—indeed, he cannot—choose good over evil. Accordingly, it takes more than just the Holy Spirit's assistance to bring a sinner to Christ; it takes a kind of regeneration in which the Spirit gives the subject a whole new nature, provided the person in question is one of those chosen in advance by God for salvation. Faith, Calvinism teaches, isn't man's contribution toward his salvation; instead, it's merely a part of God's salvation package. It's God's gift to the sinner, not the sinner's gift to God, as Calvinists interpret Ephesians 2, verses 8 and 9 ("[f]or by grace are ye saved through faith; and that not of yourselves; it is the gift of God; not of works, lest any man should boast."). The Calvinist reading of the quoted

---

[67] For details in this regard, see Bruce L. Shelley, *Church History in Plain Language*, updated 3rd Ed. (Nashville: Thomas Nelson Publishers, 2008), 237-46.

[68] For a classic translation, see John Calvin, *Institutes of the Christian Religion*, transl. by Henry Beveridge (Peabody, Mass.: Hendrickson Publishers, 2008).

[69] For extensive discussion re the five points, see Loraine Boettner, *The Reformed Doctrine of Predestination* (Phillipsburg, NJ: Presbyterian and Reformed Publishing Co., 1932), 59-201; also, David N. Steele and Curtis Thomas, *The Five Points of Calvinism* (Phillipsburg, NJ: Presbyterian and Reformed Publishing Co.,1963), 13-23.

verses makes salvation the gift when it is actually *faith* that God graciously gives us, so as to enable us to be saved by believing in Christ (John 11: 26). There would be no need to even mention faith if it were not the key to salvation.

Arminianism, named after Jacobus Arminius (1560-1609), a Dutch seminary professor whose teachings were posthumously offered in opposition to Calvin's beginning in 1610, takes a different position with respect to man's depravity and its effect. The Arminian view is that although man's nature was definitely changed for the worse by Adam's fall, man was not left in a state of total spiritual helplessness. Man does still have the ability to on his own do the right thing, to choose good over evil, life over death. Man was, after all, created in God's image (Genesis 1:27), and not all of the God likeness was wiped away at the fall. Since man has the ability to make the right choice with the assistance (but not under the domination) of the Holy Spirit, he can actually change the future. This he can't do under Calvinism, which sees a future that has been *foreordained* by God. Foreknowing is the same as foreordaining, in the Calvinist view. As Calvinist apologist Loraine Boettner states the case, "Foreordination renders the events certain, while foreknowledge presupposes that they are certain."[70]

I see more of Arminius in God's creation than I see of Calvin. The natural world, although corrupted by the fall (Genesis 3:17-19—thorns, thistles, and suddenly it's hard work living off the land), is still something of great beauty. I marvel at the way the verdure of Stony Point folds into Henderson Bay, at the delicate wispiness of the clouds, at the gurgle of the water as it washes up on the beach. And yet I see flaws everywhere. Trees weaken and come crashing down, as did the splendid mountain ash behind The Cottage. The wind was able to take it down on a June day last year because, unbeknownst to us, it was rotten at the core. Gobies, those ugly little fish that have the appearance of a piece of excrement, infest the shallows and eat the bass eggs. We kill every one of them we hook, but we can't get rid of them.

Where does this leave us on the question of change versus stasis as the fundamental principle of reality? If predestination is true,

---

[70] Boettner, *The Reformed Doctrine*, 42.

then God's knowledge doesn't change. If God's knowledge doesn't change, then neither do the objects of His knowledge as far as appreciable progress is concerned. At a minimum, any perceived change in the world or its inhabitants would be illusory. Our human lives would merely be the playing out of an already-written script start to finish. There would be no basis for saying that there is personal freedom inasmuch as there would be no real ability on our part to affect any outcome.

Also, let's stand openness theology on its head for a moment. If we're modeled after God and we change, doesn't this necessarily mean that God changes? Doesn't our being like Him, as Genesis 1:27 says we are, imply that He has to be like us? If nature tells us things about God, as Romans 1:20 declares, a similar question is raised. The natural world tends toward both visible and hidden decay, the latter under the law of entropy. Does this, also, tell us something about God that we would not wish to hear? And is there any limit to the ways in which God can change? Doesn't the Bible say that with Him all things are possible (Matthew 19:26)? Does this statement apply only to what God can do (for example, admit a rich man to heaven)? Or does this passage speak also to the question of what God can become? Can God therefore turn out to be not what we bargained for?

If there's nothing permanent about God, how can we think that there's something about us that's permanent, something that will survive the death of our physical bodies? If we concede that God changes, we've not only made his God-ness questionable; we've also given away the primary basis for believing in our own eternal life. In addition, we're close to letting materialism win its argument with idealism. Idealism holds that ideas such as Plato's perfect rose are the only thing that's real. Materialism says that the only reality is the rose you see before you, with its browning petals and its aphids. The wilted, buggy rose has changed. It had to, since it didn't have the perfection that God has.

I'm content, at least for now, to think of myself as changeable, but I don't want God to be in the same boat. He's too perfect for that (see Matthew 5:48). With Him, there is "no variableness, neither shadow of turning" (James 1:17). How I can avoid sharing with God a craft that because of my theology may not be seaworthy is a question I'll need to return to at a later time, perhaps more than once.

# 2
# ESSENCE

*The Idea of Self*

Autumn is advancing; we're now into late September. The leaves are turning, mostly green melding into various shades of gold, but there are some rust-colored leaves mixed in. I'm seeing the autumn colors from the overlook in Westcott Beach State Park. Westcott's, as it's called, which begins a half mile or so south of the Campbell's Point property, consists of two parts: there's the beach part, where all the campsites are on the water or close to it, and there's the plateau. To get to the plateau, if you're driving there, you turn off of State Route 3 onto Chestnut Ridge Road, by the Westcott Variety Store (Jon and Lori's), which is a right turn if you're coming up from the south, then make a quick right onto the plateau campsites access road. After that, you simply follow the access road as it winds up the hill.

From the lower Westcott Beach campsites or from Campbell's Point as a starting point, you may choose to walk the road up to the plateau, as I try to every morning. By the time you get to the top, you're "sucking air," as they say. You then catch your breath as you look out over Henderson Bay and Stony Point, which are to your left, and Black River Bay and Sackets Harbor, to your right. The waters of Henderson Bay and Black River Bay are silent at this distance away from them, but you can still visually make out individual waves. On this particular morning, I see something on the horizon, in the open water beyond Point Peninsula, that I've never seen from the shore at Campbell's Point: a vessel big enough to be an ocean-going ship, or at least a Great Lakes freighter. I've gotten high enough to see over the curve of the earth, is my take on the situation.

Elevation changes your perspective on all kinds of things. From this height and angle, the three islands that in effect extend northward the land mass of Stony Point—Six Town Point, Gull, and Bass—don't appear to lie in a straight line (see Fig. 7). You'd never guess their nonalignment by looking at them from the shore down below.

There's a fourth island, connected to Stony Point by a causeway, but I don't count Association as an island. It's not just because of the causeway; Association has had too long a history of heavy human use to be included in the same category as Six Town, Gull, and Bass. For a number of years in the twentieth century, it was a place where the General Electric Company sent its employees for rest and recreation. Then it was taken over by the YMCA. I'm not sure whether Association's wooden barracks-style buildings, not in use but also not torn down, are relics from the GE era or were put up by the Y. Our Campbell's Point historian might know; I just keep forgetting to ask him. Because of the lively winds that sweep down off the bluffs that surround Henderson Bay, Association was chosen as the training site of the U.S. sailing team for the 1976 Olympics. Now it's a campground, with as many sites and slips as possible squeezed into its half-mile length.

Why am I going on in what might be seen as a somewhat negative way about Association Island? I guess it's because I feel that nature (God-or-Nature, in the parlance of Dutch philosopher Baruch Spinoza [1632-77]) speaks most clearly to seekers of truth when its message doesn't have to fight its way through too many jet skis and recreational vehicles. The jet skis swoop and wallow, tearing back and forth in bursts of noise and foam as their operators show off for people along the shoreline, our shoreline, off which their whine reverberates. They seem to be a thicker infestation on our side of Henderson Bay than out by Association Island. The RVs—visual rather than noise pollution—are a problem for places like Association Island that still have foliage and water to look at (i.e., a view), where they stand as boxy monuments to consumerism that cut off one's neighbor's view in the same way your house in the city or the suburbs does.

Pardon the digression—a brief detour on my way to thinking holistically about the business of self, which is at the heart of my inquiry about change. But I'm wrong in calling the above a digression, I now realize. Looking down on Henderson Bay from the State Park overlook and thinking the thoughts that went with this perspective wasn't really a matter of my getting sidetracked; I was just bringing in some things from the periphery of my world-view in the hope of elucidating items at the center, of thinking holistically about change and self.

Holism, as I use the term, is the theory that anything worth talking about is more than the sum of its parts. If we include the self in the worth-talking-about category of things, among the entities that can only be understood holistically, this might be seen as implying that there's something transcendent about human personality. Something that goes beyond electrochemical brain activity. Something that survives the alteration of memory by additions thereto (as we gain more and more experience) or subtraction therefrom (by Alzheimer's disease, for example). Something that's changeless and unique to each one of us. This, of course, is what Christianity and the other two major Western religions, Judaism and Islam, have been saying about personhood all along.

The opposite of holism is reductionism, the idea that things can be understood fully by studying their component parts. Thus, for example, you would be able to know who your prospective wife or husband *really* is, "deep down," if you could be privy to every bit of experience he or she has ever had. (Parenthetically, I should tell you that for purposes of this paragraph's discussion—but only for purposes of this paragraph's discussion—we're adopting the view of John Locke and other seventeenth-century empiricists that the mind at birth is a blank slate, a *tabula rasa* if you will, waiting to be written on by whatever experiences one thereafter has.) There would be nothing about the self of your prospective spouse that could not be explained in terms of experience stirring up neural activity.

We're talking not just about impulses flashing through a person's head. The impulses cause a rearrangement of charged particles in the brain of the subject. These *atoms* ("uncuttables"), which were recognized as the basis of matter by the Greek philosopher Democritus some twenty-five centuries ago, are the only existence your unholistic (i.e., reductionist) self has. It's all there, the sum total of what a person mentally and spiritually is, in the tissues of the brain. When you die, that self ceases to exist. So say the materialists. So said the great British philosopher Bertrand Russell (1872-1970), whose self disappeared some forty-plus years ago.

Russell saw Christianity's idea of a transcendent self as a negative even from a social standpoint. Traditional Christianity, because of its emphasis on individual salvation, promotes selfishness, in his view.[71] Only, I reply, if Christians shirk their moral and scriptural responsibility to obey the Lord's commands. We serve, not a faceless world-soul,[72] but a personal God, whose love has been poured out on the world through the most selfless of acts, by dying for humanity's sake (John 3:16). Our God's expectation is that we likewise will behave with generosity and love (Luke 6:31; 10:30-37; 1 John 4:21). If we do not, then our salvation is open to question, meaning that we may never have been saved in the first place (1 John 2:3-4; 3:10). We obey God out of love for Him (John 14:15), which stems from our gratitude for what He has done for us. We love Him because He first loved us, enough to die for us (1 John 4:19). This is an interaction with God (appreciation) that is different from simply trying to get something from Him (solicitation). The difference between the two scenarios is in whose love comes first. What we have in the latter case is a love transference response on the part of Christians, which I have difficulty seeing as self-centeredness.

---

[71] Bertrand Russell, *Why I Am Not a Christian and Other Essays on Religion and Related Subjects* (New York: Simon & Schuster, 1957), 34, 72-73.

[72] We'll talk about nineteenth-century American transcendentalism later.

# ESSENCE

## *Eastern Thinking*

For an alternative to Christianity's alleged selfishness, Russell might have turned to "transpersonal" psychology (TP), which was a late-1960s attempt to do away with individual personhood by expanding the boundaries of self to include all of humanity. Consciousness extended beyond the individual human ego, according to TP.[73] Thus supposedly brought together were Eastern mysticism, with its holistic approach, and Western rationalism, with its assumption that reason is the fundamental source of knowledge. Not surprisingly, then, the ideational mix included Swiss psychologist Carl Jung's theory that there is a "collective unconscious," a store of archetypes, such as the expectation of being "mothered" in times of stress, which is shared by all human beings and is accessible through mysticism.[74] Transpersonal psychology also reflected the suppositions of American psychologists Abraham Maslow and Rollo May, who stressed "human potential."[75] At the top of Maslow's famous pyramid, his "hierarchy of inborn needs," ironically, was *self*-actualization.[76]

If we were Buddhists, we wouldn't be worried about expanding personal identity; indeed, Buddhist thought runs in the direction of dispensing with personhood. A key principle of Buddhism, it turns out, is that the self is not permanent. To this extent, Buddhists agree with materialists such as Bertrand Russell. The Buddhist principle of *anatman*, or "no-self," is in contrast to the Hindu concept of *atman*, of a soul that continues to exist and moves from host to host through a cycle of birth, death, and rebirth (*samsara*), which may include incarnations as plants or animals.[77]

Under the Buddhist "no-self" principle, life at any point is a temporary gathering together of cosmic elements (*skandhas*) that are continually in flux, constantly swirling around, like the atoms

---

[73] See Rohmann, *Ideas*, 408.

[74] Stokes, *Essential Thinkers*, 140-41; Rohmann, *Ideas*, 213.

[75] Rohmann, *Ideas*, 186, 408.

[76] Ibid., 186.

[77] Rohmann, *Ideas*, 48.

that make up the universe.[78] Our thoughts and feelings come and go all the time, and our selves don't exist any more than do the temporary impressions that pass through our heads. The idea that we have a self, an identity that continues, is an illusion that keeps us from becoming enlightened, Buddhists say, and one's lack of enlightenment is the reason we see repetition of the cycle of *samsara* (birth, death, and rebirth, but no animal or plant lives mixed in).[79] With enlightenment comes not only the end of the cycle of reincarnation, but also the final shedding of your *karma*, the spiritual baggage that you acquired during unenlightened past lives, and the attainment of *nirvana*, a constant awareness of the oneness of all life.[80]

After seeing the scheme laid out above, I wonder if I'm missing something. Buddhism teaches that we have no self, that the idea of self is a delusion, and yet we're reincarnated, again and again, until we get the "no-self" thing right. Who or what is getting reincarnated? Doesn't reincarnation ipso facto assume a self? Or maybe it's just that until we attain enlightenment we're living selfishly (no play on words intended) and therefore in a deluded manner inasmuch as we believe there's a surviving self, but that doesn't seem quite right either. And if even the illusion of self is no longer there when we achieve enlightenment, who or what is left that can enjoy the pure spiritual state of nirvana? It can't literally be true in every sense that you have no self. Even if you gain the sought-after freedom from delusion and all attachments, there has to be an entity to which that freedom refers, doesn't there?

Paradoxically, the Mahayana school of Buddhism recognizes the ideal of the Bodhisattva, a highly evolved person who vows to help others achieve Buddhahood before himself entering *nirvana*.[81] Even here, selfness enters the picture. If there is no separate individual self, the idea of someone entering *nirvana* alone doesn't

---

[78] Ibid.

[79] Ibid.

[80] See ibid.; also, Stephens, *Philosopher's Notebook*, 77. For a more detailed thumbnail sketch of Buddhism, see Fritjof Capra, *The Tao of Physics: An Exploration of the Parallels between Modern Physics and Eastern Mysticism*, 35th Anniversary Ed. (Boston: Shambhala Publications, 2010), 93-99.

[81] Rohmann, *Ideas*, 49.

make sense, Fritjof Capra observes.[82] How can there be aloneness if there is no person or thing that can be in a state of separation from other persons or things?

If the above understanding of Eastern thinking on my part is correct, there's no critical point on which either Buddhism or Hinduism can be reconciled to Christian teaching. The anonymous author of the epistle to the Hebrews writes that it is "appointed unto men once to die, but after this the judgment" (Hebrews 9:27). Death, in this statement, has a finality that is denied to the person subject to reincarnation, who is condemned to an indeterminate succession of returns until he works out his *karma*, regardless of whether his soul is conceived of as an illusion (the Buddhist view) or migratory (the Hindu view). The deceased Western person, contrariwise, proceeds through judgment to his final spiritual destination; nothing is postponed. See, in this regard, Romans 14:10 (judgment by Christ); 2 Corinthians 5:10 (unavoidable); 2 Timothy 4:8 (unto reward as well as punishment).

Christianity's denouement, because it involves either reward or punishment administered by an arms-length God, is different from what either Buddhism or Hinduism counsels its adherents to expect. The Buddhist, when (if) he reaches *nirvana*, is supposedly united with the universe, not reconciled to a personal God.[83] The Hindu goal is *moksha*, the union of one's soul with *brahman*, the Absolute underlying all gods.[84] This, also, is not the same as standing before a separate, personal God, who wants to bless the faithful and will indeed do so. As the apostle Paul tells the Corinthians:

> Eye hath not seen, nor ear heard, neither have entered into the heart of man, the things which God hath prepared for them that love him.[85]

From the above, it's clear that not only the nature of human personhood, but also the very existence of any entity that can be called a

---

[82] See Capra, *Tao*, 98.

[83] Rohmann, *Ideas*, 48.

[84] Rohmann, *Ideas*, 175.

[85] 1 Corinthians 2:9.

supreme deity, is at issue when Eastern and Western thinking are compared. This is something more than a contest between the God we want and the God who is, the kind of debate which throughout much of history has sapped the strength of Western religion.

Maybe we should have begun our discussion by attempting to frame a definition of "self" that meets our need to explain eternal life from a Christian perspective. Even if the Eastern "no self" and "absorbed self" ideas are laid aside, however, there's still major disagreement as to what the self is. Sometimes the definition that's given is a functional one: what the self does. It knows, remembers, desires, suffers, thinks, and acts. Such a definition, without more, unfortunately allows the self to be thought of in strictly material terms. It's possible for a person to do the above things as a result of autonomic nervous system prompts or external environmental influences, over neither of which would he or she have any control.

But what if we say that the self is *the something that does* the knowing, remembering, desiring, suffering, thinking, and acting? Essentially, give the mental or physical activity in question an individual person to be attached to. The predicate (verb and modifiers) gets a subject (noun and modifiers). It's no longer pure, unalloyed action or mental activity. This would seem to give us a definition of self that allows the individual to do what he or she does as a matter of conscious choice. Something besides sensory input, internal or external, is running the show. In saying that the mental function of the true self can't always be controlled by what happens outside the person or ruled by the state of one's digestion, we're almost certainly assuming the existence of a self that transcends the physical, as I see it. This would seem to be reason for us to consider a definition of self that says that it's really your immortal soul. That being said, we're now forced to examine in detail the teachings of empiricists John Locke and David Hume.

*Locke and Hume*

The concept of personal identity (selfhood) includes the question of how it happens that we are what we are. Do we each have a unique soul from the start or do we each acquire what distinguishes us

## ESSENCE

from others through a special set of experiences that's replicated for no one else in the world? English philosopher John Locke (1632-1704), in his *Essay Concerning Human Understanding*, opts for the latter.[86] We find Locke saying that the identity of a being or object consists solely in those things that are attributable to or discernible by sense experience. There are no innate principles, according to Locke—we bring no ideas with us, not even ones concerning morality and justice, when we enter the world. This means that we *learn* the concepts of personal identity, worship, and God.

Locke's view that there are no innate ideas raises an unsettling question: if man has no innate sense of his being a unique soul and of there being a God who should be honored and obeyed, how can he be held accountable for not showing respect for God's commands? First of all, who is it that gets held accountable? Is it the collection of sense impressions that we call John Jones? The content of empirical experience will vary from person to person. The person named John Jones may not have heard that it's not nice to murder people and may not have read the statute law declaring this. His experience to date might effectively place him in a moral milieu where the rules concerning homicide are different from what we citizens of the West are accustomed to. How, then, can we punish this man, or even force him into compulsory rehabilitation, if he in his mind lives in a world where killing is not wrong?

But isn't there natural law that comes into play here? you ask. Can't we say that Mr. Jones should have known better than to gun down his fellow Postal Service worker inasmuch as there's a universal law, which everyone knows about, that forbids such conduct? It's not that easy, I caution. The existence of the law that you refer to is dependent on the existence of a lawgiver, someone or something other than the state, is it not? That would be God. If a person lacks an understanding that there is a God who exists and who expects certain things from us, how can we charge that person with the knowledge of a moral law? Endemic moral confusion being the case in our society, we would seem to be looking at a situation like that of the nation Israel at the conclusion of the

---

[86] See John Locke, *An Essay Concerning Human Understanding*, Great Books in Philosophy Ed. (Amherst, N.Y.: Prometheus Books, 1995), 12-59.

book of Judges, where "every man did that which was right in his own eyes" (Judges 21:25). We're talking about not only spiritual but also social anarchy. The two go together.

Locke professes belief in a supreme being on the strength of deductive reasoning, seemingly his only concession to the Renaissance god Reason.[87] That puts him ahead of fellow empiricist David Hume, a Scotsman who lived some hundred years after Locke (1711-76). Hume's philosophy is essentially Locke's without the belief in a deistic God—i.e., in a minimalist God established through reason rather than revelation and not involved in any direct way in the lives of His creatures. To believe in God, one would need to believe in causation, Hume indicates, and this would be misplaced faith, since we never receive knowledge through our sense impressions of the force that compels one event to follow another. We experience only separate events that happen to occur in succession, although some sequences are more likely to happen than others. Therefore, there being no basis on which it may be said that there is cause and effect, Locke's prime mover is not a reality as far as Hume is concerned.

The above conclusion, which is essentially that we have no way apart from experience to infer a connection between a cause and its purported effect and that therefore any attempt to posit anything beyond the odds that one event will follow another is spurious, raises another question: Do Western religious believers know that God is the cause of the world merely by reason of custom? Is it simply a matter of persons of faith being accustomed to seeing the physical complexity of the universe and its inhabitants and thinking, logically, that things this amazing could only have come from an all-powerful and all-wise Creator? Is logic the thing that ties cause and effect together in this instance? Is logic something that's lacking in the competing big bang/evolution model?

Given Hume's atheism, it's not surprising to find him rejecting any concept of selfhood. Personal identity by any definition is out of the question. "'Tis absurd," he says, "to imagine the senses can

---

[87] Ibid., 527-36.

ever distinguish betwixt ourselves and external objects."[88] Inability to know that you have a personal identity is the rule even if you look only within yourself—to your feelings—for proof of your selfhood. Hume explains:

> If any impression gives rise to the idea of self, that impression must continue invariably the same, thro' the whole course of our lives; since self is supposed to exist after that manner. But there is no impression constant and invariable. Pain and pleasure, grief and joy, passions and sensations succeed each other, and never all exist at the same time. It cannot, therefore, be from any of these impressions or from any other that the idea of self is deriv'd; and consequently there is no such idea.[89]

Hume counsels that in any event we should be looking to philosophy rather than to religion ("superstition," he calls it) for guidance concerning matters such as selfhood because philosophy is less likely to "disturb us in the conduct of our lives and actions" than religion.[90] He states, "(g)enerally speaking, the errors in religion are dangerous; those in philosophy only ridiculous."[91] Regardless of whether atheism is seen as a religious or a philosophical choice, however, it can hardly be denied that it is potentially dangerous to your health for eternity to bet against God. Wouldn't it be better, if you can do so sincerely, to lay down Pascal's wager (believing in God as insurance)?[92]

---

[88] David Hume, *A Treatise of Human Nature*, 2nd ed. (Oxford, U.K.: Oxford University Press, 1987), 190.

[89] Ibid., 251-52.

[90] Ibid., 271-72.

[91] Ibid., 272.

[92] French philosopher/mathematician/ physicist Blaise Pascal (1623-62) argued that to believe in God or not is in effect to wager that He does or does not exist. If God exists, you've secured eternal life by believing in Him or damnation by denying His existence. If He doesn't exist, you've lost nothing by either accepting or rejecting Him. Hence, the wise gambler will opt to believe in God, according to Pascal. See, for further discussion, Stephens, *Philosopher's Notebook*, 86-87

What happens when the body is no longer able to receive sense impressions in the normal way? In Locke's formulation, the test for continuation of selfhood is continuity of memory and consciousness. But what is the memory or consciousness that continues? Is it the sum of what you remember or are aware of at the moment of death? Or is it everything that was in your head at one time or another during your life? The latter seems the more likely, but we're still talking about a snapshot, only with more detail, are we not? Again, as with all questions regarding death, we're reduced to speculating.

Additionally, the "portable memory" view of selfhood transcending death doesn't seem to take into account the possibility of learning new things in the life to come. The same problem you face in trying to pin down the identity of someone still living. If, in the example given earlier, you want to know who your chosen mate is "deep down" (maybe you don't; maybe she's just a "trophy" wife or he's merely a "meal ticket"), then, under the determinist model previously described, you can in theory gain this knowledge by finding out how all the atoms in his or her brain line up.

But here you run into the problem that your knowledge is valid only until your prospective spouse has had further experience. What you're looking at is a snapshot that's out of date as soon as it's taken. What would amount instead to a continually running movie, assembled moment by moment from the constantly changing data in your mate's brain, would presumably be available; however, this would likely provide only the sketchiest answer to the question, "Who am I marrying?" Your identity at the moment of death and thereafter would seem to be in a similar state of crisis, because of your continuing to acquire experience in an interactive heaven (or hell). The question now: "Who is being judged?"

At this point in the discussion, I can hear Bertrand Russell urging from the grave that we shouldn't be worrying about what experience we'll take with us into eternity, because death is both a cessation and a negation of experience.[93] Actually, if we apply Russell's understanding of life and death, the good professor would be incapable of urging anything from "the other side," since implicit in

---

[93] See Russell, *Why I Am Not a Christian*, 88-93.

his idea of death as a descent into nonbeing is the view, expressed by Jehovah's Witnesses, that the dead "cannot do anything or feel anything."[94] Cults expert Walter Martin, rightly as far as I'm concerned, does not agree with the JW interpretation of Scripture.[95]

Dr. Martin maintains instead that there is "sufficient evidence to show that man has an eternal soul and will abide somewhere, either in conscious joy or sorrow eternally."[96] Logically, you can't be nowhere, because then you would be experiencing nothingness, which is an impossibility, existentialist dread of nothingness notwithstanding. If you can't experience nothingness, then you can't *be* nothing, and vice versa. Knowing and being have a symbiotic relationship, it might be said.

Assuming you're unable to know nothingness, there's only one thing that you can experience: something. Since your god is a deity who has rules of accountability with respect to his subjects' behavior, what you will be experiencing after death is retribution or reward in heaven or hell. You will experience things in heaven or hell because neither of these places is nothingness. The treatment you will receive, good or bad, will reflect your accountability to your god. Christianity, Judaism, and Islam agree on this, if nothing else. Bertrand Russell's view is therefore in opposition to many centuries of Western religious thought as well as illogical and counter intuitive

One's conception of self is rooted in personal memory of past experiences, Paul Davies concludes, appearing to this extent to agree with Locke.[97] Thus, Davies says, it is " not at all clear that someone with no memory can be thought of as a person at all."[98] My reaction: If we hold to the "grid" theory of learning (experience enables you to grasp later experience), can a fetus ever be said to have sufficient memory to establish personhood? Could Locke's

---

[94] From the Watchtower Bible and Tract Society beliefs summary contained in Walter Martin, Ravi Zacharias general editor, *The Kingdom of the Cults* (Bloomington, Minn.: Bethany House Publishers, 2003), at 71.

[95] See Martin, *Kingdom of the Cults*, 128-32.

[96] Ibid., 131.

[97] Davies, *God and the New Physics*, 90.

[98] Ibid., 91.

thinking be applied to favor abortion as against the pro-life position? Does this also mean that an Alzheimer's patient, as he or she progresses in the disease, becomes less and less a person? And what about the trauma-induced amnesia victim who is unable to remember who he or she is or recall details of his or her pre-trauma life? If our amnesiac can still perform basic life functions (eating, drinking, and the like), may we ever correctly say that he or she is something less than human? But what if the patient is unable, for example, to self-hydrate or receive nourishment without the use of a feeding tube but is in some other way(s) responsive to stimuli?

If, taking an even more extreme hypothetical, we have a patient who is comatose and has been so over a prolonged period of time, may we at some point consider this poor unfortunate a nonperson? My question assumes no legal qualifier. Thus: In the absence of a "living will" or some other judicially enforceable expression of a patient's wishes, may we justify withdrawal of life support on the reasoning that such a patient no longer has a human self?

Davies carries the inquiry into the hereafter. If the soul depends on the brain for memory storage, he asks, how can the soul remember anything after the death of the body? And if it can't remember anything, how can we attribute a post-mortem personal identity to it?[99] Or, Davies muses, are we to suppose that the soul (which Davies apparently equates to the self) has a sort of non-material back-up memory system that functions in parallel with the brain but can equally well cope on its own?[100] This, it seems, is as close as physicist Davies comes to concluding that there is an immortal soul on which the emblems of a person's selfhood are printed. Concerning Locke's criterion of consciousness, Davies notes that many philosophers have defined consciousness—the foundation of selfhood, as I see it—as the awareness of self, one's knowing that he or she knows, perceiving that he or she perceives.[101] Davies then goes so far as to state that the property of self-awareness "is holistic, and cannot be traced to specific electrochemical mechanisms in the

---

[99] Ibid., 91-92.

[100] Ibid., 92.

[101] Ibid.

brain;"[102] but he still appears unwilling to acknowledge that there is something nonmaterial about man that survives death.

*Hegel and Selah*

German philosopher Georg W. F. Hegel (1770-1831) grounded human identity in social activity: our consciousness of self arises from our contact with others, he said.[103] This theory is not inconsistent with Hegel's interactive view of history, in which any assertible proposition (the *thesis*) will necessarily be opposed by an equally assertible proposition (the *antithesis*), and the two opposing ideas will then be reconciled by a proposition that represents a higher level of truth (the *synthesis*), which in turn will become a new thesis that starts the dialectic process all over again.[104]

Hegel's theory of personhood is potentially problematic, however, in that it seems to assume that human consciousness itself is never fixed but instead is continually changing. As noted with respect to the problem of being able to know who your mate is "deep down," each new day's contact with others will add something to a person's store of identity-establishing experience. Is there nonetheless a core self that does not change as a result of human interaction? Or is a new-and-improved or perhaps not-so-improved self constantly being generated?

Also, what about those unfortunate individuals who are physically or mentally *unable* to interact socially? A case that took place in Rochester several summers ago evokes the reaction on my part that lack of interaction is not lack of identity. The facts of this case, which follow, provide a basis for further examination of the question of self, I think. My Rochester readers will recognize the case, as it received extensive coverage in the local media.[105]

Strong Memorial Hospital, which includes Golisano Children's Hospital, is located on the south side of Rochester near the Erie

---

[102] Ibid., 92-93.

[103] Rohmann, *Ideas*, 192.

[104] See Rohmann, *Ideas*, 171; also, Stokes, *Essential Thinkers*, 103.

[105] See, e.g., Nestor Ramos, "Lost in the Canal," *Rochester Democrat and Chronicle* (October 8, 2012), 1A, 9A.

Canal, a waterway that spans upstate New York from Buffalo to Albany and is much favored for recreation because of its accessibility to pleasure boaters and the paved walking/bike path that follows it for most of its length. As in many areas that have a children's hospital, there is a Ronald McDonald House adjacent to the hospital, where parents and other family members may stay while children are receiving treatment. The Ronald McDonald House that housed the family in this case is located a hundred yards or so from the Canal.

The Florida family in question was therefore familiar with the canal path, which they had used repeatedly as a setting in which they could unwind during trips to Rochester to obtain eye treatments for a child born blind. On a bright August day in 2012, the father, Jon, was on the canal path with the blind child, eight-year-old Sam, and another child, Selah,[106] a developmentally-delayed seven-year-old girl whom Jon and his wife Yvonne had adopted out of a Ukrainian mental institution. Sam and Selah were being pushed in a double stroller. At one point Jon stopped to check his cell phone. In order to shield the display from the sun, he momentarily turned away from the stroller, which then rolled into the canal with Sam and Selah strapped in it. Jon dove into the canal, was almost immediately assisted by other canal path walkers, but was unable to bring the stroller to the surface of the murky waters quickly enough to prevent Selah from swallowing a critical amount of water and suffering temporary stoppage of her heart.

As of the time of the October follow-up media reporting relative to the above incident,[107] local medical providers were saying that Selah still had no apparent mental function above the brain stem, although she could breathe on her own and her eye pupils were responsive to light. She was therefore not "brain dead" in medical terms, but was in what doctors term "a persistent vegetative state." In spite of medical diagnoses giving little if any hope

---

[106] The word "Selah," as used in the Psalms, signifies a pause for reflection (Henry M. Morris, *The Henry Morris Study Bible* [Green Forest, Ark.: Master Books, 1995, 2006, 2012], n. to Ps. 3:2). The subject family was said to be "deeply religious" (Ramos, "Lost," 1A).

[107] See Ramos, "Lost," 1A, 9A.

of recovery, Jon and Yvonne were stating clearly that they would not consent to the removal of Selah's feeding tube. To thus starve or dehydrate her to death would be for him and his wife to break the promise made by them when they adopted Selah, Jon explained.

To withdraw life support and allow nature to take its course in a case such as Selah's would constitute euthanasia of the passive variety. As I've already noted, part of the argument for euthanasia in cases of apparent permanent unconsciousness is that there's no longer any self; thus, you're not killing anyone. But if there's an afterlife, then we have an identity while we're living that afterlife, I must assume. Where did that identity come from? Obvious answer: it survived death. But the proponents of plug pulling in cases such as the above ask us to believe that personhood, the existence of a self, would have ceased before any decision for passive euthanasia based on supposed permanent unconsciousness, before any determination to allow starvation and dehydration to work their final solution. Thus, caregivers, if they profess belief in life after death, are asked to be a party to the inconsistency of saying that there's no longer any self in the here and now but there's still a self that passes into the next life. The atheist won't be troubled by this conceptual dissonance, seemingly, although he or she may experience anxiety regarding the prospect of being responsible for another's death. It's the relative or other concerned person who is attempting to rationalize euthanasia as a means of allowing the subject to pass into an afterlife who may find himself or herself on the horns of a logical dilemma if he or she analyzes the "no self" argument carefully. Or am I putting too fine a point on this?

The above analysis fully supports the position of Jon and Yvonne. These parents were clearly still seeing their daughter as a person, to whom they had a continuing caregiving obligation. It was not, as I see it, a matter of their simply being attached to an ideal (to be "faithful and true") rather than to an individual person. Nothing in the newspaper reporting suggests that. And seven-year-old Selah would presumably never have had a chance during her until-then relatively short life to meaningfully or legally express a wish that she not be kept alive artificially. A child of tender years is in my opinion incapable legally of directing that the plug be pulled under any circumstances, a position with which I think most other legal professionals would agree. It would be a different story with respect to an adult for whom euthanasia is proposed if an enforceable health care proxy or living will were in place, but I doubt that a court would authorize euthanasia of a child even on the basis of a document executed by both the child and the child's parents.

It should be noted that by December of the same year, according to a *Democrat and Chronicle* follow-up article, Selah had recovered sufficiently that she could be transferred to a hospital in Jacksonville with a plan to send her home several weeks after that. She was to continue with rehabilitation and physical therapy, according to her mother. A pediatrician at Golisano Children's Hospital was saying that Selah had shown "remarkable progress."[108] So much for pulling the plug.

*Determinism*

Back to the abstract. Is self the same as character? I'm tempted to say that self exists while character develops, or that self encloses while character disposes, but maybe it's a bit more complicated than that. Looking at the two concepts from a lawyer's perspective, we can see that the judicial system continues to punish acts that flow from what could be called a defect of character. But where does defective character come from? A hard determinist, someone who utterly refuses to recognize that actions can result from the exercise of free will, would argue that there should be no legal responsibility for any crime because the deed in question will always have been the product of a chain of cause and effect that could not have been otherwise.

At the beginning of the nineteenth century, French astronomer and mathematician Pierre Laplace (1749-1827) even went so far as to suggest that there must be *a set of scientific laws* that would allow us to predict everything that will ever happen, including all human behavior, if only we could know the complete state of the universe

---

[108] Victoria Freile, "Small steps home for daughter Selah," *Rochester Democrat and Chronicle* (December 12, 2012), 3B, 6B.

# ESSENCE

(what all its atoms are doing) at any given time. Presumably, the predictable everything would include how any particular person's character will develop—i.e., what that person will be like by the time the opportunity to commit a crime arises. Albert Einstein, a determinist, took a practical approach to the problem of criminal culpability. "(P)hilosophically a murderer is not responsible for his crime," he said, "but I prefer not to take tea with him."[109] With all due respect to Messrs. Laplace and Einstein, I can't accept the above world-view. It's counterintuitive as far as I'm concerned, and for now that's enough. I'm hopeful that a more intelligible response to exculpatory determinism will crystallize for me as I move forward with this work.

If character doesn't exist as a physical given, in the sense of being a caused arrangement of the atoms of the brain, where *does* it come from? Searching Scripture for instruction on this question, we find David acknowledging before the Lord:

> My substance was not hid from thee, when I was made in secret, and curiously wrought in the lower parts of the earth. Thine eyes did see my substance, yet being imperfect; and in thy book all my members were written, which in continuance were fashioned, when as yet there was none of them. (Psalm 139:15-16)

"Substance" is the key word here. God sees one's "substance" before the physical features of that person appear, before there is even an embryo ("my members when as yet there was none of them."). "Substance" precedes the physical features that are formed as first the embryo and then the fetus develops in the continuum of life.[110]

There is therefore a pre-existing self, in which the kernel of character, each person's essence, is contained. This is not necessarily the determiner of behavior. One can behave against character, and we do so whenever we act in a godly manner, whenever we follow the example set by Jesus, who is said to have lived a perfect

---

[109] Denis Brian, *Einstein: A Life* (New York: John Wiley & Sons, 1996), 185.

[110] See Morris, *Study Bible*, nn. to Ps. 139:15-16.

life (2 Corinthians 5:21; Hebrews 4:15). Man is at heart "deceitful and desperately wicked" (Jeremiah 17: 9), but that doesn't mean that he has to act in a way that reflects his depravity, nor does it make God the author of evil for having created him. As I've previously urged, God has given man the ability to choose good over evil; He has not left him in a state of spiritual helplessness as the Calvinists say He has.

Very well, you reply; God has given man the means to get out of his morass, but why would He create someone with a corruptible essence in the first place? C. S. Lewis's answer, given earlier to the question of why God would allow disobedience, is that allowing man free will is essential if man is to be able to give back to God a "love worth having."[111] The above fact, that God's obvious desire to have a relationship of *mutual* love with His children requires that His children be free to love Him or not love Him, belies any accusation that we're begging the question in positing human free will. We don't need to assume without evidence the existence of human free will when it's so easily inferred from the all-encompassing nature of God's love toward His creatures (see John 3:16). Love on the part of God the Father, so deep that it sent God the Son to the cross, would certainly be justification for the Father's expecting and making possible meaningful love from His children in return.

The freedom described above, which includes the ability to act against character, is what allows us to skirt what's been called the "Basic Argument" for free will impossibilism.[112] As I understand the Basic Argument, it runs as follows (my paraphrase): (1) Actions or decisions reflect the character of the actor, the way he or she is. (2) In order for me to be morally responsible for any action or decision on my part, I would have to be the cause of the way I am. (3) But I can't be the cause of the way I am, because I would have needed at some point to decide to be this way, and I would have had to make that decision on the basis of the character I had at that time, for which I would not have been responsible because actions or decisions invariably flow from how we are. As suggested in the

---

[111] Lewis, *Mere Christianity*, 47-48, 183.

[112] See, in this regard, Shaun Nichols, *Great Philosophical Debates: Free Will and Determinism* (DVD) (Chantilly, Va.: The Teaching Company, 2008), Lecture 10.

preceding paragraph, the effect of God's love, which He wants us to reciprocate freely, is to release us from the trap of the Basic Argument. God's nature as a heavenly Father desiring meaningful love from us trumps man's natural tendency to act like the "desperately wicked" creature that he is (Jeremiah 17:9)

Even more constricting than the straitjacket imposed by the Basic Argument, it seems to me, is the rigid notion of personal identity propounded by Gottfried Wilhelm von Leibniz (1646-1716), a German philosopher, writer, and mathematician of the seventeenth and eighteenth centuries. Leibniz rejected his fellow rationalist Spinoza's view that there's only one substance (God or Nature) and held instead that there is an infinity of individual substances that are like Pythagorean geometry points in that they're neither extended nor composed of any matter.[113]

The above-described entities, which Leibniz called "monads," were said by Leibniz to be irreducible, indestructible, independent concentrations of force each of which was a microcosm mirroring the universe as a whole.[114] Their "microcosm" property meant that the monads were self-contained and self-motivated. They were, as Leibniz put it, "windowless:" Everything that was true of a monad was contained *within* it, so that there was no need for it to enter into a cause-and-effect relationship with any other monad.[115] In every true proposition, Leibniz further explained, the predicate was contained within the subject, so that every truth was *necessarily* true and everything happened the way it did because it had to.[116]

The above rule of necessity carried over into Leibniz's cosmology. Since God could have created any logically consistent universe, the one He chose to make must have been the "best of all possible worlds," Leibniz reasoned.[117] He peopled that world with persons who could not have been other than who they were. Julius

---

[113] See Stokes, *Essential Thinkers*, 81.

[114] Rohmann, *Ideas*, 228.

[115] Stokes, *Essential Thinkers*, 81.

[116] Ibid.

[117] Rohmann, *Ideas*, 228-29.

Caesar could not have not been an emperor of Rome.[118] Adolph Hitler could not have not been a cruel fascist dictator. This would make God, as the creator of Leibniz's monads, the originator of evil as well as of good.

Applying to Jesus the idea of the predicate's being contained in the subject could in certain situations be seen as denying our Lord the autonomy one normally expects a member of the Trinity to have. An illustration of how this might work is presented in connection with the shortest sentence in the King James Bible, "Jesus wept" (John 11:35). As previously noted, Jesus was fully man and therefore presumably vulnerable emotionally (see John 1:14; also Hebrews 4:15). As someone who loved the recently deceased Lazarus and his sisters Mary and Martha (John 11:5), Jesus would therefore probably have found it difficult not to cry upon accompanying the sisters to Lazarus's burial place. In the present analysis, however, the Master's weeping does not need to be predicated on His humanity or His love for Lazarus and his sisters. Rather, it can be based on His being God the Son and therefore sharing in God the Father's nature as love itself (John 10:30; 14:9; 1 John 4:16). It is God's unconditional love that causes Him to grieve when ills befall His children, although circumstances may be the trigger (see Genesis 6:6). Under Leibniz's rule of necessity, Jesus could not have not wept. We therefore find the Creator of the universe trapped in an identity cocoon (an arresting if not somewhat disturbing thought).

Leibniz's conclusion that of necessity this is the "best of all possible worlds" was satirized by French philosopher and social critic Voltaire (Francois-Marie Arouet) (1694-1778) in the novel *Candide* (1759), which concludes with a statement by the (some would say) excessively optimistic Dr. Pangloss that if the chain of disasters that he recounts hadn't happened, Candide would not now be eating candied citrons and pistachios raised in his and Pangloss's garden. Somehow this doesn't quite equate to the statement of the apostle Paul, at Romans 8:28, that "all things work together for good to them that love God, to them who are called according to his purpose."

---

[118] Stokes, *Essential Thinkers*, 81.

## ESSENCE

*Pan(en)theism*

We're now into October. A chilly day today, with intermittent clouds and sun. I've chosen to stay inside and organize the 5-by-8-inch cards on which I summarize my research and impressions. Contrary to habit, I'm sitting at the landward end of our long dining room table, so that if I stare straight ahead, what I see through the window at the far end of the room is the lake. Indeed, nothing *but* the lake. Neither our front lawn nor the beach is visible from where I sit. Thus, even though our cottage is on ground that's five or six feet above the water level, I have the feeling that the waves are lapping against our foundation. I would even have the sensation that our cottage is floating on the water, I'm sure, were it not for my next-door neighbor's massive cottage, which looms in the corner of my eye (it's visible through another window, a few feet to my right) and acts as an anchor for my perception. The separation between The Cottage and the lake disappears when I stare straight ahead, however.

My disorientation at this moment is not unlike what I've experienced on the Maid of the Mist. The Maid of the Mist is a craft that carries passengers up to within two hundred feet or so of the foot of the Horseshoe Falls at Niagara. When the boat begins to get close to the cataract, you can feel (but can scarcely hear because of the roar of the falls) the captain gunning the engine to keep the Maid from getting swept downstream. At the point of nearest approach, the waters churn all around the boat, in swirling vortexes that seem somehow at odds with the water falling from above, which is taking the straight downward path dictated for it by gravity. Which is the ruling principle here, you wonder—fluid dynamics or gravity? Whichever it is, you've become one with a great mass of water, as with the surf seemingly lapping at The Cottage's foundation, or so it seems.

The above descriptions, of ordinarily discrete things forcefully running together, are a perfect segue to talking about pantheism and panentheism. *The Encyclopedia of Religion and Ethics* defines *pantheism* etymologically as "the view that all is God, and that God is all."[119] *Panentheism*, looking strictly at the Greek roots, means simply, "all is in God."[120] The difference is inclusiveness: under *absolute pantheism*, there is no part of God that's outside the universe, while with panentheism we find all the world in God but a part of God outside of time and space. The world, or universe, includes all of humankind; thus, we're part of God and He's part of us no matter which of these two systems we're under.

Only it isn't quite as simple as the above description might make it seem. Absolute pantheism, in which God and nature are identical and the universe is merely the way God appears, is but one of several varieties of pantheism. *Modal pantheism* sees God and the universe as one infinite substance and views all things in the universe as finite modes of that substance. By way of contrast, God is merely part of the universe in *dynamic pantheism* but is its immanent (within space and time) force or "soul." God nonetheless regains His full magnitude in *emanational pantheism* and *panentheism* (the latter is sometimes defined as a type of pantheism), in both of which the universe is merely part of Him and He's transcendent (outside space and time). As mentioned earlier, panentheism, at least in the form presented by Alfred North Whitehead, views the universe as the finite, developing part of God's bipolar nature.[121]

Modal pantheism is what we see in the philosophy of Baruch Spinoza (1632-77), whose single substance comprising the world is said by him to be neither physical nor mental (Spinoza's way, perhaps, of avoiding entanglement in the longstanding materialism/idealism debate). The substance of all things, God or Nature, has infinitely many attributes, Spinoza says, but human beings, because they're finite, can only perceive two of them, *extension* (matter) and *thought* (mind).[122] Unlike seventeenth-century philosopher/mathematician Rene Descartes (1596-1650), who viewed mind

---

[119] A. E. Garvie, Pantheism article author, *The Encyclopedia of Religion and Ethics* (New York: Charles Scribner's Sons, 1917), 609.

[120] Hartshorne, Pantheism and Panentheism, *Encyclopedia of Religion*, 165 (emphasis supplied).

[121] The above distinctions are found in Prof. Cooper's pantheism article, *Encyclopedia Americana* vol. 21, at 363.

[122] Stokes, *Essential Thinkers*, 79.

# ESSENCE

and body as separate and distinct, Spinoza argues that mind and body are just different ways of conceiving or expressing the same reality.[123]

Since everything in the universe is part of God, everything that happens is an expression of the divine nature, in Spinoza's view.[124] This has two effects. First, it removes any element of free will from human thoughts and actions, since it's God who is thinking and acting. Second, it means that there's no such thing as sin or evil. That which seems to be evil looks that way only because human beings lack the perspective of the bigger picture, of the chain of causation that makes all events a part of God as Spinoza conceives Him and therefore necessarily good.[125]

Spinoza further holds that since all individual things—objects, creatures, ideas—are nothing more than transient modes of substance, there can be no personal immortality.[126] Death is the passing away of a transient mode; only substance in its totality is eternal.[127] Stating Spinoza's underlying assumption another way, each individual is a temporary localized concentration of the attributes of reality, really a quasi-individual, since the only person or thing that has reality is the universe *in toto*.[128]

What's the use, then, of doing what we think of as living? To become free, Spinoza answers.[129] To become free, a person must understand by rational reflection the extended causal chain that links everything together.[130] This still doesn't get you out from under Spinoza's determinism; it just frees you from ignorance of your true nature.[131] My response: What then? You still don't exist as an autonomous being. You're right where the Buddha left you.

---

[123] Ibid.

[124] Ibid.

[125] Ibid.

[126] Rohmann, *Ideas*, 375.

[127] Ibid.

[128] Stokes, *Essential Thinkers*, 79.

[129] Ibid.

[130] Ibid.

[131] Ibid.

I guess Spinoza never heard the Christian version of the story: our *names* (emblems of self) are written for eternity in the Lamb's book of life (Revelation 13:8).

For me, pantheistic thinking such as that outlined above is not a satisfying answer to the question of what one's self is. Nothing is gained, either, by switching one's thinking from absolute pantheism as a world view to the seemingly less radical panentheism. There's still no evil (since everything that happens is of God) and still no free will (since God rather than the individual is doing the doing). If there's no freedom of choice, how can anyone actually choose Jesus as savior? And yet God's written Word directs us, not to passively accept a predetermined fate, but to believe in Him as a condition for inheriting eternal life (see, e.g., John 3:16; 14:1; Acts 2:21). And if there's no sin the penalty and practice of which you need to be saved from, why would you need to trust Jesus for anything? The purpose of Christ's coming, we Bible believers vigorously assert, was not simply so that He could be in the world as a good moral example, but also to free mankind from the consequences of sin (see Romans 6:23; 2 Corinthians 5:20-21; Colossians 1:12, 14; 1 Timothy 1:15). Grace first, guidance second.

There's also the small matter of cosmology. The pantheist, of necessity, would make the universe beginningless. If the universe didn't always exist, he says, then God didn't always exist. By the same token, if the universe ends, God ends. In either instance, God is not God by any sensible reckoning. To maintain his position, the pantheist must deny not only the Bible (Genesis 1:1, for starters), but also modern science (the "big bang" and the second law of thermodynamics, the latter of which we know for certain to be true).

By saying that God and the universe are the same or partially the same, the pantheist and panentheist in addition make God a blurry everythingness rather than a person with thoughts and attributes. Having taken away God's personhood, the pan(en)theist would also take away man's. "Made in the image of"—the universe? Compare Genesis 1:27 ("So God created man in his own image,"). Thus, pan(en)theism imposes a concept of personality that does not amount to selfhood by any biblical definition but is more like the Buddhist "no-self."

# ESSENCE

Do pantheism and panentheism share anything with *animism*? Animism is the belief that all things have within them a soul or spirit. It can take a number of forms, to include believing in a "world soul" pervading creation (a core New England transcendentalism tenet), attributing consciousness to natural phenomena such as wind and rain, assigning human characteristics to animals and objects (anthropomorphizing), and believing that a soul is what makes animals and plants alive.[132] Clearly, the pans and animism occupy some common ground. In both pan(en)theism and animism, you *worship* nature. Thus, nature, even inanimate nature, must be thought to have something spiritual about it.

It's not always easy, however, to make the distinction between giving attention to nature as an object of worship and allowing God to speak to you *through* nature. God may speak to you *through* nature based on nothing more than well-formed metaphors that your natural surroundings evoke. When your encounter with nature is said to involve mystical experience, it is especially difficult to avoid pan(en)theism, since, as William James has suggested, true mystic states involve our becoming one with the One and becoming aware of our oneness with the One.[133]

Worth noting in the above regard is a statement that C. S. Lewis has offered which might be viewed as ambivalent on the question of whether stones and trees have soul or spirit while at the same time could be seen as a rejection of pan(en)theism (everything is God) or of monism (everything is one substance). Analogizing the relationship of the universe's Director and the component parts of the cosmos to a system of mail delivery through which the Director gives instructions by letter to everyone and everything, Lewis says:

> The only packet I am allowed to open is Man. When I do, especially when I open that particular man called Myself, I find that I do not exist on my own, that I am under a law; that somebody or something wants me to behave in a certain way. I do not, of

---

[132] Rohmann, *Ideas*, 19-20.

[133] William James, *The Varieties of Religious Experience* (New York: Barnes & Noble Classics, 2004; orig. publ. 1902), 363.

course, think that if I could get inside a stone or a tree I should find exactly the same thing, just as I do not think all the other people in the street get the same letters as I do.[134]

A possible interpretation of this statement is simply that we human beings only know what we know about God and His wishes for the universe in terms of our own experiences and consciousness, and can hope for no broader knowledge than that. This aside, one is still left wondering whether in Lewis's world-view there is anything animate about the inanimate.

*Existentialism*

There is one thing a person seeking to know more about selfhood may be ready to accept even now: the assumption that you can't be the author of your essence. If "essence" is defined by Christianity as the "you" that was known to God since before the day of your conception (Psalm 139:1-4, 13-16), then on that count alone Christianity is in conflict not only with openness theology (God finds out about you as your life progresses), but also with existentialism.

The existentialists were a group of French and German philosophers mainly of the twentieth century, whose system was anticipated nearly a century earlier by the writings of a Dane who was at odds with his Lutheran state church, Soren Kierkegaard (1813-55).[135] The salient features of existentialism were (1) the inference that the universe is absurd, without apparent meaning or purpose, (2) atheism in the version propounded by the Frenchman Jean-Paul Sartre (1905-80), as well as his circle of friends, (3) the conclusion that "authenticity," the individual conducting himself solely as he sees fit, answering to no one but himself, is the only possible response to the world's absurdity, (4) accordingly, freedom of choice in all matters, including personal morality, (5) dread of nothingness, which is said to await all persons at the end of their

---

[134] Lewis, *Mere Christianity*, 25.

[135] See Stokes, *Essential Thinkers*, 145.

earthly lives, and (6) most importantly for the present discussion, personal essence resulting from the choices made by the individual rather than its being imparted by God or coming from any other external source.[136]

The name "existentialism" was attached to the aforesaid movement by the German philosopher Karl Jaspers in response to Sartre's proclamation that existence precedes essence.[137] In existentialist thinking, you first exist—i.e., have a place in time and space—and only after that is your personal identity created, by you, based on your actions (you *are* what you *do*). The "existence first" dictum was illustrated in the teaching of the longtime companion of Jean-Paul Sartre, feminist icon Simone de Beauvoir, who famously stated, "One is not born, but rather becomes, a woman."[138]

One of the writers most frequently associated with the existentialism movement was French-Algerian author/playwright/philosopher Albert Camus (1913-60). The main contribution of this thinker to existentialism—which movement he denied having any part in—consisted of his development of the theme of absurdity. In his essay "The Myth of Sisyphus" (1942), Camus attempted to answer the question, how can you find meaning in a meaningless universe?[139] Sisyphus was a character in Greek legend who was condemned for eternity to have to push a large stone up a hill only to see it roll back down, causing him to again have to labor up the hill with the stone, a cycle repeated endlessly.

The understanding of Camus regarding the above situation was that it showed that work is futile; in the end we will have nothing to show for our labors. Indeed, Camus argued, anything we achieve that happens to last, that at least survives us as individuals, will go up in smoke when the solar system comes to an end.

---

[136] See Rohmann, *Ideas*, 127-28, 172-73, 222-23, 280-81, 352; also Stokes, *Essential Thinkers*, 144, 146-47, 150-51, 153, 155, 157, 212.

[137] See Jean-Paul Sartre, *Being and Nothingness: A Phenomenological Essay of Ontology*, transl. by Hazel E. Barnes, English Eds. 1956, 1984 (New York, Washington Square Press), 603; see also ibid., 45, 69, 72, 802.

[138] Simone de Beauvoir, *The Second Sex*, transl. by Constance Borde and Sheila Malovany-Chevallier (New York: Alfred A. Knopf, 2010), 283.

[139] See Rohmann, *Ideas*, 128; also Stokes, *Essential Thinkers*, 155; and Stephens, *Philosopher's Notebook*, 154.

Is life worth living under such circumstances? he asked. It has to be *made* so, he reasoned, because death is unacceptable, a capitulation to the absurdity of life. Thus, suicide is ruled out, Camus concluded. Sisyphus's salvation therefore became that of continuing to commit himself to his task, regardless of how fruitless this might be. Resisting one's crushing fate is what gives a person identity, Camus concluded.[140]

The commitment strategy backfired on German philosopher Martin Heidegger (1889-1976), a contemporary of Sartre and Camus. Heidegger defined "commitment" as attachment to one's national culture; thus, he chose to support Hitler and nazism during the 1930s, a decision that did lasting damage to his reputation.[141] As a result, he spent the last few years of his life in self-imposed exile in the Black Forest.

Talk of commitment evokes the case of Kierkegaard, the sole Christian in the existentialist pantheon, who insisted that religious belief and duty can only be the product of a "leap of faith," an example of such a "leap" being Abraham's willingness to sacrifice his son Isaac based on his belief that this was the will of God (Genesis 22:1-13). True faith, Kierkegaard said, has nothing to do with rationalistic proofs of the existence of God offered by "pencil pushers."[142] Instead of being preoccupied with metaphysical schemes, he urged, adherents to Christianity should simply and passionately believe. He explained:

> *An objective uncertainty, held fast through appropriation with the most passionate inwardness, is the truth*, the highest truth there is for an *existing* person (emphasis Kierkegaard's).
>
> But the definition of truth stated above is a paraphrasing of faith. Without risk, no faith. If I am able to apprehend God objectively, I do not have

---

[140] See ibid.

[141] See Stokes, *Essential Thinkers*, 150-51.

[142] See "Concluding Unscientific Postscript" (1946), essay in Howard V. Hong and Edna H. Hong, eds., *The Essential Kierkegaard* (Princeton, N.J.: Princeton Univ. Press, 2000), 204; also Stokes, *Essential Thinkers*, 145.

faith; but because I cannot do this, I *must* have faith (emphasis supplied).[143]

The above is consistent with the New Testament letter to the Hebrews, which declares that faith "is the substance of things hoped for, the evidence of things not seen" (Hebrews 11:1). The author of the epistle to the Hebrews continues, "But without faith it is impossible to please [God]; for he that cometh to God must believe that he is a rewarder of them that diligently seek him" (Hebrews 11:6). Like Kierkegaard, the author of Hebrews sees saving faith as subjective in nature.

*Whence Cometh Life*

As stated above, existentialism is atheistic in its twentieth-century version. The nonbeliever, having rejected the idea that life requires life as a source, must answer the question of how human or other living essence could be created by an impersonal process. The problem goes beyond the reach of the argument for intelligent design (ID), which is essentially an assertion that the odds are overwhelmingly against a universe or region thereof that's physically able to support intelligent life and that therefore the cosmos must reflect the creative work of a being with a mind.

Materialists have sought to nullify ID by theorizing an infinity of universes (thus, it's said, under the law of averages the combination of physical factors needed for life had to have happened somewhere). They've also referred to the anthropic principle (AP), the "weak" version of which states that the constants of nature must be tuned to allow for intelligent life, so as to give us sentient beings a "goldilocks zone" in which to live; otherwise, there wouldn't be anyone here to talk about the anthropic or any other principle.[144] We're here, the "weak" AP asserts, not necessarily because God or

---

[143] Hong and Hong, eds., *Essential Kierkegaard*, 207.

[144] See Michio Kaku, *Parallel Worlds: A Journey Through Creation, Higher Dimensions, and the Future of the Cosmos* (New York: Random House, 2005), 241-49; Hawking, *A Brief History*, 128-29; Rohmann, *Ideas*, 21; Davies, *God and the New Physics*, 173-74.

some other intelligent creative force put us here, but because consciousness doesn't give us any choice but to be here.

Hearing the AP stated as above reminds me of the cheerleader-led singing at the end of our bus rides to and from high school basketball games, to the tune of *Auld Lang Syne:* "We're here because we're here because we're here because we're here. We're here because we're here because we're here because we're here."

I find the "infinite universes" argument amazing in a perverse kind of way: men are willing to believe in an infinity of universes but not in an infinite God. Moreover, to be able to say that there are or have been universes beyond numbering, we must first theorize the existence of matter sufficient to constitute them, and yet multiple-universes advocates have not told us where such building material might have come from.

Nor do we derive any help, with respect to the enigma of life's origin, from the theory of biological evolution, which presupposes life in some form—perhaps at the bacterial level—and then has living things developing greater and greater complexity through the process of natural selection (survival of those species that have the genetic characteristics most compatible with survival), until we arrive at Albert Einstein and Ray Charles. How could entry-level species have existed so as to provide a starting point for evolution? Science has never explained this.

Science's inability to explain how life arose has not been for lack of trying. In 1953, Stanley Miller and Harold Urey of the University of Chicago attempted to simulate the conditions that were believed to have prevailed on the primeval earth—an atmosphere of methane, ammonia, and hydrogen, combined with a pool of water and a thunderstorm (the latter mimicked by an electrical discharge). A few days after their perfect storm, the experimenters found that the water had turned a red color and contained some of the compounds found in living things, including amino acids, but nothing more.[145]

Paul Davies speculates that a primeval "soup" like that concocted by Miller and Urey could possibly have been driven into a

---

[145] Davies, *God and the New Physics*, 68-69.

sequence of self-organizing reactions by some external influence such as the sun, but cautions that

> it would be wrong to suppose that we have anything like an understanding of the steps intermediate between the Miller-Urey experiment and full-blown replicating molecules. *The origin of life remains a mystery*, and contentious even among scientists (emphasis mine).[146]

In more recent times, members of the intellectual community have not been content merely to speculate on the processes that might have led to life after the supposed zapping of the primeval soup; they have gone farther and made it clear that they are unwilling to leave the door open for any thought that life is a holistic concept and therefore cannot be understood solely in terms of electrochemical reactions in the human brain and body. They have chosen instead to define life in strictly material terms. Thus, led by "Four Horsemen" Richard Dawkins, Christopher Hitchens (now deceased), Sam Harris, and Daniel Dennett, the "New Atheists" have stridently urged that there is no God, there is only the natural, and have proceeded on the assumption that it is their duty to disabuse the population at large of the notion that God exists.[147]

The New Atheists rationalize their fervor in preaching an anti-gospel with the assertion that religions such as Christianity "tyrannize" their adherents. It's said that the offending religions deny believers personal freedom not only by imposing doctrine, but also through communicated expectations that the faithful will obey certain rules of conduct—no extramarital sex, no same-sex contact, and abstinence or restraint as regards alcohol, for example, complaints that the atheist wing of existentialism also appears to have

---

[146] Ibid.,69.

[147] See, generally, Dave Hunt, *Cosmos, Creator, and Human Destiny: Answering Darwin, Dawkins, and the New Atheists* (Bend., Ore.: The Berean Call, 2010). Regarding the passing of Christopher Hitchens—salvation status unknown—and the journey of his brother Peter to faith, see Sarah Palin, *Good Tidings and Great Joy: Protecting the Heart of Christmas* (New York: HarperCollins, 2013), 46-47.

had against religion. Apparently seeking to prevail by nomenclature, the New Atheists refer to themselves as "the brights," implying that those of us who believe in God are dimwits.[148]

Although it may be something of a detour from examining the question of how life in general arose, I feel compelled to respond to the New Atheists' contention that religion "tyrannize(s)." What the New Atheists fail to recognize or acknowledge is that their own personal freedom, also, may well be circumscribed. Lusts of the flesh assert their claims to the soul of the unbeliever, it may be assumed, just as happens with believers.

On a more subtle level, the emotional need for peer acceptance likely prevents the individual "bright" from doing or saying anything that could be seen as evidencing disloyalty to the pack. The practicing atheist is thus under a pressure to conform that is every bit as real as, if not more than, that which is said to be experienced by a serious religious believer. The difference is that, at least where the religion at issue is Bible-based Christianity, the God-believer's mandate to submit to authority comes ultimately from a benevolent deity who, if the witness of the Word is believed, demonstrably has the best interest of the believer at heart. The atheist, on the other hand, having rejected God's providence, has no one to look out for him or her.

Since not wanting to submit to the authority of religion is an important motivator for atheists, it would seem disingenuous for them to claim, as they often do, that anti-Christian fervency on their part is born simply out of a desire to prevent the rest of mankind from being deceived regarding the existence of God.[149] As so framed, atheist militancy is made to look like compassion. To thus characterize New Atheist behavior would in most cases be to attempt to create a false impression, however, since the bent of the iconoclast typically is to destroy the faith of the believer without providing anything in its place. I can't have a savior, the militant atheist seems to be saying, so I don't want you to have one. Having wrought destruction such as that implied above, the New Atheist

---

[148] Hunt, *Cosmos, Creator*, 9, 36.

[149] See, in this regard, Ben Stein's interview of Richard Dawkins in *Expelled: No Intelligence Allowed* (DVD) (Universal City, Calif.: Premise Media Corp., 2008).

fails even to offer a cohesive scientific explanation for the life that is said not to have come from God.

Where do the New Atheists say that life came from? Ultimately from the "big bang," comes the answer, but we can't tell you much more than that. The big bang theory, still the cornerstone of secularist cosmology but under increasing attack, states that both space and time began with the explosion of an infinitely dense point of matter, which is said to have set in motion a chain of events that ultimately led to the appearance of life.[150] The details of exactly how all this happened, to include what set off the big bang, remain obscure notwithstanding the success of recent attempts to find the Higgs boson, a subatomic particle that supposedly could only have been created if the big bang theory were true.[151]

Researcher-lecturer the late Dave Hunt observed that atheist embrace of the big bang hypothesis has a motivation more tactical than technical. In his 2010 book *Cosmos, Creator, and Human Destiny*, he stated:

> The so-called Big Bang is the atheist's last hope to explain everything without God. It is also the *sine qua non* of theistic evolution.[152] For both atheistic and theistic evolution, the Big Bang, with its lengthy cooling-down period, provides a rationale for the billions of years needed for life to have arisen out of the completely sterilized (by the extreme heat produced by the "big bang") universe so that evolution and natural selection could begin and eventually produce the forms of life we have today. Hoping to hide behind billions of years,

---

[150] For a summary of the reasons for not believing the big bang theory, see the "Big Bomb" section of my Paradigm Paralysis chapter, infra.

[151] See James Goodman, "Big Bang boon," *Rochester Democrat and Chronicle* (July 3, 2012), 1A, 6A; also the Associated Press, "Physicists exult over particle discovery," *Rochester Democrat and Chronicle* (July 5, 2012), 3A; and John Heilprin, "Physicists say they have found a 'God particle,'" *Rochester Democrat and Chronicle* (March 15, 2013), 1A, 6A.

[152] "Theistic evolution," a doctrine held by some religions, including Roman Catholicism, assumes that God used evolution to produce humanity (my footnote).

evolutionists ignore the law of biogenesis (that life can only come from life).[153]

"Billions of years," as a foundation for biological evolution and atheism (if you wait long enough for life and then more advanced forms of life to appear on their own, they will come) has been challenged by a number of scientists and writers, professionals with far more expertise in this area than I possess. You'll meet some of these people as you make your way through this book. As things stand now, I'm hoping to get into the "young earth" literature more in depth over the coming winter, when I'm not focused as I am at the moment on getting a different season's version of Mother Nature to speak to me and do so directly. Research into Nature's workings goes on for me all year long; it simply moves indoors and becomes more bookish when the weather turns cold.

"Billions of years," even if somehow shown, would not allow evolutionists and atheists to ignore the failure of the Miller-Urey experiment and of similar efforts to craft a materialistic explanation of life.[154] Nonetheless, these groups, led by the New Atheists, continue to maintain that somehow the right particles of matter came together in the right environment and produced the electrochemical reaction that materialists see life as. From there it was "onward and upward," they theorize, until there was human consciousness. It's commonly agreed in the philosophy and science communities that what distinguishes human cognizance is the awareness of self (animals react; humans are aware they're reacting). For atheists and evolutionists, however, human consciousness is nothing more than a highly developed animal sentience, possessed by "an ape that has graduated" (my phraseology).

How was the gap between animal awareness (one-dimensional) and human awareness (two-dimensional) bridged? This is not explained by the New Atheists, any more than is the leap from inanimate matter to an animate state. The New Atheists are confident, nonetheless, that human consciousness—indeed, the

---

[153] Hunt, *Cosmos, Creator*, 224 (parentheticals mine).

[154] See, in this regard, Mark Eastman, *Creation by Design* (Costa Mesa, Calif.: The Word for Today, 1996), 9-21.

human self—is strictly a material phenomenon, which happened and continues to happen on its own. An immaterial soul, a part of the human essence that transcends the life of the physical body, is therefore an impossibility, they say.

The atheist/materialist view that the world and its inhabitants are self-existent also raises the question, what was the source of the physical laws that would have applied when the interaction of electrical charges and chemicals that Miller and Urey attempted to duplicate was happening on the earth? Physical events, even spontaneous ones, require physical laws that allow them to happen, do they not? Also, one may ask, where did the impetus for the alleged life development process, the nudge that got things started, come from?

To the first question above, the New Atheists answer that there is and always has been energy—self-existent, self-organizing energy, which, they say, is the principle behind all things.[155] Energy provides the laws, in other words. If energy is without beginning or end, however, this violates the second law of thermodynamics, which states that the amount of energy available to do work in the universe is continually decreasing. Also, while matter is convertible to energy and energy to matter, this presupposes that one or the other is already present. The New Atheists contend that their god energy was present before there was any matter, but they don't tell us how it got there.

For their response to the second question above, the no-God apologists cling to the theory of abiogenesis, the idea that living things can arise spontaneously from inorganic matter, which is the opposite of biogenesis (life only from life). Abiogenesis has been discredited in the scientific community generally.[156] If the New Atheists' theory of life fails, clearly so does their materialistic view of self, which leaves them nothing to turn to but a vague holism or (we may hope) a belief in the supernatural. Quaere whether their materialism would leave room for holism in any form.

---

[155] See, e.g., Hunt, *Cosmos, Creator*, 114, 144, 146, 357.

[156] Ibid., 106-07, 112.

## Dogs and Computers

Life implies sentience, which in turn implies a consciousness upon which facts or feelings may be impressed.[157] French philosopher and mathematician Rene Descartes had some thoughts on sentience that I would suspect are shared by many persons today and that continue to be a subject of debate even now. I think of Descartes' "animal-machine" doctrine as I watch a golden retriever lope down the beach. Campbell's Pointers love dogs, unless they don't own one. One of the prerogatives of private ownership of a recreational area such as the Point is being able to have your pet with you. For Descartes, mind and matter were separate varieties of substance, "mind" being described by him as that which can think and reason, and "matter" as that which simply occupies space. Animals fell into the latter category in Descartes' thinking, as purely mechanical objects incapable of any kind of thought.[158] Descartes would thus have rejected the position of modern animal-rights advocates that animals, at least higher-order ones, are entitled to certain basic rights by virtue of being sentient, social creatures who are able to feel emotion as well as physical pain.[159]

If animals are in the game, why not computers? Can a machine be intelligent? Can it have "consciousness" or "mind"? In the 1930s, English mathematician Alan Turing proposed a simple test, which is summarized by scholar Philip Stokes in *Philosophy: 100 Essential Thinkers* (2002), for determining whether computers can think.[160] The so-called "imitation game" suggested by Turing requires three players. As initially explained to the participants, Player A's role is that of an interrogator whose mission is to guess the sexes of the other two players in the game, one of whom is a man, the other a woman. The three players, as Turing set up the game, would be in separate rooms and send and receive questions and answers by teletype. Player B's assignment would be to answer

---

[157] See definitions of "sentient" and "conscious," *The Merriam-Webster Thesaurus* (Springfield, Mass.: Merriam-Webster Inc., 2005), 120, 538.

[158] See Rohmann, *Ideas*, 19, 99.

[159] See ibid., 19.

[160] See Stokes, *Essential Thinkers*, 199.

questions in such a way as to keep the interrogator from correctly guessing his or her gender, in effect to imitate the opposite gender. Player C, on the other hand, would be instructed to help the interrogator guess his or her gender correctly. The interrogator would not know at the outset which player is trying to help him and which is trying to deceive him.

A technological twist was then secretly added by Turing to his hypothetical game: a machine (not yet built but theorized as being capable of obstructive human-like responses) was now to take the place of Player B. The question would thus become that of whether the interrogator could correctly guess who was trying to help him and who was trying to deceive him more or less often than he would have if Player B and Player C were both human. In other words, could the machine fool the interrogator to the same degree that a person could? The answer to this question, Turing maintained, would resolve the issue of whether a machine can think, the assumption being that any machine sophisticated enough to effectively replace a person in a game such as the one described above must possess the same intelligence as a human being.

As the cited *Essential Thinkers* article points out, Turing's game raises some issues that may be of interest to philosophers. In particular, is the ability to imitate (in this case, to mimic someone of a sexual identity the imitator doesn't have) a sufficient basis for concluding that the imitator can think? A child can imitate the behavior of an adult, but that doesn't make him or her an adult; neither is original-version Player B, if he is a man and fools the interrogator into believing he is a woman, thereby made a woman. So why should we suppose that a computer, which merely imitates the behavior of a thinking person, is really thinking?

A chimpanzee, also, can imitate. But that doesn't make him able to think, any more than the dog who mindlessly barks at everyone who walks past his property. In both cases, it's instinct rather than intuition or intelligent understanding that's at work, as best I can judge the matter. But perhaps I'm just trying to nail down the difference between simians and us humans so as to negate the evolution thesis. I'll admit that theological politics can enter into a discussion such as this one.

As of the time of this writing, no machine has been able to pass the Turing test, at least none that this author is aware of.[161] Should such a supercomputer be constructed or discovered, this would direct our attention once again to the anthropic principle, that old bugaboo for atheists. The "strong" version of the anthropic principle holds that the physical constants of our universe, which allow for life and intelligence, are so finely tuned that they could only reflect the existence of a multiplicity of universes or be the work of an intelligent designer of some kind. The "weak" AP version, mentioned earlier, acknowledges the fine tuning but leaves open the question of who or what is responsible.

Whichever AP version one entertains, there remains the bothersome reality that one day our sun will die and with it the earth, and that it is unlikely that by that time mankind will have been able to find suitable alternative living arrangements. Bible-believing Christians don't see the above as a problem, since they hold with certainty that long before any natural end of the solar system the Lord will return to Earth and set in motion the chain of events prophesied in Scripture, which will culminate in "a new heaven and a new earth" as well as the final triumph of righteousness (Isaiah 11:1-16; 65:17-25; Daniel 12:1-3; Matthew 24:4-31; 25:31-46; Luke 1:31-33; Acts 2:29-30; 1 Corinthians 15:20-28, 51-55; 2 Corinthians 5:10; Titus 2:13; 2 Peter 3:10; Revelation 19:11-21; 20:1-15; 21:1-27; 22:1-17).

Anticipating the end of usable energy, secular thinkers have written man but not intelligence out of the script. The "anthropic principle" article in Chris Rohmann's *A World of Ideas* notes:

> Some scientists have proposed an even stronger (than the "strong" AP) principle, known as the *final anthropic principle,* which states that once intelligent life (or intelligent information processing) has come into existence, it will never pass away. Since the sun will eventually die, and with it all organic terrestrial life, this principle implies that "strong artificial intelligence"—intelligent, non-carbon-based,

---

[161] See, in this regard, Rohmann, *Ideas*, 28-29.

self-reproducing entities—will be developed before
then (first parenthetical added).[162]

The certainty of the drying up of the "goldilocks zone" does more than just give occasion for the thoughts of mortal man to turn to the possibility of eternity with God in another realm. The temporariness of the coalition of "happy accidents" necessary to sustain our life here means that there will come a time when there will be no human observer to give the universe existence by perceiving it. Under Bishop George Berkeley's *esse est percipi* ("to be is to be perceived") and other solipsistic systems, it would then be necessary for God to exist, at least as an observer, in order to allow for the existence of the universe in any form. Artificial intelligence, created by man and therefore inferior to him ontologically, would not qualify as an observer any more than it would qualify as human.

*A Philosophical Recap*

At this point, fellow traveler, as we continue to think about thinking, it might be helpful for us to capsulize the major "ism(s)" of philosophy over the centuries. This is, after all, a book written with the objective of giving Christians enough information that they'll be confident in their faith and maybe even share it with "educated" persons. Therefore a little additional discussion concerning some of the things that we've talked about so far would probably be helpful.

*Empiricism*, in its various forms, holds that all knowledge derives ultimately from experience—i.e., from direct observation of physical things and analysis of one's observations. The faith of the early Christians was empirically-based. In his first epistle, the apostle John proclaims to his readers "(t)hat which we have heard, have seen with our eyes, have looked upon, and our hands have handled, of the Word of life" (1 John 1:1). He refers to the incarnate God the Son[163] and to the events detailed in the eyewitness accounts contained in the gospels of Matthew, Mark, and Luke, as well as in his own gospel.

---

[162] Ibid., 21.

[163] Cf., with regard to Christ as the Word, John 1:1, 4.

Empiricism stands in contrast to *rationalism*, which says that reason, consisting of your either (1) directly grasping first principles or (2) developing an understanding of things through logical argument, is a more dependable path to knowledge than experience or observation. With both empiricism and rationalism, the focus is *epistemological*, on the "how" of knowing, how you know that something is true. *Materialism*, which declares that matter is the only thing that exists—i.e., that everything, including human thought, is exclusively physical in nature—is *ontological* in its emphasis, meaning that it focuses on the nature of existence or being as such.

Materialism, it seems to me, is at the heart of some of the problems people have with relationships in our postmodern world. Romantic love is seen as created by looks and line and maintained by an assemblage of sex tricks such as those detailed in *Cosmopolitan* magazine. God's wishes and one's personal character are not part of the equation. You never get beneath the surface of your partner, therefore, and because of aging and other factors the surface changes. You wind up, as a result, with a mate that you don't know either on the outside or on the inside.

Materialism disagrees ontologically with both *dualism*, which distinguishes between mind and matter, and *idealism*, which sees reality as fundamentally mental or spiritual in nature. For the materialist, mind is just an aspect of matter, as previously noted. There is no world beyond the sensory world. The apparent-to-the-senses world is genuine and the "world beyond" is a delusion. With both dualism and idealism, there's room to rise in one's thinking to the contemplation of God and an afterlife, since each allows for a human self that may be eternal.

With *panentheism* and *pantheism*, there's no upward thought path to follow, since you're either part of God or co-extensive with Him. You can't hike up to the top of the state park and look down on the harbor, since you *are* the harbor. *Monism* has the same effect in a backwards sort of way (there's only one substance, God or Nature and we're just one of the forms that substance takes). *Solipsism*, which announces that one's self is all that can be definitely known to exist—reality is all in your head, you might say—further complicates the business of thinking about the self and its eternal destination.

*Transcendentalism*

The New England transcendentalists, a nineteenth-century phenomenon, were pan-something, it's clear. "Transcendentalist" is probably a misnomer as applied to members of this group, since, purely speaking, transcendentalism is the philosophical belief in a reality that can't be known through everyday experience or normal reason. The ultimate reality is "out there" and not easily grasped. The New England transcendentalists believed not only that they could grasp but also that they *were* the animating principle of the universe. They enjoyed, they thought, a oneness with God and nature and the rest of mankind through the "Oversoul." Their self-appointed leader, essayist-lecturer-poet Ralph Waldo Emerson, described the "Oversoul" in terms that suggest an amorphous entity something like Spinoza's monist substance. It was, Emerson said, the "divine spirit or mind" that was present in each and every man and in all of nature.[164] In some of his less abstruse writing, Emerson states of the "Oversoul":

> The Supreme Critic on the errors of the past and present, and the only prophet of that which must be, is that great nature in which we rest as the earth lies in the soft arms of the atmosphere; that Unity, that Over-Soul, within which every man's particular being is contained and made one with all other; that common heart of which all sincere conversation is the worship, to which all right action is submission; that overpowering reality which confutes our tricks and talents, and constrains every one to pass for what he is, and to speak from his character and not his tongue, and which evermore tends to pass into our thought and hand and become wisdom and

---

[164] See generally Emerson's essay "The Oversoul," in Larzer Ziff, ed., *Ralph Waldo Emerson: Nature and Selected Essays* (New York: Penguin Books, 1982), 205-24.

> virtue and power and beauty. We live in succession, in division, in parts, in particles. Meantime within man is the soul of the whole; the wise silence; the eternal beauty, to which every part and particle is equally related; the eternal ONE.[165]

Elsewhere, Emerson says of the "ONE" that

> the larger experience of man discovers the identical nature appearing through them all (i.e., all persons). Persons themselves acquaint us with the impersonal. In all conversation between two persons, tacit reference is made, as to a third party, to a common nature. That third party or common nature is not social; it is impersonal; is God.[166]

This is clearly not the God of the Bible, whose holiness stands in stark contrast to man's sinfulness and therefore bespeaks man's separation from Him. And by giving humanity an identity that is the same as or overlaps God's, Emerson removes the need for men to be reconciled to God through Christ's atoning sacrifice. The suffering and death of Christ on the cross are meaningless if God and man are one, and therefore are ipso facto not estranged. Insult is added to theological injury when Emerson informs us:

> To truth, justice, love, the attributes of the soul, the idea of immutableness is essentially associated. Jesus, living in these moral sentiments, heedless of sensual fortunes, heeding only the manifestations of these, never made the separation of the idea of duration from the essence of these attributes, nor uttered a syllable concerning the duration of the soul. It was left to his disciples to sever duration from the moral

---

[165] Ibid., 206-07.

[166] Ibid., 212.

## ESSENCE

elements, and to teach the immortality of the soul as a doctrine, and to maintain it by evidences.[167]

Emerson apparently never read much Scripture. If he did, what would he have made of Matthew 10:28 ("And fear not them which kill the body, but are not able to kill the soul")? Or Mark 8:36 ("For what shall it profit a man, if he shall gain the whole world, and lose his own soul?")? Can anyone doubt that personal identity persists in the hereafter upon reading Luke 10:20 ("rejoice, because your names are written in heaven.")? Or Luke 23:43 ("to day shalt thou be with me in paradise.")? The above words, reported by the Gospel writers, were all spoken by none other than Jesus himself.

While Emerson sees mankind as united with God and Nature through the "Oversoul," he affirms the existence of individual souls. "The good, by affinity, seek the good; the vile, by affinity, the vile," he declares.[168] "Thus of their own volition, souls proceed into heaven, into hell."[169] Emerson then goes on to say that the facts that truth ultimately prevails and that goodness admits one to heaven have always suggested to man "that the world is not the product of manifold (i.e., many kinds of) power, but of one will, of one mind; and that one mind is everywhere"[170] Such oneness of the divine and human minds seems to preclude personal freedom, regardless of what Emerson says about "volition." Choosing not to act within God's will, presumably the reason for one's going to hell, assumes that the actor and God are separate, does it not? How can you rebel against someone or something that your existence is contained within? I, for one, find the whole transcendentalist oversoul/soul structure confusing and inconsistent.

The crisis that the "Oversoul" idea creates as far as personal selfhood is concerned is not resolved by statements elsewhere in the transcendentalist manifesto. For one thing, Emerson gives

---

[167] Ibid., 216.

[168] Ralph Waldo Emerson, "Divinity School Address," in Lawrence Buell, ed., *The American Transcendentalists: Essential Writings* (New York: Random House Inc., 2006 Modern Library ed.), 132.

[169] Ibid.

[170] Ibid (parenthetical supplied).

indications that he believes in heaven. The human mind, he says, is more real than the world, which is "a divine dream, from which we may presently awake to the glories and certainties of day."[171] "The foundations of man are not in matter, but in spirit," he declares.[172] "(T)he element of spirit is eternity," Emerson adds, and he goes on to state: "A man is a god in ruins. When men are innocent, life shall be longer and pass into the immortal, as gently as we awake from dreams."[173]

Are some people, specifically those who have awakened from the dream that is the world, already in heaven? If so, how did they get there? If determinism is the operative principle (the yes or no of which is by no means clear in Emerson's writing), then why all the talk about the soul by an act of volition being able to admit itself to heaven or hell? And if man can will himself into heaven or hell, exactly how is that accomplished? Do you simply either "seek the good" or "seek the vile"? Emerson's qualification that you do whichever you do "by affinity" further complicates matters. It implies that you're naturally inclined toward whatever you do, does it not? This would seem to at least amount to a "soft" determinism.

Henry David Thoreau, transcendentalism's number two man, brings more poesy than clarity to the discussion regarding souls. *A Week on the Concord and Merrimack Rivers* is his account of a boat trip taken with his brother John, and along with *Zen and the Art of Motorcycle Maintenance* is a model for the present book. In one of his many digressions, in the context of his "Tuesday" chapter, Thoreau tells of having hiked up Saddleback Mountain, outside Williamstown, Massachusetts. Freshmen at Williams College still walk up to the top of Greylock (formerly Saddleback) as one of their orientation activities. Indeed, it was during such a hike that my son Jeremy, perhaps for the first time, saw the inbreeding of the intellectual elite, when one of his fellow hikers asked the group, "Is there anyone here who isn't a legacy?"[174] Thoreau, in his digressive narrative, has just spent the night camping out on the top of

---

[171] Emerson, "Nature," in *Nature and Selected Essays*, 72.

[172] Ibid., 77.

[173] Ibid.

[174] A "legacy" is the child of a parent who attended the same college.

the mountain. He describes what he sees when he awakens in the morning as follows:

> As the light increased I discovered around me an ocean of mist, which reached up by chance exactly to the base of the (College observatory) tower, and shut out every vestige of the earth, while I was left floating on this fragment of the wreck of a world, on my carved plank in cloudland;[175] a situation which required no aid from the imagination to render it impressive. As the light in the east steadily increased, it revealed to me more clearly the new world into which I had risen in the night, the new terra firma perchance of my future life. The earth beneath had passed away like the phantom of a shadow, and this new platform was gained. As I had climbed above storm and cloud, so by successive days' journeys I might reach the region of eternal day beyond the tapering shadow of the earth;... The inhabitants of earth behold commonly but the dark and shadowy side of heaven's pavement;. But my muse would fail to convey an impression of the gorgeous tapestry by which I was surrounded,...[176]

In short, what we seem to have in transcendentalism is the idea of a self that's not individual—it can't be, if everybody is part of the Oversoul—but still in a sense floats in its own boat on the river of life, pulled by the current toward a vaguely-defined God and heaven. Contradiction multiplied on contradiction, as in other transcendentalist attempts at metaphysics.

The search for metaphysical truth through philosophy may be over, in any event. Rene' Descartes, in what has come to be known as the seventeenth century's *"epistemological turn"* of philosophy,

---

[175] Thoreau, having brought no bedroll with him, had found some planks to sleep on and under.

[176] Henry David Thoreau, *A Week on the Concord and Merrimack Rivers* (Orleans, Mass.: Parnassus Imprints, Inc., 1987), 231-33, parenthetical added.

raised the issue of doubt, the question of *how* you know something is true. No more free rides for widely accepted assumptions. The aforementioned "how" of knowing rather than the "what" now became the focus (although, it should be noted, Descartes appears to have entertained with very little skeptical questioning the assumptions that God and demons exist).

The "*linguistic turn*" of philosophy in the nineteenth and twentieth centuries reinforces the conclusion that philosophy is no longer about ultimate truth. Stephen Hawking, writing in 1988, offered the following summary at pages 190 and 191 of his *Brief History*:

> Up to now, most scientists have been too occupied with the development of new theories that describe *what* the universe is to ask the question *why*. On the other hand, the people whose business it is to ask *why*, the philosophers, have not been able to keep up with the advance of scientific theories. In the eighteenth century, philosophers considered the whole of human knowledge, including science, to be their field, and discussed questions such as: did the universe have a beginning? However, in the nineteenth and twentieth centuries, science became too technical and mathematical for the philosophers, or anyone except for a few specialists. Philosophers reduced the scope of their inquiries so much that (Austrian Ludwig) Wittgenstein, the most famous philosopher of this century, said, "The sole remaining task of philosophy is the analysis of language (parenthetical mine)."

So whither is a truth-seeker to turn? How is one to establish that there is transcendence and that it has a name?

*Back to Impermanence*

A day or two later; I haven't recorded the exact date in my notes, but I never do write down dates and sometimes don't even note the day of the week. Obviously it's a weekday, since I indicate in my notes that hardly anyone is out on the beach. It's warmer today than the day mentioned above when I chose to stay inside and play with my note cards, so I'm sitting outside in a folding chair on the front stoop of the Cottage, staring at the lustrous mid-morning surf. My frustration with philosophy, with the whole metaphysical enterprise, has caused whatever intensity I had up to this point to be transmuted into apathy or at least confusion. Where do I go from here? I ask myself in my present/future tense as researcher and writer, a time orientation which you'll recall was adopted during the early going in this book to help you as the reader experience more of what I felt in the course of my still-not-concluded quest for truth.

The row-on-row march of the waves, which tumble onto the beach in explosions of foam, makes me think of La Jolla, a seaside town north of San Diego that I visited while our Family Court conference group was waiting for airline flights to resume after the September 11, 2001 attack on the United States. Like many communities in southern California, La Jolla sits atop a bluff overlooking the Pacific. Down the cliff from the main part of the town, the surf has carved out a grotto. Visitors can descend some steps that lead down to the cave. You then traverse a walkway which suspends you above the churning waters that are the floor of the cavern. The waters are relentless in their motion. At the same time, the presence of the grotto informs us that rocks likewise have a dynamism. They're not as permanent as we'd like to think they are. They erode. They crack. They even melt.

Also impermanent are the businesses—clothing and jewelry boutiques, art and curio shops, and the like—that sustain the town above the grotto. Someday there will be a new tenant or owner for each of these main-drag storefronts, if the host building still exists at that point. There will be new dreams for these prime locations. Reveries of richness or at least of solvency. Meanwhile, the homeless sleep in the park nearby, on the benches and under the bushes that this oasis provides for their comfort. The would-be prosperous and the de facto poor are united in a dreamy transcendental temporariness.

## WATER

At Campbell's Point and in the geographical area that surrounds it, the water doesn't produce the same effect as at La Jolla. We don't have the sheer rock face that you find in southern California, not even on Stony Point, so there's nothing in our part of Ontario that the water could act on to make a grotto. Here, the waves simply re-arrange the sand. What was a sand bar one day is no longer a sand bar the next. Nearby, a new shallows has appeared at the water's edge. The sand beyond the water's edge isn't stable under your feet. If you stand where the waves can wash over them, your feet will quickly begin to sink into the wet sand, which is constantly being displaced from under you by the in-rushing water. At such times, there comes to mind the refrain of a hymn learned by me in my youth but thereafter forgotten until we started coming to The Cottage:

> On Christ, the solid Rock, I stand;
> All other ground is sinking sand,
> All other ground is sinking sand.[177]

Reassuring for a Christian, certainly, but this still doesn't satisfy my need to understand the metaphysics of it all, to know *why* we can say with confidence that there's anything of permanence, including an immortal self. Oh, the tyranny of the intellect! Why can't I just stand on the Rock and let it go at that!

It's just as well that metaphysically-oriented philosophy has bitten the dust, I tell myself. In his letter to the Colossian church, as I've already noted, the apostle Paul warned: "Beware lest any man spoil you through philosophy and vain deceit,"[178] He explained that to lead men captive (i.e., to "spoil" them) through rationalization and misrepresentation would be "after the tradition of men, after the rudiments of the world, and not after Christ."[179] In commenting on this verse, Dr. Henry Morris states that Paul's inspired words must be read as applying to all philosophy, not just the Epicurean

---

[177] Edward Mote (words), William B. Bradbury (music), "The Solid Rock," in *Majesty Hymns* (Greenville, S.C.: Majesty Music, 1997), no. 411.

[178] Colossians 2:8.

[179] Ibid.

## ESSENCE

and Stoic systems that dominated thinking in the Roman world.[180] As hard a pill as this may be for some to swallow, the divine revelation recorded in the pages of God's written Word must in all instances be trusted over extrabiblical "reason."

---

[180] Morris, *Study Bible*, n. to Col. 2:8.

# 3
# THE REGION OF AWE

*Drawn to The Islands*

It's now a Sunday, later but still early in October. A nice afternoon and so I've decided to explore Pillar Point, which, as I think I mentioned before, is on the other side of Black River Bay from us and from Sackets Harbor. What I'm doing today other than sightseeing, and probably will be doing in days to come, is trying to be Robert Pirsig, Henry David Thoreau, and C. S. Lewis all rolled into one. Specifically, I'll be traveling around the eastern Ontario area and seeing what metaphysical or spiritual truth I can find either above or beneath the surface of things in nature.

Since a lot of my time while I'm out exploring will be spent parked by the side of the road making notes, Sue doesn't want to tag along. She knows how involved I get with things during my voyages of discovery, and has, I think, come to realize that it's best to just let me do what I do at my own pace rather than her getting involved and trying to make me live my life—at least the professional part of it—according to her rhythms.

Before we proceed farther, I should tell you that I'll be making a number of turns on my drives around this end of the Lake, so you may soon be experiencing dizziness if you try to visualize every change of direction on the mind's-eye map I'm trying to create. All you need to know, to avoid tumbling into befuddlement, is that various capes and peninsulas circle (or, rather, half-circle) the smaller land masses that I identify herein as The Islands—Bass, Gull, Six-Town Point, and (I reluctantly include) Association. If you want a more complete picture of this topography, I suggest you consult Figure 6 (Henderson Bay Area) and Figure 7 (The Islands), both found at the beginning of this book. Compass orientations, except as they show a focus on The Islands, are unimportant under my present plan.

To get to Pillar Point from Campbell's Point, you begin by taking Route 3 north to Route 180, which takes you into the village of Dexter. Dexter is where the Black River loses its navigability by passing over a weir. Once you cross the Black River and turn left onto the Pillar Point road coming up out of Dexter, you can look off to the left and see, a mile or so in the distance, the point at which the river widens to become Black River Bay. The road soon dips down and crosses the Perch River, which is narrow and marshy, then rises again and comes up to a T-intersection at which I turn left. I then go past the Rustic Golf Course and turn left again, onto South Shore Road. At last I'm traveling in the direction I wanted to go, which I know from having studied my map earlier this afternoon. I'm headed toward The Islands.

Across Black River Bay, which is on my left, I now see Sackets Harbor. The masts of the yachts at Navy Point are visible, as are the buildings of Madison Barracks, which looks like a college campus and presides with dignity over the Bay below.[181] Partially obscuring the Sackets shoreline and separated from it by a few hundred feet of water, I see the leafy corpulence of Horse Island, which played a role in the Battle of Sackets Harbor during the War of 1812, as a staging area for an ultimately unsuccessful British attack.[182]

The only battles that get fought at Sackets Harbor these days are between individual property owners or are of the

---

[181] The rowhouse- and dormitory-style buildings of the Barracks, some red brick, some gray limestone, but all permanent, have now been converted from their former military use to condominiums.

[182] See Patrick A. Wilder, *The Battle of Sackett's Harbour: 1813* (Baltimore: The Nautical and Aviation Publishing Co. of America, 1994), 57, 85-123.

cottager-versus-municipality variety. Regarding the former kind of conflict, I think of the antiques dealer whose building was situated between my in-laws' winter home and the harbor and who refused to trim back a row of evergreens on the back edge of his property, so that eventually Sue's parents' (and later their successors') view of the harbor was almost completely blocked. The latter type of clash, between landowners and government, has also involved our family—as, for example, when the municipality raised our school taxes nearly two thousand dollars in one year, even though we and most of the other Campbell's Pointers had no children attending Sackets Harbor schools. But I digress.

As I drive farther on the South Shore road, catching glimpses of the Bay and Lake through the willow trees that line the shore on my left, The Islands finally come into view. Bass Island, the nearest, is easily distinguishable by its scraggly but tall main tree, which stands out against the visually fuzzier shoreline on which Campbell's Point is found. To the left of Bass and farther in the distance is the brushy outline of Gull Island, deforested a few years ago in a fruitless attempt to get rid of the cormorants that were nesting there. Beyond Bass and visible to its right are the full-vegetation outlines of Six-Town Point and Association Islands. From this angle and at this distance, it's hard to tell that they're separate islands.

Although at this time I'm doing no more than fifteen miles an hour, I slow even more in the hope of catching a glimpse of the Lime Barrel Shoal light buoy, but see nothing that looks like it. The flash of the Lime Barrel light, which when seen from The Cottage is farther out toward the main body of the Lake than The Islands, is visible almost any night, weather conditions permitting. At the time of my father-in-law's passing, as I was preparing to eulogize him at his memorial service, I found myself thinking of his steadiness through many life changes, including paralysis from an auto/trailer accident, divorces suffered by two of his four children, and the death of several close friends. His imperturbableness was not unlike the dependability of the Lime Barrel light.

Up to this point, South Shore Road has been a more or less uninterrupted succession of summer cottages and docks interspersed with occasional year-round homes (there are "School Bus Stop Ahead" signs, I note), with the majority of the houses being on the inland side of the road. South Shore Road now swings away from the shoreline and crosses overgrown farmland. A sign announces that lakefront property (presumably undeveloped) is available at Everleigh Point.

The road rejoins the lake shore shortly before a narrow bridge that crosses the mouth of the Sherwin River, and then we're back to the earlier pattern of cottages, year-round homes, and docks. The road now curves to the right, and as I look out on the water I realize, from the relative lack of land on my right or left and the angle of the sun, that I've reached the tip of Pillar Point. I spot a grassy area between the road and the water, pull off into it, and get out of the car.

Finding no convenient way to descend the embankment before me to the rocky beach, I remain on the grass, surveying the water from left to right as I attempt to get my bearings. At last I see The Islands, so far distant now that it's difficult to separate them visually from the land mass behind them, the shoreline that includes Campbell's Point. I find this unsettling. My world, formerly defined by the land and water that I can see looking west from Campbell's Point, has shrunk. Satisfying my curiosity concerning The Islands by looking at them from the Lake side, I no longer see The Islands against the background of the Lake's infinity. This would not have happened, I tell myself, if I had been able to journey to The Islands by boat and from there look out toward the main body of the Lake; however, spinal arthritis has set in since I rowed out to Bass with my son the swimmer, and I'm still looking for a dependable, reasonably-priced motor for the boat that I bought to replace the skiff that Sue's parents had but sold.

Viewing The Islands from Campbell's Point or anywhere on the water between Campbell's Point and The Islands means that you see topography beyond The Islands that's very, very distant but still obscures a considerable amount of water horizon (see this book's cover). You're similarly denied the bigger world view — nothing but lake between you and the curve of the earth — when you're standing on Pillar Point looking southeast in the direction of The Cottage.

I feel reassured when I look due south, nevertheless. Between the silhouettes of Stony Point and Stony Island, the latter of which is to the right of Stony Point in my present line of vision, I see shimmering open water, waves catching the sunlight as far as the eye can see. I'm looking out into the main part of the Lake. The same is true when my eyes scan the area between the outlines of Stony Island and Galloo Island. I'm saved! My world expands again! There's still a real horizon, which for me at this moment, as I now realize, symbolizes the infinitude of God. I need that contact with infinity, with open water that disappears from my view because of its incalculable distance from me. It's no good if the shimmering infinitude is invisible by reason of its being obscured by intervening land features.

My reverie is now interrupted. A voice from behind me says, "May I help you?" I turn and see a middle-aged man in a gray windbreaker standing in the yard of a house on the other (inland) side of the road. He's staring at me with a pained expression—as if, by the planting of my feet where they are, I've overturned his entire lifestyle. "You're standing in my front yard," he announces.

"I didn't know that; I thought this was just a little park area," I explain. "No problem," I assure him.

As I get back into the car and begin driving up the South Shore road in the direction from which I came, I feel a heightened awareness of just how hard it is to get your arms around this body of water, this Ontario. Every square inch of shoreline except for the state parks, which are overcrowded, is privately owned. The same as it was with Otisco, only there are no parks there. It occurs to me, on later reflection after I'm back at The Cottage and sitting out on the back patio, that this is symptomatic of the way churches are now denying access to the truth, the living water (John 4: 10-14), by either standardizing and sanitizing it or stepping off into the occult in ways that are beyond the range of ordinary knowledge or understanding. They're the property owners along the shore, if you will, and they're keeping us common folks away from God.

I still haven't been able to spot the Lime Barrel Shoal light buoy.

# THE REGION OF AWE

*Point Peninsula*

Monday. Another step is taken to explore my world. Another chance for you to see how I've tried to work this nature mysticism gig so as to learn hidden things about the character of reality. I'm going to Point Peninsula, the more distant of the two major land masses that I can see from my park bench on Campbell's Point. On the way, I take a detour to Storrs Point, which is just outside Sackets Harbor, beyond Madison Barracks and in the direction of Dexter.

The paved lead-in to Storrs takes me past farmland. Then the highway becomes a one-and-a-half-lane dirt road that winds through woods. The cottages along this stretch, which are scattered among tall trees, seem more like summer camp cabins than seasonal homes. Their rectangular one-story construction helps create this impression. Visible on my left is Black River Bay, considerably narrower at this point than its mile-or-so width down by the Barracks. I reach the heavily wooded tip of Storrs Point, beyond which is perhaps a quarter mile of water and beyond that a marsh. I later learn, from looking at my U.S. Geological Survey maps, that the water is Muskalonge Bay, which runs east off Black River Bay.

I get out of the car to look around Storrs Point a bit more, but there isn't much to see. The only word that enters my mind is seclusion. The people in this neck of the woods are here because they want to be left alone, is my surmise. They want to be able to get their boats into the water from their camps ("camp" is the Central New York word for cottage, as I think I mentioned before) and in this way become part of the local boating and fishing community, but other than that they desire no connection with the rest of the world, or so it seems. Do they want to get out into the main part of the Lake and try to reach—or see what's over—the horizon? Probably not, is my conclusion, which is not to say that none of them are inclined toward God if you choose to interpret this situation metaphorically.

A second detour that I take on my planned trip to Point Peninsula is from the village of Chaumont (pronounced Shuh-mo') south onto Point Salubrious (wonderful name!). Cottages, docks, some farmland, much the same as what I saw on Pillar Point the day before. The open water of the east end of Lake Ontario is now to the south rather than the southwest of me, but my view of this water is largely blocked by the tips of Pillar Point and Point Peninsula (on my left and right respectively), both of which protrude farther out into the Lake than Point Salubrious, and also by Cherry Island (straight ahead). There's no way I can see The Islands, either. They're behind Pillar Point.

A third detour is to Three Mile Point. Again I make a loop to the south, see what there is to see (nothing noteworthy), and then loop back to Route 12E, the state road that eventually leads west to Cape Vincent, from which you can catch a ten-car ferry to Wolfe Island, and from there a larger transport to Kingston. The same complaints apply to Three Mile Point as with respect to Point Salubrious: no window to the main part of the Lake and no Islands visible.

Finally, I turn off Route 12E onto County Road 57 and begin to make my way south on Point Peninsula. This imposing expanse of land, which encloses Chaumont Bay on the south, is shaped like an inverted pterodactyl neck and head when you look at it on the map (and even when you don't [joke!]). After crossing the isthmus that connects the creature's neck and head, I make a right turn and drive northeast to Long Point State Park, which would correspond to the bill of the pterodactyl, to scout the place out for my Rochester daughter-in-law's family, who are inveterate campers.

From Long Point State Park, after almost hitting a deer, the third one to cross the road in front of me today, I head south, pass through the incorporated village of Point Peninsula, and suddenly find myself farther south than the end of Pillar Point, which is now across the wide expanse of choppy water seen over my left shoulder. I look to the east and see, miles distant but still visible even through today's haze, Bass Island! Considerable ways to the right of Bass, but still plainly visible, are the outlines of Six Town Point and Association Islands! Gull Island, because of its low profile, can't be definitely seen, although I know it's there. I still see no Lime Barrel Shoal light buoy, no symbol of my father-in-law's stabilizing presence.

I've lost track of time, I realize. There's no way I'm going to get back to The Cottage by our normal five-thirty supper time, and

I've forgotten to bring my cell phone, so there's no way for me to let Sue know I'll be late. I begin to retrace my steps of earlier in the afternoon. When I get away from the shore, I'm back on fifty-five-mile-an-hour road that runs through farm land. Good! Now I can put the pedal to the metal, as they say. Many of the farms that I'm whizzing past look abandoned, and from the architecture of their houses and the condition of their barns and other outbuildings, I can tell that these homesteads predate most of the lakefront cottages and year-round homes I'm seeing. Occasionally I pass a farm whose house is next to the water. A conjunction of two cultures that otherwise seem to remain separate.

When I arrive back at The Cottage at 6:15, I find Sue in a venting mode. She reminds me that she has repeatedly told me to take my phone with me when I go out. "Don't ever do this to me again!" she inveighs. I'll be lucky if this project doesn't conclude with me in divorce court as a respondent, I think to myself. But then my longsuffering wife becomes calm and pleasant. She seems to understand that what I'm involved in is business that's very important to me. I remove my mobile phone from my briefcase, which I neglected to take with me on this particular day of roaming Ontario, and place it in the glove compartment of the car to assure that there will be no future instances of my being incommunicado while out on The Quest.

*Stony Point*

Tuesday the same week. I'm taking a day trip in the opposite direction from my forays of the previous two days. Again, gentle reader, you're coming along with me on a trip that you may never have occasion to take other than in your imagination. I'm venturing to a part of the Ontario lakeshore that lies to the south of Campbell's Point and that you see only from a great distance when you come up Route 3, which is our normal approach to The Cottage as we travel here from Rochester. I refer to Stony Point, the headland that nestles Henderson Bay.

The settlement known as Henderson Harbor, not surprisingly, is on Henderson Bay, which opens north toward The Islands and

Pillar Point. When you break over the crest of the plateau to the east of Henderson Bay and Henderson Harbor, driving north on Route 3, you can see The Islands in the bay below and also the massive promontory of Stony Point, which is due south of Point Peninsula, from which it's separated by an expanse of water that appears to be several miles wide.

On the other side of Stony Point, acting as barriers between it and the sometimes tempestuous main body of the Lake, are Stony Island and Galloo Island, Stony being the only one of these outer islands that's visible from our cottage. The political powers-that-be have been eyeing Galloo as a possible location for a wind farm (picture huge rotors beating the sky above a pristine island, in order to produce a tiny fraction of the region's power). The Islands that have heretofore so much been the focus of my attention—Bass, Gull, and Six-Town Point—are within the sheltering embrace of Stony Point on the south and the points Pillar and Peninsula on the north, and are thus somewhat protected from the fury of the Lake when storms come in from the south or west.

The drive in to Henderson Harbor from Route 3 takes me along County Road 123, which services the many impressive homes on the east side of Henderson Bay and is more like a private drive than a public highway. There are no summer-camp-like cabins such as I saw at Storrs Point. The homes on the bay side of this drive are close to and only slightly downhill from the road, allowing me to take in their pleasing architectural details. Similar homes on the non-Bay side of County 123 sit above the road and stand out cleanly against the dark green evergreens on the steep wooded hillside behind them. This stretch of the drive in to Henderson Harbor defines upscale.

I then come to what little commercial area there is at Henderson Harbor. Some motels and restaurants, apparently geared to fishermen, and a boat yard. The road now climbs a hill to meet State Road 178, which I turn right onto and follow as it descends and crosses the flats at the south end of Henderson Bay, where there's another marina. When I'm nearly to the top of the hill on the other side of the flats, I turn right onto Snowshoe Road, which I know leads to Snowshoe Bay.

The dwellings on the west side of Henderson Bay have more the character of cottages—i.e., are smaller and less architecturally cohesive—and sit farther below road level than any of the elegant waterfront homes on the east side of the Bay. I'm often looking down at rooftops that are as much as a hundred feet below me, on homes that must be reached by steeply inclined driveways. The road on which I'm traveling, one that no doubt was carved out of the rocky spine of Stony Point with some difficulty, has no shoulders; nonetheless, it's wide enough for cars to pass. This is not an appropriate time for a driver to be rubbernecking, however.

Snowshoe Road now curves to the right and descends sharply. Ahead of me is a narrow bridge that leads onto Hovey Island, a tiny chunk of land that is a steppingstone to Association Island, which is connected to Hovey by a causeway. A sign next to the bridge says, "Association Island RV Resort Registrants Only." I take this as an invitation to stop at the Resort office and tell whoever's there that I'd like to look around. I'm interested in Association Island for itself, I can honestly say; it's a place that I haven't been to until now, and it has the history that I've already noted. So I'm not just trying to sneak another peek at The Islands.

I feel keen anticipation as I start across the causeway to Association. When I first explored Stony Point in the early 1980s, the roadway out to the island was still awaiting reconstruction after having been washed out some years earlier (a life story similar to that of the Otisco causeway, but with a supposed happy ending for "civilization").

I'm off the causeway now and stop at the campground office. The manager provides me with brochures and opens the gate for me. As I follow the gravel drive past the RV sites, I see weathered wooden buildings that either have been purposely left as reminders of the history of the island or haven't been torn down by Association's present owners because of the money and trouble involved, most likely the latter. One of the buildings, which is barn-shaped, has the look of a community center. The structure that stirs my imagination the most, however, is a long, one-story building with alternating doors and windows that give it the appearance of a poor man's motel or a migrant worker camp, I'm not sure which.

What place these buildings should occupy in my thinking, other than as what they are, is not clear to me at this point.

I reach the far (north) end of Association Island. Ahead, not more than half a mile distant, is Six-Town Point Island. Beyond Six-Town Point are the remaining two islands of the group that I've been calling The Islands. Gull, to the right of Six-Town in my line of vision, displays its characteristic brushy vegetation and guano-whitened rock apron. Gull is the only island that could possibly be where this one is, based on available maps. Buried deeper in the day's haze, between Six-Town and Gull islands, which are closer, is Bass Island. Bass can be identified not just by map location, but also by the semi-skeletal tree that towers above the other vegetation and an abandoned stone house on the island. I believe I mentioned the tree at the beginning of my account of circling The Islands. This great gray ghost was probably close to being dead when my son Geof swam out to Bass. Age looking down on youth.

One last stop I have to make today. After getting off Association Island and following Snowshoe Road back to Route 178, I head southwest across a more barren part of Stony Point. Gone are the tall evergreens and assorted leaf-shedding trees that stood over the cottages encircling Henderson Bay. In their place, climbing out of a tangle of grasses, are wiry bushes and low evergreens, cedars of some kind. I pass through what has been designated by the State as a "Unique Area," where there are hiking trails and no homes or farms. Eventually, I begin to see cottages again, part of a landscape punctuated by trees and bushes that have a more windswept look than the greenery on the other side of Stony Point.

Finally, I'm at the end of the paved road, and the Stony Point lighthouse stands before me, against a background of crashing surf. The historic masonry tower has painted white stucco as its exterior surface. The keeper's house, which hugs the tower, is now protected from the elements by white nylon siding. I notice that there's no light in the masonry tower. The actual Stony Point light, I find, is on a steel frame tower several hundred feet north of the masonry lighthouse. The lighthouse is now merely someone's picturesque home, although it flies the United States and New York State flags.

# THE REGION OF AWE

The surf is higher, the waves more aggressive, than was the case at Campbell's Point when I left to come over here. This was to be expected, since the Stony Point light looks out to windward over almost totally open water. This is what Stony Point protects us from. The horizon that I see as I gaze southwesterly is broken only by the Ontario eastern shore on the left and Galloo Island on my extreme right. Galloo comes farther south than Stony Island, my auto club map tells me.

When I get back to The Cottage, I look at my most detailed map of the east end of Ontario, published by the National Oceanic and Atmospheric Administration. The Lime Barrel Shoal light buoy, it turns out, was behind Six-Town-Point Island when I was standing on Association Island. Otherwise, I might have been able to see it. But perhaps I'll never be able to see it, unless I get a boat big enough to allow me to brave the choppier waters on the outer side of The Islands.

*Additional Perspectives*

The question of why I would be taking the tack of seeking out any location from which there's a view of The Islands—Bass, Gull, and Six-Town Point, mainly—hasn't been answered, I realize. And what did I find so fascinating about the way other chunks of land hover around The Islands? In my Introduction, I said with optimism that I would be seeking God-provided insights regarding spiritual and metaphysical reality. I stated that I intended to bring together science, philosophy, my observations of nature, and the teaching of the Bible to give us (you the reader and me the writer) the benefit of a Chautauqua-like experience. You may count the above-described multi-day voyage of exploration toward, but not as the totality of, the promised nature observation.

To say that what I've been doing is part of a Chautauqua somehow doesn't make our adventure sound vital enough, however. This isn't just a seminar that's been hatched out of someone's felt need for continuing education. We're doing something more integral, aren't we? We're trying to understand who God is, and what His relationship to the world is. We're trying to comprehend how we can be part of His plan for eternity. We're trying to be less at variance with Him.

Does the above mission statement mean that, laying aside prior resolve to the contrary, we're wading into the waters of mysticism? When I announced to you at the beginning of the current chapter that I was going to spend some time cruising around our end of Lake Ontario in search of truth, I didn't use the word "mysticism." It could have been implied, nonetheless, that the activity I was contemplating was a kind of nature mysticism, inasmuch as I was expressing hope that God would somehow speak to me through the land and the water. That hasn't happened so far, at least not in the way I thought it would. It seems that staring at scenery through a car window, even if it's accompanied by stops to get out and look around, is not enough to put someone in direct contact with the Almighty, any more than is sitting in a lawn chair on the front stoop at Campbell's Point with the Lake before you. There is, nonetheless, some value in treating things in nature as metaphors for spiritual or metaphysical reality, I think. As I see it, this was what C. S. Lewis was doing, even if an observer of Lewis (a mystic, perhaps?) called it something else.

Also of questionable usefulness for the purpose of directly contacting God are the techniques I've categorized in my Preface as "mindless mysticism." These methods, as I noted, are "mindless" in that they're not connected in any way to the working of the intellect. Some would argue that unreason is a good thing in this context, that we're wasting our time when we attempt to rationally comprehend a God who is clearly beyond our power of comprehension. I disagree, obviously. I'm not ready at this juncture to cast my lot in with Richard Foster and company, so we'll just have to continue for now to chase after the truth by means other than the so-called spiritual disciplines. Is there an approach that's halfway between the cerebral (mind full of ideas) and the contemplative (mind emptied of ideas)? I wonder.

Philosophy, to the extent it asks what the self is (still our central object of investigation), feeds into something beyond mere rational inquiry. We can't just reason our way to an understanding of self, it appears; what we're dealing with is too slippery a concept to allow that, and more is needed than to simply pronounce "self" a holistic concept. At some juncture, if we're to proceed further with our investigation, *intuition* must be brought into play. In earlier discussion, I've spoken repeatedly of intuition as a source of knowledge or understanding.

Webster's Unabridged defines "intuition," *inter alia*, as "direct perception of truth, fact, etc., independent of any reasoning process; immediate apprehension."[183] This sounds like mysticism, which also bypasses the intellect; however, intuition is not specifically spiritual, whereas mysticism seems to always be found in the context of a religion.[184] For this reason, mysticism is usually specific to the religious culture of the individual mystic,[185] a condition that would seem to augur bias in the mystic's view of the truth. Better, however, that the subject approach the mystical encounter with ideas than that he or she, going into the encounter, be bereft of them because of "spiritual formation" purgation. The latter situation would seem to invite activity of a mind-filling nature on the part of Satan, the enemy of our souls.

Other than being united with God and man through the Oversoul, gaining new views of the theological landscape was what New England transcendentalism was all about, or so its adherents thought. The transcendentalists were not comfortable with Calvinism, which they saw as imposing unnecessarily harsh doctrine, and they were also uneasy with the rationalism of Unitarianism. These were belief systems that relied on second-hand information and handed-down tenets, the transcendentalists said, whereas by virtue of their pantheism, their unity with all people and all things, the transcendentalists were (they said) able to understand spiritual truths on their own, intuitively, without the assistance of the Bible or church tradition or even reason. Emerson spoke of "the infinitude of private man,"[186] the divinity that was said to be latent

---

[183] *Random House Webster's Unabridged Dictionary*, 2nd Ed. (New York: Random House Reference, 1987), 1002.

[184] See Rohmann, *Ideas*, 270.

[185] Ibid.

[186] Buell, *American Transcendentalists*, 208.

in every person. God's perspective was man's perspective, supposedly, when God was approached directly.

The approach to truth necessarily had to rely on multiple views, Emerson observed. The frame-by-frame-change principle of motion pictures seems to have been anticipated by Emerson when he wrote:

> Nature is made to conspire with spirit to emancipate us. Certain mechanical changes, a small alteration in our local position apprizes us of a dualism. We are strangely affected by seeing the shore from a moving ship, from a balloon, or through the tints of an unusual sky. The least change in our point of view, gives the whole world a pictorial air. A man who seldom rides, needs only to get into a coach and traverse his own town, to turn the street into a puppet-show. The men, the women—talking, running, bartering, fighting—the earnest mechanic, the lounger, the beggar, the boys, the dogs, are unrealized at once, or, at least, wholly detached from all relation to the observer, and seen as apparent, not substantial beings. What new thoughts are suggested by seeing a face of country quite familiar, in the rapid movement of the rail-road car![187]

In the "Friday" chapter of *A Week on the Concord and Merrimack Rivers*, Henry David Thoreau similarly comments on the need for varying perspectives, stating, "The most familiar sheet of water viewed from a new hilltop, yields a novel and unexpected pleasure." Thoreau, *A Week*, 436.

Hilltop views sometimes evoke feelings other than pure pleasure. My thoughts return to early December 1963. My college friend Woody had gotten his hands on what he described as a "very old" shotgun and was, he said, anxious to try it out. So we drove the forty or so miles from Hobart to my boyhood home, stayed with Mother and Dad overnight, and spent a chilly Saturday tramping

---

[187] Emerson, "Nature," in *American Transcendentalists*, 54-55.

through my Aunt Jo's woods, which were part of a farm perched high on a hill overlooking Otisco Lake. The causeway that I mentioned earlier is visible if you stand in the middle of Slate Hill Road, which ran past my aunt's house.

The temperature that day was in the low twenties and the ground was frozen solid, feeling unyieldingly rough under my boots (and, I suspect, Woody's as well). We didn't find anything that was big enough or close enough to shoot at, but that didn't bother me, and it didn't seem to matter to Woody either. We hadn't come for prey. The important thing was to be out in the wilds, away from the Hobart campus, which still had a pall hanging over it. John F. Kennedy had been assassinated two weeks earlier.

Kennedy was younger than U.S. Presidents usually are, prompting thoughts in early-twenties folk such as ourselves of our mortality and how hopelessly short life is. I wasn't saved at that point and neither was Woody that I know of, although as a freshman he had talked of going into the Episcopal ministry. So the cold, hard ground on which we trod, snapping twigs and pushing aside leafless branches, was a reminder of death, which we were trying to banish from our thoughts. I think that Woody and I both realized by the end of the early-darkness-shortened afternoon that the hunting trip hadn't had the salutary effect that we'd hoped for.

Death, at least impending death, had already made an appearance before me in a large field near the woods that Woody and I made our way through, a field situated on the lake side of the woods and so sloped that you could see both the eastern and western shores of Otisco Lake from the upper part of the field. In the summer of 1960, just before I started college, I helped my cousin Floyd (Aunt Jo's son and only child) and two of his boys with picking stones out of the above-described field, which was a yearly ritual. Floyd was a number of years older than I was, as were all my cousins. We would dig stones the size of softballs out of the loose dirt created by the plowing of the field and throw them onto a concave wooden sled, called a stone boat, which was hauled behind a tractor.

As we worked, Floyd, normally robust but now thinner and more sallow than before, jokingly claimed that he could still service his wife, even though he had lost both testicles to cancer, as we all knew. False bravado, affected mainly for show? A sense of humor so tenacious that even "the big C" couldn't make it go away? Or did he believe that he had knowledge (that he would survive this health crisis, specifically) that the rest of us lacked? I never was able to sort this out. Floyd was dead a few months later, as I learned in a phone call from Mother during my freshman year at Hobart.

By the time of my hunting expedition with Woody, the Thompson family farm had ceased being a working farm; the land was now either fallow or rented out to "bigger" local farmers. Aunt Jo nonetheless still needed help with various things—repair of household appliances, grass mowing, snow plowing, and maintenance on her aging Chevy coupe, for example. In the meantime, she was supporting herself by cooking for, chauffeuring, and otherwise being at the beck and call of several eccentric but relatively solvent older couples in the Navarino area and surrounding communities.

Taking care of needs that Aunt Jo had such as those mentioned above, as well as looking in on her every day, particularly during the winter, fell on Floyd's three sons and their wives after Floyd's death, and also on Floyd's widow, Levy (Levina). Floyd and Levy had lived in a house a quarter mile up the road, and after Floyd's passing Levy remained in the house and commuted from there to her job as a nurse in a hospital in Syracuse, some fifteen miles away. Two of Floyd's boys eventually started an auto repair business that was located in a barn-like steel building behind the family home, to which they daily betook themselves from their homes one and two hills away. The third son, an X-ray technician who worked in the same Syracuse hospital as his mother (Community General), eventually built a house directly across the road from Aunt Jo's, on the edge of the field from which we had plucked stones years before.

Thus, there was no shortage of family support for Aunt Jo. Having grown up in the country, I had seen this kind of pattern repeated again and again: family members in adult life tended to live close to each other, often less than a mile apart. North of Navarino, it was the Searles, west of Navarino it was the Ramsdens, and off to the south it was the Cases. Clannishness, perhaps, but life was thus invested with a feeling of context, a sense of having a place in which you felt comfortable living and dying because almost everyone else in your extended family would live and die there. The old urban ethnic neighborhood but with more elbow room, you might say.

Why am I taking time to describe the social topography of Navarino for you? Why would you want detail such as the above concerning my roots? Read on.

*Mysticism and Metaphor*

As must be unmistakably clear by now, I'm feeling that God is drawing me not only into metaphor, but beyond metaphor into mysticism itself. Mysticism is defined, in the words of Emory University professor Luke Timothy Johnson, as "the individual search for unmediated contact with ultimate power."[188] You skirt the Bible and enter directly into what C. S. Lewis called "the region of awe."[189] You throw aside the curtain behind which is the Holy of Holies, God's inner sanctum.[190]

More specifically, what I'm envisioning is what has been called "extrovertive" mysticism, defined as that in which the mystic finds the One within the multiplicity of the things of nature.[191] Pantheism would not be a characteristic of this kind of mysticism, in which there is, as Christian educator Georgia Harkness describes it, "a vivid sense of the divine immanence in all things rather than their identity with God."[192]

---

[188] Luke Timothy Johnson, *Mystical Tradition: Judaism, Christianity, and Islam* (DVD) (Chantilly, Va.: The Teaching Company, 2008), lecture 1.

[189] See Devin Brown, *A Life Observed: A Spiritual Biography of C. S. Lewis* (Grand Rapids, Mich.: Brazos Press, 2013), 130. Lewis applied the phrase to what for him was "evidence of something supernatural beyond himself." Ibid.

[190] See, in this regard, Exodus 26:31; Matthew 27:50-51; Hebrews 9:1-8; 10:20. The curtain referred to above is the veil that separated the "holy place" in the Hebrew tabernacle and temple from the "most holy place," which was to be entered only by the high priest and was considered the very presence of God. Matthew records that at the instant when the life passed from the body of the crucified Jesus, this curtain was torn in two, signifying that through Christ's sacrifice believers were now able to gain admittance into the presence of God.

[191] See Georgia Harkness, *Mysticism: Its Meaning and Message* (Nashville: Abingdon Press, 1973), 60.

[192] Ibid.

Alternatively, as I've just suggested, God may be inviting me to simply seek out well-framed metaphors in nature that direct me to spiritual truth. A metaphor is something that's an emblem, representation, or symbol of something else. God is not literally a massive stone enclosure, but saying that He's "a mighty fortress" points to certain of His qualities and therefore describes Him.

Regardless of whether God's plan for my illumination is mysticism of some description or simply toying with correspondences (see further discussion infra), I think I made a wise choice in fleeing Navarino, as pretty as the countryside is, since I'm reasonably certain that had I stayed there I would have found myself starving for the resources and support needed to sustain a project such as this book. As I now realize and explain in the paragraph following, taking a fresh look at life would have been a problem. Fortunately, instead of staying home I married an urbane girl from New Jersey and have ever since lived with her in city or suburb including during our four-year stopover in the restored colonial city of Williamsburg.

In the Navarino kind of situation, for a mystic wannabe who has never known anything as a home but the world he grew up in, the question would arise: is this person communing with God or with the ghosts of ancestors and neighbors? I'm talking not about spiritism, but rather about paradigms that are reinforced for you as an adult by seeing only the kind of people that you knew as a child and adolescent, who in turn happen to be people who think and act in the same ways as their forebears. All of this taking place against the backdrop of a landscape that never changes. How can you get oxygen in that kind of setting! How can you hear directly from God! Maybe you can, but I wonder.

In his "Nature" essay, Ralph Waldo Emerson explained the metaphorical approach using the idea of "correspondence." The physical world, he said, is a reflection of the spiritual—specifically, of the Creator's mind—and is therefore *symbolic* of the divine, so that there is a one-to-one correspondence between natural laws and spiritual laws. In the "Language" chapter of "Nature," Emerson elaborated on the above idea, saying that the whole of nature symbolizes spiritual reality and offers insight into the universal, and once again claimed that there is a correspondence between moral and material laws.[193] This doesn't sound like Dr. Johnson's "unmediated contact" definition of *mysticism*, suffice it to say. We're definitely talking metaphorically.

In other instances, the mysticism-metaphor call is not made so easily. As I've observed many times on the beach at Campbell's Point, a dying campfire, one that's down to coals, "knows" when you add wood to revive it. Instantly, before the new pieces have had a chance to catch fire, the embers begin to glow more brightly. Does this reflect that there's some kind of soul in the fire or the wood? Does it show the truth of animistic belief? This reminds me of the Einstein-Podolsky-Rosen quantum physics experiment in which distant paired electrons are seen as communicating with each other. At a minimum, the encouraged reviving campfire seems to show that the world is not as impersonal as the materialists would have us believe. The world has personality if you look for it, a rudimentary cognition of some kind, and in this way reflects a Person behind it, it seems.

Things get even more complicated when you consider the teachings of Mary Hesse, Professor of the Philosophy of Science at the University of Cambridge, U.K.. Referring *inter alia* to papers by Max Black entitled "Metaphor" and "Models and Archetypes,"[194] Prof. Hesse opines that *scientific theories* are models, analogies, or metaphors rather than direct representations of reality.[195] The subject of the metaphor and the metaphor itself can never be separated, she says, since in explaining any scientific phenomenon we are *of necessity* engaging in a metaphoric description of that thing.[196] Not only does the metaphor necessarily become the theory; the theory also becomes the metaphor.[197]

The next logical step, it seems to me, is to say that Hesse's supermetaphor may refer to something *unscientific*. Why not agree

---

[193] See Emerson, "Nature," in *American Transcendentalists*, 48-55.

[194] These two papers are contained within *Models and Metaphors* (1962, Ithaca College Press).

[195] See Mary Hesse, "The Explanatory Function of Metaphor," in *Revolutions and Reconstructions in the Philosophy of Science* (1980, Indiana University Press).

[196] Ibid.

[197] Ibid.

that there are metaphysical or theological statements that can similarly be said to be metaphors and thus give these statements, also, a status co-equal with scientific theory? The above seems to be an approach that might get some of the more resistant non-religious thinkers over or around the wall of positivism (the philosophical position that the only genuine knowledge is that which can be obtained using the methods of science). Only when the conceptual barriers to religious belief are cogently challenged—by what passes for logical and persuasive argument in the secular world— can some people ever believe, it seems.

A word of warning, however: don't take the above approach as a suggestion on my part that God is nothing more than a metaphor. I'm merely trying to show that statements about God logically deserve the same respect as is accorded to scientific theory under thinking such as Mary Hesse's. Of course, I'm leaving the Holy Spirit out of the discussion at this point because I'm talking to secularists. Once secularism is overcome, by a "throw caution to the winds" decision to believe the Gospel, the Spirit appears on the scene and provides persuasion far superior to human logic or science.

C. S. Lewis explores the question of correspondence between natural things and spiritual things in his essay, "Transposition."[198] In Platonic fashion, Lewis observes that pictorial and other representations of things acquire meaning by virtue of their being part of the *real* world, by reason of their reflecting archetypes.[199] We therefore need to approach our transpositions (identifications of correspondences) "from above," he says.[200] This is essentially an endorsement of mysticism, as I see it, since in Lewis's analysis a small entity—a line drawing purporting to show light patterns, for example—can be fully understood only if the viewer is directly acquainted with something larger within which the smaller thing is contained— in this case, the earth's light sources and the visual effects they produce. The viewer, in other words, needs to have seen the sun and

---

[198] C. S. Lewis, "Transposition," in *The Weight of Glory and Other Addresses* (New York: Macmillan Publishing Co., 1949, 1980 rev. ed.), 54-73.

[199] Ibid., 61-62, 68-69.

[200] Ibid., 64-65, 72-73.

the moon and the world bathed in their light. This translates into us as humans needing to appropriate God's understanding of things.

Only by adopting God's thinking as our own will we be able to overcome personal perceptual limitations, such as might be evidenced by one's mistaking physical sensations for emotion, lust for love, politics for mere economics (James Carville: "It's the economy, stupid!"), or consciousness for nothing more than the activity of atoms in the brain. Otherwise, as Lewis points out, we become dog-like mentally.[201] Dogs generally don't understand what it means when humans point their fingers. Point to a morsel of food on the floor and the dog, instead of looking at the floor, will sniff your finger.[202] This is because a finger is nothing more than a finger to the dog; his world, as Lewis characterizes it, is "all fact and no meaning."[203] God, if we're successful in seeking to know His mind, allows us to rise above the dog's level. As light itself, He overcomes the darkness of our understanding (John 8:12; 2 Corinthians 4:6; 1 John 1:5).

Does mysticism therefore eschew facts? In a manner of speaking, yes. It casts aside what has been established empirically and seeks new facts, by encouraging experience that goes beyond "normal" experience. Its aim, we're told, is to uncover something that's "more real" than ordinary sensory input. In the materialist view of the world, contrastingly, ordinary sensory input is the sum total of our thought, although that input may be arranged and rearranged in the mind this way and that. For the contrary position, we're referred back to Plato, whose philosophy encouraged thinkers who would come after him to seek knowledge of the "ideal" that underlies what we see on the surface of things.[204] The "ideal" knowledge here referred to would be of Plato's forms or archetypes, mentioned earlier, which Plato illustrated by means of his parable of the cave. Being chained facing the back wall of Plato's cave is then a metaphorical description of everyday

---

[201] Ibid., 71.

[202] Ibid.

[203] Ibid.

[204] See the discussion at the end of the "Dancing Shadows" section of my Fluid Reality chapter, supra.

experience, which is thought to be all there is unless you're the rebel prisoner who escapes into the light of day and discovers that there's a better, "more real" world outside the cave.

*Scripture and Spirit*

Resort to analogy, as C. S. Lewis pointed out, is nothing more than an attempt to explain the unknown in terms of the known.[205] For example, we may start with a supposedly mystically-revealed understanding of God, one that might be said to require high-level comprehension, and then attempt, for whatever reason(s), to transpose it down into a lower or simpler one.[206] There will always be some people who insist that the simpler view is the only real one, Lewis indicates, and these are the persons who will be factually oriented and therefore dog-like.[207] But here we come face-to-face with the question, posed by mystics, of whether the Bible, by itself, is adequate as a guide for faith and practice. Is the Bible "all fact and no meaning" and for that reason not exactly what we need?

The Bible is a book of facts, the biblicist states, and is complete in the information it provides. God spoke and acted in specified ways, and his words and actions were straightforwardly recorded in writing in Hebrew, Greek, and Aramaic, later translated into Latin, English, and many other languages. The Bible's account of things is to be trusted because Scripture, all of it, is "given by inspiration of God,"[208] who "cannot lie."[209] We can rely on God's Word not only to give us a cosmology, but also for moral guidance: it is "a lamp unto (the psalmist's) feet, and a light unto (his) path,"[210] and is "profitable for doctrine, for reproof, for correction, for instruction in righteousness."[211] What need have we, then, to venture outside

---

[205] See Lewis, "Transposition," in *Weight of Glory*, 65-66; also Downing, *Awe*, 29.

[206] See Lewis, "Transposition," in *Weight of Glory*, 64-65; also Downing, *Awe*, 29.

[207] See Lewis, "Transposition," in *Weight of Glory*, 71-72; also Downing, *Awe*, 29.

[208] 2 Timothy 3:16.

[209] Titus 1:2.

[210] Psalm 119:105.

[211] 2 Timothy 3:16.

the Bible? Why do we require any other means of ordering our understanding of God's world and will?

The mystic might answer that ultimately it is the Spirit of God, not the Book, that "will guide you into all truth,"[212] since the Bible merely records what the Spirit has said. The counter-reply that is most often heard from the lips of the nascent mystic is subtly different, however. It's noted that the Bible in a number of places encourages the idea of God, as one or more of his three Persons, *actually living in* the believer. Cited in this regard are 1 Corinthians 3:6 ("Know ye not that ye are the temple of God and that the Spirit of God dwelleth in you?"), Galatians 2:20 ("I am crucified with Christ: nevertheless I live; yet not I, but Christ liveth in me"), and 2 Peter 1:4 ("Whereby are given unto us exceeding great and precious promises, that by these ye might be partakers of the divine nature,"). Verses such as these are interpreted by mystics as indicating that God *shares essence* with His earthly subjects. However, they could just as easily be construed as nothing more than a reference to the Godhead and a human soul inhabiting the same frame. Being in a house with another doesn't necessarily imply that there's a oneness of you and that person. Being a "partaker of the divine nature" in all likelihood refers simply to being "born again."[213]

That the mystically-minded sometimes leave the Holy Spirit out of their lexicon is not surprising. Not wishing to be separated from God the Father by an intervening advocate, Jesus (1 John 2:1), they decline to acknowledge the role of the Spirit for a similar reason: they believe that the immediacy of their relationship with Father God would be diminished should they concede that the assistance of the Spirit is needed in drawing men to the Father. And—what is critical for our discussion here—they also exclude the historic Bible from the game, reciting the mantra of "later revelation."

When God's written Word is properly handled, the goal is to find its true meaning rather than to become able to fancifully think of oneself as drawing closer to God, as mystics often do. The only reliable guide to the meaning of the Bible—i.e., to the truth—is

---

[212] John 16:13.

[213] See Morris, *Study Bible*, n. to 2 Peter 1:4.

Scripture literally interpreted.[214] Seek the truth about God first, and then you cannot help loving Him for who He is and what He has done, for the freedom that comes with knowing His truth.[215] Biblical literalism and biblical completism have to go hand in hand, it should be clear. Start putting words in God's mouth that He supposedly spoke after the canon of Scripture was complete and you change the meaning of what was already there in the canon. Thus, biblicists are right in taking the stance that the Bible cannot be added to by mystical experience. The revelation contained in the Word of God is frozen in accordance with proscriptive statements by the apostle John and Jesus himself (Revelation 22:18-19; Matthew 5:18).[216]

Against the rejection of "recent revelation" is raised the "inadequacy of language" argument. The Sapir-Whorf hypothesis in linguistics is the theory that language determines the way we see the world.[217] This hypothesis is named for American linguistic anthropologists Edward Sapir (1884-1939) and Benjamin Lee Whorf (1897-1941), whose studies of the American Indian, Eskimo, and Hopi languages disclosed structures and vocabularies that were not always parallel to those found in European languages.[218] In Hopi, for example, there were no Indo-European past, present, or future tense markers, a reality found to be connected to a cultural perception that time is not a dimension but rather a condition of unchanging, certain, or uncertain knowledge. The idea that thought is structured by language raises the possibility that mystics and pentecostals are right. Perhaps there is communication that we need to have with or from God that exceeds the ability of our language to express or relate.

Pentecostals (of which I was once one, more or less—I never could get the hang of speaking in tongues, as I told you) are of particular interest as regards the language issue. The pentecostal/charismatic solution to the problem would seem to be given legitimacy

---

[214] See Morris, *Study Bible*, v.

[215] See John 8:32.

[216] See further in this regard Morris, *Study Bible*, nn. to Rev. 22:18, 19, Matt. 5:18.

[217] See Rohmann, *Ideas*, 351-52.

[218] Ibid.

by verses such as Romans 8: 26, which says that the Spirit of God "maketh intercession for us with groanings that cannot be uttered;" however, Luke's exciting account of the coming of the Holy Spirit in Acts 2 is counterbalanced by Paul's cautions to the Corinthian church regarding tongues (1 Corinthians 14:7-25).

It's tempting to suggest that sociological factors are the key to understanding tongues speaking as a solution to the "inadequacy of language" problem. Pentecostals have often been characterized—unfairly, in my experience—as hillbilly snake handlers and other socially-deprived folk looking for miracles to lift them out of their physical or financial problems. Assuming for argument's sake that the above characterization is correct in some cases, it may be that there is a linguistic poverty on the part of pentecostals that makes them more receptive than others to tongues as a way of communicating with a God whose vocabulary and syntax are still sometimes viewed as limited by Elizabethan standards—i.e., who speaks King James English. This by itself does not invalidate their attempt to establish a better communication with God, however.

Whereas the aspiring mystic and his pentecostal cousin try to get to God by skirting the restrictions of language, the postmodernist tries to get around the idea that there even is a God and uses, as an excuse for his nihilism, the lack of fixed meanings for the words by which God would be described. This could be seen as the flip side of Anselm's claim that God exists because no person or thing greater than He is can be described. In Anselm's ontological argument, the inadequacy of words is a result not of the instability of their meanings, but rather of the total stability or excellence of God. If, as postmodernists urge, words have no fixed meaning—i.e., don't attach to anything certain "out there"—how can we begin to think about eternal verities such as God? goes the postmodernist argument. Such reasoning can be applied in reverse: If there's no divinity underlying the world, there's nothing of substance to which human understanding can attach. We can't know anything because there's nothing to know.

Also, if we're congenitally unable to use our words in crafting a statement of truth, how can we ever hope to establish the boundaries of self? That would be a logical first step toward understanding our place in the universe and our eternal destiny, would it not? But postmodernism won't let us arrive at even that point. How, then, can we hope to attain true knowledge of *anything* as long as we're suffering from the postmodern condition? Derrida, Foucault, and company answer that we can't, but that response somehow doesn't satisfy. We'll talk more about postmodernism in my Identity and Orientation chapter, infra.

*Maps and Mariners*

A Friday, later in October, two days before we're scheduled to leave The Cottage for the winter. Though I'm retired and neither Sue nor I has any obligations in Rochester, our departure date is set because that's the last day the Campbell's Point water will be on. The water pipes and valves of our community system are close to the surface of the ground and therefore susceptible to freezing, which is the reason for the on-or-about-October-fifteenth shut-off date. Some of the forty-nine families on the Point have their own wells, enabling them to ride out the winters, but we're not one of them. The water comes back on around April fifteenth (an email is sent to everyone announcing this); thus, our season is six months minus July and August, when we rent The Cottage out to help meet expenses. Our summer home is only a spring and fall home, for economic reasons.

I'm standing before the wall map in the dining room, studying it as I often do. This map, published by the National Oceanic and Atmospheric Administration (NOAA), is thirty by thirty-six inches borders to borders and shows a geographic area of slightly more than two thousand square miles. Kingston, Ontario, appears at the top of the map, slightly left of center, and Stony Point—the same Stony Point I explored a few days ago—is in the lower right hand part of the map. As might be expected with a U.S. Department of Commerce map, information concerning water depths in the Lake and surrounding waters is assiduously and plentifully provided, and underwater contours of the various land masses are shown, along with light houses and light buoys. NOAA wants to make sure you don't run aground.

# THE REGION OF AWE

Of interest to me on this particular day is the chain of islands and shoals that runs more or less due west from what to us is the far side of Stony Point. The names of the islands and shoals are picturesque—(from east to west) Stony Island, Little Galloo Island, Galloo Shoal, Yorkshire Island, Main Duck Island, Psyche Shoal (I particularly like that one), Harris Shoal, and, at the western edge of the map, a group of islands identified as the False Ducks islands.

A friend from Kingston, Michael, has sailed out to the False Ducks islands. We know Mike, a physics professor retired from teaching at Queens University, through his sister Jenny, who was Sue's roommate during the college year Sue spent in England. Jenny has an intriguing conversion story in that, as she tells it, she was saved when she went into a cathedral to say "thank you" to "someone or something that[ she] didn't believe in" and at that time distinctly heard a voice ask her if she was going to go on with her life in the manner she'd been living it up until then (as one of the founding members of the University of Exeter's skeptics society, *inter alia*). She answered "no," whereupon the voice said, "follow me, then," and it was after this that her life changed entirely.

Jenny flies over from England to visit Mike every summer, and Sue and I drive up to Kingston to see both of them, usually for a day, while Jenny is there. Mike, with Jenny, has visited us at Campbell's Point by car, and has talked of sailing his boat over to Association Island, a distance of some forty or fifty miles of open water. Like the sail to the False Ducks, this would be an impressive feat, I think, for a man in his seventies in a twenty-foot boat sailing single-handed; nonetheless, Mike is dismissive of the thought that there's something epic about such a jaunt.

I wonder what it would be like to be out on the water and be unable to see land, navigating strictly by compass, sextant, or what-have-you. I keep forgetting to ask Mike what device or method he used to find his way on his long sail to the False Ducks. Nowadays, I know, he has GPS (Global Positioning System), which ties him electronically to a network of satellites. I try to imagine looking around and seeing nothing but heaving waves. On a cloudy day or with the sun directly overhead, would I be hopelessly disoriented as the waves tossed our little boat around?

"Our little boat" these days is the twelve-foot aluminum Starcraft of indeterminate age that I bought from an auto repair guy in Cortland, whose number I got from a Penny Saver ad. "That's a long way to come for a boat," he said, referring to my having come from the North Country to Central New York to pick the boat and its motor up. "It was what I was looking for, and all the others were snapped up by the time I called on them," I explained.

As it turned out, it was just the boat and trailer that I took back to Campbell's Point, since the 9.5-horse Mercury that our garage man was offering wouldn't start. Also, the lights on the trailer didn't work, despite a half- hour's effort on the seller's part. I hot-footed it up Interstate 81 so as to be sure to get back to the Point before dark, and was not pulled over by State or local police. I was thereafter content with the net result of the day's work. We weren't going to venture far from shore with the grandkids in the boat, and the trailer would never need to be taken off of Campbell's Point Association property.

Now I find myself wishing I had a boat with a motor, so that I could at least get to the Lake side of The Islands (Six Town Point, Gull, and Bass). Any travel farther out than the three familiar Islands[219] might be too much to undertake, given the suddenness and severity of Lake Ontario storms. I've asked the proprietor of Henderson Marine to be on the lookout for a reconditioned motor for me, but haven't heard anything back from him.

At least I have the wall map to let me know what's out there in the Lake. Appropriately, C. S. Lewis had something to say on the subject of maps in *Mere Christianity*.[220] He noted that someone who has looked at the Atlantic Ocean from one of its shores and then looks at a map of the Atlantic will be turning from something that's real to something that's less real for him. The map, nonetheless, is based on what a great many people have found out by sailing on the ocean. It will therefore be more useful in getting you to (America) (England) than what you learned gazing at the water while standing on the beach.

---

[219] As previously noted, I don't regard Association, the fourth such land mass in the Stony Point chain, as an island. I've been on it without boating to it.

[220] See Lewis, *Mere Christianity*, 154-55.

Similarly, Lewis tells the reader, Christian doctrine is based on the experience of hundreds of people who actually were in touch with God, who had "experiences compared with which any thrills or pious feelings you and I are likely to get on our own are very elementary and very confused."[221] The benefit of foundational religious experiences of others is what the Bible gives you, in other words. Nonetheless, Lewis counsels, you must *use* the map; you won't get anywhere by looking at maps without going to sea.[222]

Does "going to sea," in Lewis's analogy, mean exposing oneself to whatever spiritual weather there may be out on the deep, being ready for one's boat to be swamped by the experience should it happen that way, while still navigating according to the chart provided by God? Is a Bible-directed mysticism what we should be looking toward? Or should it be a mysticism-directed Bible—i.e., a central document with respect to which mysticism is, at the very least, a major hermeneutical tool? The answer is not always clear from Lewis's writing. Isaiah 55, at verses 6 and 7, does seemingly counsel a direct approach to God:

> Seek ye the Lord, while he may be found, call ye upon him while he is near:
>
> Let the wicked forsake his ways, and the unrighteous man his thoughts: and let him return unto the Lord, and he will have mercy upon him; and to our God, for he will abundantly pardon.

Meeting with God after repenting has a mystical sound to it, you may say, or at least there's nothing unmystical about it. But does this lead to an elevated level of understanding about spiritual things? Isaiah cautions:

> For my thoughts are not your thoughts, neither are your ways my ways, saith the Lord. For as the heavens are higher than the earth, so are my ways

---

[221] Ibid., 154.
[222] Ibid., 154-55.

higher than your ways, and my thoughts than your thoughts. (Isaiah 55:8, 9)

This would seem to be a clear rejection of mystical pantheism, which infected New England transcendentalism and now poisons New Age/Emergent Church thinking. We can expect forgiveness, God tells us through Isaiah, but not God-like status.

May we nonetheless, as I've suggested, see nature as a *metaphor* for where man stands in relation to God? We could say that the universe is pointing us toward God in much the same way as it did C. S. Lewis, but stop short of attaching the label "mystical insight" to what the water and the trees and the stars tell us.

Whatever we call nature's light, we must be sure to pass it through the filter of Scripture. Otherwise, we risk committing the error of the Emergent Church folks, who insist that their religion be a scriptureless wending of one's way, supposedly toward God but with no guarantee of arrival in His kingdom. If the latter is what mysticism is supposed to be, then we may rightly reject it out of hand, it seems to this observer.

From what I've said in my Preface and at the beginning of the present chapter (The Region of Awe), it should be apparent that the mystic's vaunted encounter with God does not necessarily have to involve nature. American philosopher and psychologist William James, in the first of his Gifford lectures on mysticism, published in 1902, began by identifying four marks of mystical experience: (1) what happens is *ineffable* (not describable in words), (2) it has a *noetic* quality (the feeling of knowing something that could not be known before), (3) there's a *transiency* to it (the experience is described as fleeting), and (4) the mystic has a feeling of *passivity* (of being not in control).[223] After quoting several accounts of alleged mystical happenings, James observed that certain aspects of nature "seem to have a peculiar power of awakening mystical moods."[224] Most of the striking cases that he had collected occurred out of doors, he said.[225]

---

[223] James, *Varieties of Religious Experience*, 328-30.

[224] Ibid., 340.

[225] Ibid.

Nonetheless, it will be noted that enhanced appreciation of nature is not included in James's marks of mystical experience, nor does it figure significantly in the approaches suggested by New Age teachers and by English mysticism authority Evelyn Underhill.[226] As I state in my Introduction, however, I'm regarding our acquisition of The Cottage as a seminal event as far as the writing of this book is concerned; thus, our focus at present must be on succeeding to God's knowledge with the help of Nature—i.e., on making nature mysticism work as a way of discovering metaphysical reality .

*Intuition and Exposition*

Returning to the debate with respect to the relative provinces of mysticism and the written Word of God: it all seems to come down to the question of whether nature is a metaphor for spiritual reality or an object in itself. If it's the latter and the nature-as-object position is taken to its extreme, you wind up with pantheism a la Spinoza (everything is *God-or-Nature* rather than God *or* nature). If, on the other hand, you say that there's nothing in your encounter with God that transcends normal experience or Scripture, then there's nothing supernatural in the intuition you seem to have gained. That doesn't sound right, either. Is there somewhere a middle ground, a milieu in which one may spiritually comprehend God and be inspired by Him without essentially becoming God? The character of the aspiring mystic's "intuition" is key, I would think. In this regard, William James states:

> Mystical truth resembles the knowledge given to us in sensations more than that given by conceptual thought It is a commonplace of metaphysics that God's knowledge cannot be discursive but must be intuitive, that is, must be construed more after the pattern of what in ourselves is called immediate feeling, than after that of proposition and judgment.

---

[226] See, e.g., Foster, *Celebration of Discipline*, 1-201; and see Evelyn Underhill, *Mysticism: A Study in the Nature and Development of Spiritual Consciousness* (Mineola, N.Y.: Dover Publications, 2002), 167-443.

> But our immediate feelings have no content but what the five senses supply; and we have seen and shall see again that mystics may emphatically deny that the senses play any part in the very highest type of knowledge which their transports yield.[227]

If the mystical experience is primarily sensual, do we then have any hope of intellectually understanding our existence? A tenet of postmodernism that I neglected to mention earlier is that there is no grand account of history that explains everything by referring to a single theory or principle.[228] Grasping such a metanarrative would entail primarily cerebral rather than carnal experience, I'm assuming. If the search for a grand narrative is abandoned, does this push modern philosophy farther in the direction of seeing mystic experience as sensual rather than intellectual?

The question of whether a mystical encounter is a sense experience or an intellectual experience, like the "Bible-versus-beyond-the-Bible" issue, has long been debated. David Downing (*Into the Region of Awe*) notes that C. S. Lewis defined mysticism as a "direct experience of God, immediate as a taste or color."[229] But taste and color must be apprehended by the physical senses. With what sense, Downing asks, is an invisible deity to be understood?[230] And if words, the currency of the intellect, are inadequate to describe a direct sense experience such as color or the aroma of coffee, how can they be of much use in portraying God?[231] Nonetheless, William James insists:

> The kinds of truth communicable in mystical ways, whether these be sensible or insensible, are various.

---

[227] James, *Varieties of Religious Experience*, 351.

[228] See Rohmann, *Ideas*, 310.

[229] Downing, *Awe*, 19.

[230] Ibid.

[231] Ibid For a discussion of the implications of having sensual knowledge versus the supposedly "higher" mystical or metaphysical knowledge, see David K. Clark and Norman L. Geisler, *Apologetics in the New Age: A Christian Critique of Pantheism* (Grand Rapids, Mich.: Baker Book House, 1990), esp. 160-65, 182.

> Some of them relate to this world, (including) *the sudden understanding of texts* (emphasis and parenthetical supplied)[232]

James continues with the statement that "the most important revelations are theological or metaphysical."[233] It's not clear, however, whether he's referring to truth directly revealed by God or truth discovered by applying divinely-approved hermeneutics.

Regarding the interpretation of Scripture, a preliminary question is that of whether the reading of a sacred text can open the door to truth beyond its words. Can the reader or a listener thereby be put in a condition of receptivity to *unexpected* truth from God? In this instance, the focus would be not on the document in question (the Bible) but on its Author. The document or text would be merely a pass-through. The Book rather than the brook, in this case, is the means of access to the Beyond, but while what happens transcends hermeneutics, it couldn't happen without The Book.

As I write the above, I worry that I may sound like I'm advocating the practice of *lectio divina*, also known as "sacred reading," "divine reading," or "spiritual reading," a mystical practice that is becoming increasingly popular with evangelicals.[234] With *lectio* as with contemplative prayer, the key supposedly is to not let your intellect interfere with the enterprise of hearing from God.[235] The proper technique is to begin (in the *lectio* stage of a four-stage progression that is also known as *lectio*) with a chosen passage of Scripture, read it aloud two or three times, pick out from this passage a particularly interesting word, and repeat this word over and over again without trying to figure out its meaning or why it was given.[236] By the time he or she reaches the so-called *meditatio* stage of this exercise, the attention of the participant should be turned

---

[232] James, *Varieties of Religious Experience*, 355.

[233] Ibid.

[234] See Gilley, *Out of Formation*, 63-80.

[235] Ibid., 73, 79.

[236] Ibid., 74-75.

inward, away from the Word of God and to subjective thought that is being interpreted as coming from the Lord.[237]

The *oratio* and *contemplatio* stages of *lectio* follow. These represent the achievement of the actual goal of *lectio*, which is not to bring about more complete understanding of the Bible but rather to deliver the believer to a contemplative state consisting of a heightened awareness of God and feelings of love and nearness toward Him.[238] This is basically the same result as would be sought through contemplative prayer, in which the methodology is that of emptying oneself mentally, maintaining silence before the Lord, and being receptive to His communication and leading.

It should come as no surprise that some people would prefer to not involve the Bible at all in the mystical experience other than by acknowledging that a monotheistic god of some description exists. No journey into mysticism can proceed except from the starting point of belief in a single powerful deity, it seems clear, at least none embarked on under the Western world-view (Christianity, Judaism, and Islam). But other than that, it's all up for grabs. The teachings of Rick Warren and others prominent in the New Age/Purpose Driven/Emergent Church movement clearly demonstrate this. Pantheism ("As above, so below") is embraced without reservation.[239] The Bible is consistently devalued and marginalized.[240]

The lack of New Age interest in Scripture goes beyond the longstanding mystic preference for "being" over "seeing,"[241] which might be described as seeking a spiritual bubble bath and nothing more from one's "unmediated contact" with God. The danger should be apparent: when you don't keep the Bible at your side at all times during your spiritual wanderings, you become susceptible

---

[237] Ibid., 75-76.

[238] Ibid., 77.

[239] See, e.g., Warren B. Smith, *A "Wonderful" Deception: The Further New Age Implications of the Emerging Purpose Driven Movement* (Silverton, Ore.: Lighthouse Trails Publishing, 2009), 18-20, 43-46, 165-67.

[240] See, e.g., ibid., 100-101, 104-6, 125-26, 187.

[241] See Downing, *Awe*, 147-48.

to odd ideas such as world peace under the umbrella of mysticism.[242] It needs to be understood, in this regard, that the Bible is more than your helpful home companion. It's an end in itself, along with God the Son. "Word" is applied to both Christ and His gospel in Scripture (see, e.g., John 1:1, 4; 15:20; Acts 13:49; 2 Corinthians 2:17; Colossians 3:16; Revelation 19:13). Know the Word, and you know the Word. Strike out on your own, without the Word, and you know nothing. I know that much.

*Chemical Mysticism*

Some who are wading through this discourse may be wondering whether chemicals are of any value in facilitating mystical experience. Drag your lawn chair out onto the beach, Campbell's Pointer, with a double Scotch in hand. Do it after dark so the jet skis won't drown out your thoughts, or rather God's thoughts. Listen to the waves breaking on the shore. Look up at the sky. Pour yourself another double if you think you need it. William James concluded that alcohol has "power to stimulate the mystical faculties of human nature," as "the great exciter of the *yes* function in man."[243] Nitrous oxide (laughing gas), also, can give you a mystical high, according to James, but what you're left with after you come to your senses makes no sense.[244] James tried various other chemicals.[245] He vomited for twenty-four hours after using peyote, but had no visions.[246] He seems to have viewed alcohol the most favorably of any of the substances he experimented with.[247] David Downing comments that after James's early experiments with mind-altering drugs, "the

---

[242] See Smith, *"Wonderful" Deception*, 155, 167-71. The truth is that there will be no peace until Christ, the Prince of Peace, returns (Isa. 9: 6-7; 11: 1-16; 65: 17-25; Luke 1: 32-33; Acts 1: 9-11; Rev. 2: 27; 12: 5; 20: 1-6; 21: 1-6).

[243] James, *Varieties of Religious Experience*, 334.

[244] Ibid., 335.

[245] Downing, *Awe*, 155.

[246] Ibid.

[247] See James, *Varieties of Religious Experience*, 334-38.

boundaries between research and recreation became increasingly blurred."[248]

A substance can be psychotropic (perception-altering) without being psychedelic (perception-creating; hallucinogenic). One of the earliest "researcher(s)" to advocate psychedelics as an aid to mystical insight, Downing notes, was Aldous Huxley. In *The Doors of Perception* (1954), Huxley described his sensations while under the influence of mescaline.[249] He reported that in his altered state he saw "what Adam had seen on the morning of his creation—the miracle, moment by moment, of naked existence."[250] Downing's summary comment concerning *Doors* is that in general Huxley's account "sounds less like a mystical revelation than a 1960s fraternity party."[251]

Even more unkind was the critique that came from British historian of religion R. C. Zaehner, who refused to credit Aldous Huxley's claim of having had a mystical experience and flatly rejected Huxley's suggestion that mescaline be made a part of Christian worship.[252] These and other negative reviews notwithstanding, *Doors* became a counterculture classic in the 1960s and 1970s and was followed by the writings of other mind-expansion gurus, including Carlos Castaneda (*The Teachings of Don Juan: A Jaqui Way of Knowledge*) and Timothy Leary ("Turn On, Tune In, Drop Out"). Anyone who was an adult or near-adult during the 60s and 70s will likely find it difficult to suppress a smile when the above names are mentioned; nonetheless, what these guys were urging—"break(ing) through to the other side" through drugs—was thought of as serious business at the time.

---

[248] Downing, *Awe*, 155.

[249] This was the grandson of Thomas Huxley, "Darwin's bulldog." His *Doors of Perception* inspired the name of a 1960s rock group, The Doors, who likewise used drugs as a portal to "another world." For further biography, see David W. Cloud, *Seeing the Non-Existent: Evolution's Myths and Hoaxes* (Port Huron, Mich.: Way of Life Literature, 2011), 106-7.

[250] Aldous Huxley, *The Doors of Perception and Heaven and Hell* (New York: HarperCollins Publishers, 1954, 1955-56), 17.

[251] Downing, *Awe*, 155.

[252] R.C. Zaehner, *Mysticism Sacred and Profane: An Inquiry into some Varieties of Praeternatural Experience* (Oxford, U.K.: Oxford University Press, 1957), 19-20.

# THE REGION OF AWE

As a college English major during the 1960s, I may have been helped scholastically by the mind-expansion craze. I had a professor, one of those unpublished types often found on the faculties of small liberal arts colleges, from whom I had been unable to get an "A" on a paper. With a due date of the following day bearing down on me, I sat down with my Royal typewriter and a couple of six-packs of Ballentine's Ale and began to write. Predictably, the paper soon took shape as a stream-of-consciousness piece. I passed out (went alcohol-assisted unconscious) as I was about to begin page seven. I later awoke, bleary-eyed finished my writing, and staggered off to class. A few days later I received my masterpiece back with an "A" written on the first page, along with a brief comment that I can't exactly remember now. My professor had apparently viewed my work as the product of a mystic experience.

Oxford scholar Zaehner, in critiquing *The Doors of Perception*, further suggests that mysticism of the kind engaged in by Huxley will do nothing to improve your character.[253] Mysticism, if it's worth anything, should lead to altered *traits*, not just altered states, as a paraphrase of Huston Smith puts it.[254] Profs. Smith and Zaehner find support for their position in the writings of the apostle Paul, who stated, in his first letter to the Corinthian church:

> (T)hough I have the gift of prophecy and understand all mysteries, and all knowledge; and though I have all faith, so that I could remove mountains, and have not charity, I am nothing. (1 Corinthians 13:2)

Henry Morris urges that the word "charity" should be here read as "a generous and unselfish concern for others."[255] Unfortunately, "charity" as thus defined often seemed to be lacking in the large pentecostal church to which Sue and I belonged in the closing years of the twentieth century. Call the church office in search of someone to help you with a personal problem and you find that all the pastors are off on a spiritual sightseeing trip in California or

---

[253] Zaehner, *Mysticism Sacred and Profane*, 3, 93, 104-5.

[254] Downing, *Awe*, 156, 190.

[255] Morris, *Study Bible*, n. to 1 Cor. 13:1.

some other remote place, "getting blessed." Come into the church sanctuary with your aged mother on a Sunday morning and you find that seats are being staked out, by people throwing their coats and Bibles down on them, faster than Mother can reach any of them even with you dragging her along. Park your car in the church's lot in time to let you get in and get a good seat for the service and you'll come back afterward to find that you've been blocked in by latecomers.

The above is symptomatic. A shortcoming of contemporary Christian, in my opinion, is its focus on the believer's experiencing a personal "high" rather than transcending the limits of self by reaching out to others in altruistic love.[256] Which is more Christlike? we need to ask. As I'll develop at greater length later, the self-centeredness is especially evidenced in the present reversion to mysticism.

"Altered traits?" I'm not sure alcohol ever made me any better as a person, even when I was consciously seeking God, so I would have to agree with Drs. Smith and Zaehner that there's no reason to believe that substances that make you high will lift you high.

I have no basis, either, for concluding that drugs can produce an enhanced appreciation of one's natural surroundings. R. C. Zaehner notes that experimenter Huxley reported a heightening and deepening of normal color perception.[257] I never experienced anything like that in my collegiate research with beer and ale, and I never took any of the more exotic substances that Mssrs. Huxley, Castaneda, and Leary fooled around with, so I can't comment on their mind-expansion claims other than I have above. If anything, as I found on many an occasion before I reached drinking age,[258] an awareness of God seemed to arise from the *unaltered* (sober) version of nature. God didn't later use alcohol to show me more of nature, that I can see. And God wasn't nature Himself. As an underachieving undergraduate, I didn't even know who Spinoza was.

---

[256] Regarding the increasing self-absorption of modern man generally, see David Brooks, *The Road to Character* (New York: Random House, 2015), esp. the block quote at 10.

[257] Zaehner, *Mysticism Sacred and Profane*, xiii, 4-5, 12.

[258] The drinking age in New York at that time was eighteen.

# THE REGION OF AWE

*Summer Camp*

"Nature," for me, has more often than not involved water, as you may have guessed. Enter two more of the Finger Lakes, Cayuga and Owasco (Fig. 8).

Camp Gregory was buried deep in woods that sloped down to Cayuga Lake, and was what you would call a rough-hewn complex (bare wood inside and on the outside of all buildings). The camp was open to groups of all denominations. The summer after my sophomore year in high school, our Methodist Youth Fellowship rented Gregory for three days over a weekend. We took our own kitchen workers, older ladies from the Navarino church. The Youth Fellowship girls had little in the way of active chaperonage, however, since the cooks mainly stayed in a structure behind the dining hall. Between the boys' cabin and the girls' there was a deep gully, but that didn't deter panty raids by the boys any more than did the theoretical chaperons. An adult counselor stayed with the boys, but only for part of the night.

The above aside, spiritual things did happen at Camp Gregory that weekend. Both mornings, campers were directed to each find a place where they could be alone with God and spend an hour there. The idea was not specifically to seek mystic solitude—indeed, there was little awareness in the mainline churches at that time of mysticism and the spiritual disciplines thought to support it.[259]

I chose the lake shore. Only one other youth made the same choice, and his "spot" was approximately fifty yards up the shore from mine. I could see him by turning my head slightly, which I did from time to time to check on whether he was doing anything more spiritual than I was. My head wasn't bowed at any time, nor was his that I could see, so there was no prayer going on, at least not in the customary fashion. We were each meditating in his own way, I told myself.

Cayuga, one of the larger of the Finger Lakes, was quite turbulent both mornings we were there; waves broke on the stony shore with force sufficient to throw spray three or four feet in the air, and

---

[259] See, with respect to the discipline of solitude, Foster, *Celebration of Discipline*, 96-109; also Gilley, *Out of Formation*, 81-100.

the wind blew droplets of lake water against my face. I found the roar of the surf and the misty wind reassuring. There had to be a God, I thought. From what other source could things this awe-inspiring come! I hoped to hear from Him during my time by the lake, but He would have to make the first move. He never did, except to announce His presence through the surf. Perhaps I should have seen that as enough.

The next summer, also for the first time during my Youth Fellowship years, I attended Casowasco, a camp on Owasco Lake, a body of water that was smaller and calmer than Cayuga (see Fig. 8). Casowasco had buildings, especially the main house, that were more finished and gracious in appearance than Camp Gregory's, and also had shaded lawns that were bounded on the west by tennis courts and on the east by the lake and a boathouse. The place, it was obvious, had been someone's estate at one time.

The boathouse was of considerable interest to the young men and women who came to Casowasco. It was a storied place for romance, although camp staff at the beginning of the six-day session warned against its use for that purpose. The danger of amour was multiplied by Casowasco's campers being from churches that were scattered all over central, southern tier, and western New York. Many of the teens were their church's sole representative; thus, there was a Las Vegas-like anonymity[260] that didn't exist at Camp Gregory. I was the only person from Navarino Methodist.

I should add that I was painfully shy when I went to Casowasco that summer. I had no idea that I would become the subject of the attentions of a young lady from Branchport, a village on the north end of the west fork of Keuka Lake (my wife's college was at Keuka Park, on the east fork, as I noted earlier). A high school cheerleader and very flirtatious. We quickly found ourselves paired off and hanging out with two other couples. A fat boy among the other campers, seeing something happening that did not involve him, complained at one of the teaching sessions mid-week that the only reason anybody came to Casowasco was to "see what the other sex look(ed) like."

---

[260] "What happens here, stays here."

Thursday night, after the evening chapel service, a number of the kids that had become friendly were standing around in groups in a dimly-lit area under a grove of trees between the chapel and the main house. Suddenly I found myself in the embrace of my cheerleader, who planted what the world would call "a big wet one" on me. She kissed me. My first kiss, and *she* kissed *me*. What followed in the remaining two days of the camping session was hand holding and discreet hugging and kissing as the opportunity presented itself, nothing more. We never did make it to the boathouse. Donna was not up for such risky behavior, I surmised, and I wasn't sure I was.

Did anything spiritual happen at Casowasco? I liked to think that this was the case. Perhaps it was the idea that I and this raspy-voiced (from cheerleading) girl from another Methodist church in another part of the state were united by more than mouth contact, by our being in a place where devotions were read and nondoctrinal Christian songs ("Kum by yah, my Lord," "Do, Lord oh do, Lord, oh do you remember me," and the like) were sung at each meal.

Saturday night's lakeside service confirmed, at least in *my* mind, that we were having a "religious experience." We lit our candles, dripped the wax from them onto pieces of cardboard so as to in effect affix the candles onto flatboats, and then sent our candle boats serenely drifting out into the lake. We would never have been able to do this on choppy Cayuga, I thought. As the candles floated farther and farther away, their flames and the reflections thereof becoming tinier and tinier, I thought of the promise I had made to God a few minutes earlier, not knowing exactly who I was talking to, that I would try to find out His purpose for my life. Not a mystical encounter, but a "strong awareness of God" experience as I viewed it then and, I guess, still do.

The desire to have a "religious experience" would characterize church attenders decades later. Pastor Gary Gilley, in *This Little Church Stayed Home*, has indicated that in order to evangelize the citizens of the present postmodern age, the church must understand and respond to a mindset that wants its felt needs met and wants to have a religious experience.[261] It is Dr. Gilley's observation that the

---

[261] See Gary E. Gilley, *This Little Church Stayed Home: A faithful church in deceptive times* (Darlington, U.K.: Evangelical Press, 2006), 39.

postmodern lost "will never be reached through the offer of authoritative truth," since any claim that a belief system is in possession of absolutes is viewed with suspicion.[262] Pleasant candle boat feelings rather than hard theology are the order of the day, it seems.

*Romans 1: 20*

In my Introduction, I quoted Romans 1: 18-20 for the proposition that we have the witness of nature, by which C. S. Lewis was conspicuously influenced, to tell us about God. It now strikes me, as I sit at the dining room table of our house in the city of Rochester, watching the leaves from the maple tree in our back yard drift hesitatingly downward, bits of yellow fluttering past a cold gray sky, that Scripture is not only *explained by* nature; it also *explains* nature. Verse 20 of Romans 1 is a perfect example. Paul states of God that "the invisible things of him from the creation of the world are clearly seen, being understood by the things that are made, even his eternal power and Godhead;" The meaning of this verse on a surface level could not be clearer, it seems to me. The world and everything in it were created, "made," at a definite point, the "beginning" referred to in Genesis 1: 1, by a God possessing "eternal power." A personal God ("him"), a triune God—Father, Son, and Holy Spirit, as "Godhead" has always been understood by Christian theologians.[263] No member of the Trinity was visible on the earth as of the time of Paul's letter to the Roman church; thus, there were indeed "things of [God]"—i.e., divine attributes—that were "invisible."

Is our understanding of God, who is pointed to by the above verse, helped by considering "the things that are made"—i.e., the material objects and beings of the universe created by our Lord? Absolutely. We look around and see a universe of staggering magnitude and complexity, almost certainly a product of exercised "power." Science tells us that this universe is "running down," that the amount of energy available to accomplish useful work is constantly decreasing under the second law of thermodynamics, also

---

[262] Ibid.

[263] See Morris, *Study Bible*, nn. to Romans 1:20.

known as the law of entropy. Therefore, this universe cannot have existed from eternity, since if it had always been here, it would by now be in a state of physical equilibrium—thermodynamically dead, cold, and barren. In fact, if the universe had existed from eternity, it would always have been in a state of thermodynamic equilibrium—a logical impossibility that further illustrates how incompatible an eternal universe is with the law of entropy. Given the lack of usable energy, life could no longer be sustained, if it ever were possible (see the preceding sentence).

The cosmos thus has to have had a beginning (Genesis 1:1). And something or someone has to have preceded that beginning; otherwise, there would have been no source from which could come that which was necessary—a basis for being—in order for the beginning to happen. The so-called basis for being could have been nothing more than a physical principle, but it had to have existed; otherwise, nothing could have come into existence. To assure that the universe and its inhabitants would actually become reality, the creative something or someone could not have been subject to the disability of itself (Himself) needing a beginning; thus, the power that created the world had to have been eternal, as Romans 1:20 says it was.

Is the Trinity specifically evidenced in nature? Keep in mind that matter (convertible to or from energy), space, and time are the totality of what we find in the physical universe. Thus, by virtue of their collective completeness, these three things can be said to mirror the Godhead, which (who) similarly is unitary and all-encompassing in nature (2 Corinthians 13:14; Ephesians 4:6; 1 John 5:7).

How does Romans 1:20 *explain* nature? Answer: the verse posits the presence, since the beginning of the world, of man, who was given the ability to understand "the invisible things of God"—i.e., God's attributes and workings, including the Trinity—from "the things that are made"—i.e., from God's creation and creatures. God would have had no need to further understand Himself by looking at His creation; thus, "from the creation of the world are clearly seen" can only refer to human observation of the world. This excludes cosmologies in which man appears on the scene millions or billions of years after Earth comes into existence.

Late-appearance-of-man schemes have included the "gap" theory, which interprets verse 2 of Genesis chapter 1 as referring to the aftermath of a cataclysm that is said to have preceded man's appearance and left the already-many-geologic-ages-old earth, with its already-accumulated multiple layers of fossils, in a state of desolation.[264] There is also the "day-age" theory, under which each day of biblical creation was supposedly an age billions of years long.[265] It hardly needs stating that with both "gap" and "day-age," the name of the game was to propound an old-earth cosmology that gave evolution time to have worked its supposed wonders.

What it comes down to, then, is that nature may indeed have something to show us about ultimate reality, the explanation of the unexplained. The apostle Paul—correction, the Holy Spirit speaking by means of the apostle Paul—has encouraged me to see nature as informative, through the revelation contained in Romans 1:20. From what I've already said, I think you can fairly conclude that I'm not looking to secular philosophy for answers anymore; indeed, as I've noted, the Bible counsels against this (see, again, Colossians 2:8 and related discussion). My investigation is by no means closed with respect to mysticism as a source of truth, however, at least not yet, even if what we wind up with is metaphors rather than mystic insight. So be patient, gentle reader. The rising sun discloses itself gradually, and is more real and magnificent for revealing itself in this manner than would be the case if it simply popped up.

---

[264] See Jonathan Sarfati, *Refuting Compromise: A Biblical and Scientific Refutation of "Progressive Creationism" (Billions of Years), As Popularized by Astronomer Hugh Ross* (Green Forest, Ark.: Master Books, 2004), 135; also Morris, *Study Bible*, nn. to Gen. 1:2, Isa. 45:18; and *The Scofield Study Bible, King James Version* (Oxford, UK: Oxford University Press, 2003), nn. to Gen. 1:2, Isa. 45:18.

[265] See Sarfati, *Refuting Compromise*, 67-105, 135-36; also Morris, *Study Bible*, n. to Ex. 20:11.

# THE REGION OF AWE

*Breathing Water*

I'm now back in Rochester, on a day later on in October. A library book that I've brought with me to the Colgate Rochester Divinity School reading room, *Mysticism Sacred and Profane*, by R. C. Zaehner,[266] has caught my attention for the accounts it gives of experiences with water. Once again we're at the Lake in my imagination, but to the previously-observed fluidity of the water has now been added the dynamic of alleged mystical encounter. Zaehner refers to passages, quoted by William James and Swiss psychologist Carl Jung, in which mystics in a seaside setting wind up anthropomorphizing nature.[267] In an example from James's *The Varieties of Religious Experience*, a person identified only as Amiel records in his journal:

> 'Shall I ever again have any of those prodigious reveries which sometimes came to me in former days? One day, in youth, at sunrise, sitting in the ruins of the castle of Faucigny; and again in the mountains, under the noonday sun, above Lavey, lying at the foot of a tree and visited by three butterflies; once more at night upon the shingly shore of the Northern Ocean, my back upon the sand and my vision ranging through the milky way—such grand and spacious, immortal, cosmogonic reveries, when one reaches to the stars, when one owns the infinite! Moments divine, ecstatic hours; in which our thought flies from world to world, pierces the great enigma, breathes with a respiration broad, tranquil, and deep as the respiration of the ocean, serene and limitless as the blue firmament; instants of irresistible intuition in which one feels one's self great as the universe, and calm as a god. What hours,

---

[266] I mentioned this book earlier, you may recall, as one challenging Aldous Huxley's idea that genuine mystical experience could be had through the use of psychotropic drugs.

[267] See Zaehner, *Mysticism Sacred and Profane*, 37-40.

what memories! The vestiges they leave behind are enough to fill us with belief and enthusiasm, as if they were visits of the Holy Ghost.'[268]

It should be noted parenthetically that in mysticism God speaks to hearts, not minds.[269] The mind gets in the way, especially in contemplative prayer and sacred reading.[270] Thus, we find entries such as the above that are notable for their pure effusiveness.

A statement that is more complex but arguably more structured than Amiel's is found in the writing of German philosopher Karl Joel, quoted by Carl Jung in *Psychology of the Unconscious*:

> 'I lay on the seashore, the shining waters glittering in my dreamy eyes; at a great distance fluttered the soft breeze; throbbing, shimmering, stirring, lulling to sleep comes the wave beat to the shore—or to the ear? I know not. Distance and nearness become blurred into one; without and within glide into each other. Nearer and nearer, dearer and more homelike sounds the beating of the waves; now like a thundering pulse in my head it strikes, and now it beats over my soul, devours it, embraces it, while it itself at the same time floats out like the blue waste of waters. Yes, without and within are one. Glistening and foaming, flowing and fanning and roaring, the entire symphony of the stimuli experienced sounds in one tone, all thought becomes one, which becomes one with feeling; the world exhales in the soul and the soul dissolves in the world. Our small life is encircled by a great sleep—the sleep of our cradle, the sleep of our grave, the sleep of our home, from which we go forth in the morning, to which we again return in the evening; our life but the short journey, the interval between the emergence from

---

[268] James, *Varieties*, 341; Zaehner, *Mysticism*, 37-38.

[269] Gilley, *Out of Formation*, 69, 73.

[270] Ibid., 48, 69, 73.

the original oneness and the sinking back into it! Blue shimmers the infinite sea, wherein dreams the jelly fish of the primitive life, toward which without ceasing our thoughts hark back dimly through eons of existence. For every happening entails a change and a guarantee of the unity of life. At that moment when they are no longer blended together, in that instant man lifts his head, blind and dripping, from the depths of the stream of experience, from the oneness with the experience; at that moment of parting when the unity of life in startled surprise detaches the Change and holds it away from itself as something alien, at this moment of alienation the aspects of the experience have been substantialized into subject and object, and in that moment consciousness is born.'[271]

The environment speaks with equal elegance in a second passage taken from James's *Varieties*, quoting German idealist Malwida von Meysenbug:

'I was alone upon the seashore as all these thoughts flowed over me, liberating and reconciling; and now again, as once before in distant days in the Alps of Dauphine, I was impelled to kneel down, this time before the illimitable ocean, symbol of the Infinite. I felt that I prayed as I had never prayed before, and knew what prayer really is: to return from the solitude of individuation into the consciousness of unity with all that is, to kneel down as one that passes away, and to rise up as one imperishable. Earth, heaven, and sea resounded as in one vast world-encircling harmony. It was as if the chorus of all the great who had ever lived were about me.

---

[271] Carl G. Jung, *Psychology of the Unconscious: A Study of the Transformations and Symbolisms of the Libido*, transl. by B.M. Hinkle (New York: Dodd, Mead and Company, 1916, 1963), 360-61; Zaehner, *Mysticism*, 38-39.

> I felt myself one with them, and it appeared as if I heard their greeting: "Thou too belongest to the company of those who have overcome."[272]

Not only the vitality of the inanimate, but also the mystic's goal of unity with the Absolute and all things living, is seen in the above passages. These statements, I think, illustrate how difficult it is to draw the boundaries between metaphor, mysticism, and pantheism.

*Approaching the Absolute Systematically*

Mysticism guru Evelyn Underhill distinguishes "emanation" theory and "immanence" theory as approaches to the transcendental world.[273] The "emanation" approach consists of a step-by-step process by which the mystic labors toward a distant deity, a Godhead who is conceived as "removed by a vast distance" from the ordinary sensory world.[274] The emanationist God is "out there," emitting rays of illumination from a position that is external to the believer just as the sun is external to the earth.[275] Under the diametrically opposite "immanence" approach, the mystic strives to recognize the God that is "in me," a God who "does not hold himself aloof from an imperfect material universe, but dwells within the flux of things."[276] The "immanence" God sounds like the "openness" God mentioned in my Fluid Reality chapter, and seems also to bear a strong resemblance to the God described by Paul in his sermon on Mars Hill, who, Paul said, is "not far from every one of us,"[277] and is a God "in (whom) we live, and move, and have our being."[278] Regarding this up-close-and-personal God, Ms. Underhill issues a strongly-worded warning:

---

[272] James, *Varieties*, 341-42; Zaehner, *Mysticism*, 38.
[273] Underhill, *Mysticism*, 96-99.
[274] Ibid., 97.
[275] See ibid.
[276] Ibid., 99.
[277] Acts 17:27.
[278] Acts 17:28.

## THE REGION OF AWE

> Unless safeguarded by limiting dogmas, the theory of Immanence, taken alone, is notoriously apt to degenerate into pantheism; and into those extravagant perversions of the doctrine of "deification" in which the mystic holds his transfigured self to be identical with the indwelling God.[279]

Given this danger and assuming that pantheism should be avoided at all costs, dare we approach God at all? Underhill answers yes, provided one does so with "(a) good map, a good mystical philosophy," which, she says, will leave room for both "emanations" and "immanence" as ways of interpreting the mystic's experience.[280] The only trouble is that the features of this "map," in Ms. Underhill's formulation, are to come either from the religious community in which the mystic happens to find himself or herself or from no church tradition at all—it's not clear which—but not in either case necessarily from the Bible. While she seems to favor Roman Catholicism as a medium for mysticism, Underhill states that her recommendations are offered "(w)ithout prejudice to individual beliefs, and without offering an opinion as to the exclusive truth of any one religious system or revelation."[281] Seen from the standpoint of a Bible-believing Christian, this is a recipe for disaster.

Are the possible benefits of mysticism nonetheless such that it should be given a continued try, maybe—in my case—by doing something more than praying and reading books about it and sitting outside in the nice weather and driving around in a car? I don't know. It all depends on what the prospect of success is, I guess, which in turn depends on what I'm looking for from this process. I've been trying to see water and land from as many different angles as possible, so as to understand, as much as an unglorified

---

[279] Underhill, *Mysticism*, 99.

[280] Ibid., 103.

[281] Ibid., 104.

human being can, the meaning of everything.[282] I've been praying for light from whatever source—if not direct revelation, then a more penetrating view of my sense experience, whatever God sees fit to give me. I've been asking for something recognizable as a genuine mystical experience—something that, if it will not admit me immediately to the Throne Room, will at least give me some indication that I'm on a path that will ultimately get me there.

Keep in mind, with respect to the question of our next move, my prior rejection of what I termed the mindless mysticism of Richard Foster and company (basically, the Emergent Church's practice, which harkens back to the "desert fathers," is more Eastern than Western, and incorporates Roman Catholic custom). The Emergent Church's way of approaching God, which declines to deal with anything on an intellectual level, promises little in the way of metaphysical or theological understanding, as far as I can see. A feeling of greater closeness to God is perhaps possible. But will God then whisper in my ear the explanation of all things spiritual that my mind is presently unable to comprehend? It hasn't happened yet as I write this, gentle reader.

I hasten to add that I haven't been a spiritual thrill-seeker as some of the individuals quoted or referred to above seem to have been. Truth matters to me more than titillation. Am I wrong to be expecting an intellectual bottom line as opposed to a sensual experience from my mystical endeavor? Am I misunderstanding what mysticism is at its core? If I have gotten it straight and mysticism is a legitimate hermeneutical tool, which is what I originally intended it to be, why don't I have more to show for my efforts? Is it worth the trouble trying to proceed farther down this road?

---

[282] In his 1828 *American Dictionary of the English Language*, Noah Webster defined "glorify" as "to exalt to glory, or to celestial happiness." "Glory" was defined by Webster as "The felicity of heaven prepared for the children of God; celestial bliss." See *Noah Webster's First Edition of An American Dictionary of the English Language*, Facsimile Ed. (San Francisco: Foundation for American Christian Education, 1967, 1995). The above is the sense in which "glorify" has on occasion been used in the Bible (see, e.g., John 12:23; 13:31; 17:1, 5; Romans 8:30). Thus, an "unglorified" human being would be one who is still living this mortal life, not yet home with the Lord.

Maybe I just don't fall in the category of persons who are "able to set up direct relations with the Absolute"[283] Evelyn Underhill says I do, however. Rejecting the idea that some men have a separate "mystical sense" while others do not, she states, "every human soul has a certain latent capacity for God."[284] To the extent this statement is applied to mystical experience as opposed to merely being able to see correspondences between natural things and spiritual things, I'm beginning to have my doubts.

Underhill in her classic *Mysticism* outlines five stages in the approach to the Absolute:

(1) *Awakening to divine reality.* This involves one's becoming aware, for the first time perhaps, of God's surpassing greatness.[285] You "see (H)im as (H)e is" (1 John 3:2).

(2) *Purgation.* The supplicant becomes aware of what separates him or her from God, and makes a disciplined effort to discard habits and associations that are clearly not pleasing to God (2 Corinthians 6:14-18).[286]

(3) *Illumination.* Bit by bit, an understanding of the Absolute is gained, so that the pilgrim can be pronounced "not far from the kingdom of God" (Mark 12:34).[287] It is now possible to see with clarity (1 Corinthians 13:12).

(4) *The dark night of the soul.* This is the culmination of the purgation and illumination processes, which according to Underhill generally run together, and entails one's completely putting to death his or her old self (Galatians

---

[283] Underhill, *Mysticism*, 167.

[284] Evelyn Underhill, *The Mystics of the Church* (New York: Schocken Books, 1964), 11.

[285] Underhill, *Mysticism*, 169, 176-97; Downing, *Awe*, 83-84.

[286] Underhill, *Mysticism*, 169, 198-231; Downing, *Awe*, 84.

[287] Underhill, *Mysticism*, 169, 232-65; Downing, *Awe*, 84.

2:20).[288] Many spiritual travelers will never reach this stage, obviously.

(5) *Union*. This is the goal, the object of all the struggle involved in carrying out stages (1) through (4). The soul now rises from mystic death to what has been described as a state of being married to God.[289] In a state of joy and certainty, supposedly, the believer walks with God (Galatians 5:16: Philippians 4:7).

The mystic's journey to God was described somewhat differently by Spanish Carmelite and mystic Teresa of Avila (1515-82) in her book *Interior Castle*,[290] written some five years before her death. Teresa began writing her magnum opus on June 2, 1577, and completed it on November 29 of the same year,[291] which for a manuscript of approximately two hundred pages suggests writing done at a feverish pace. In what she wrote, Teresa described a vision in which she saw "a most beautiful crystal globe, made in the shape of a castle."[292] This castle contained seven mansions, each of which Teresa referred to in the plural ("First Mansions," "Second Mansions," and so on) because at each stage of her journey to the center of the castle (where the brilliantly illuminated King of Glory was to be found) there was the possibility of an endless variety of experiences.

In the First Mansions are the rooms of self-knowledge and humility. Both of these virtues are needed before there can be further progress into the Castle.[293]

By the time a soul arrives at the Second Mansions, as Teresa envisioned the scenario of spiritual advancement, the soul's longing

---

[288] Underhill, *Mysticism*, 169-70, 380-412; Downing, *Awe*, 84.

[289] Underhill, *Mysticism*, 170, 413-43; Downing, *Awe*, 84.

[290] See generally St. Teresa of Avila, *Interior Castle*, Image Books Ed., transl. and ed. by E. Allison Peers (Garden City, N.Y.: Doubleday & Co., 1961).

[291] Ibid., 17, 235.

[292] Ibid., 8.

[293] Shirley du Boulay, *Teresa of Avila: An Extraordinary Life* (New York: Blue Bridge Books, 1991, 2004), 228; and see Peers transl., *Interior Castle*, 11, 27-43.

for God would have crystallized into some form of conversion experience.[294] The important enterprise at this point would be for the spiritual traveler to bring his or her will into conformity with the will of God.[295]

The Third Mansions, in Teresa's description, is a place in which complacency through familiar religious observances creates the possibility of the soul's slipping back into sin.[296]

In the Fourth Mansions, human effort gives way to God's grace, through which indescribable love is said to be bestowed on the supplicant.[297]

In the Fifth Mansions, the soul is asleep, totally abandoned to God, and can neither see nor understand what is happening.[298] There is no longer any temptation for the traveler to take the initiative in prayer.[299] Nonetheless, the union has not yet reached the point of actual spiritual betrothal.[300]

In the Sixth Mansions, God confirms the betrothal of the soul to Him by sending raptures, ecstasy, and trances, which are supposedly almost continuous and toss the soul violently about.[301] In the Seventh Mansions, the union of the soul and the divinity is finally complete.[302]

In the scheme presented by Saint John of the Cross (1542-91), a contemporary of Teresa and fellow Carmelite, the upward path takes the soul up a "dark ladder," the ladder of faith.[303] Faith

---

[294] Du Boulay, *Teresa*, 228; and see Peers transl., *Interior Castle*, 11, 45-54.

[295] Du Boulay, *Teresa*, 228; and see Peers transl., *Interior Castle*, 51.

[296] Du Boulay, *Teresa*, 228-30; and see Peers transl., *Interior Castle*, 11-12, 55-69.

[297] Du Boulay, *Teresa*, 230-32; and see Peers transl., *Interior Castle*, 12, 71-94.

[298] Du Boulay, *Teresa*, 232; and see Peers transl., *Interior Castle*, 12, 95-123.

[299] Du Boulay, *Teresa*, 233; and see Peers transl., *Interior Castle*, 12, 95-123.

[300] Du Boulay, *Teresa*, 233-34; and see Peers transl., *Interior Castle*, 12-13, 125-203.

[301] Du Boulay, *Teresa*, 234-35; and see Peers transl., *Interior Castle*, 12-13, 125-203.

[302] Du Boulay, *Teresa*, 234-35; and see Peers transl., *Interior Castle*, 205-235.

[303] See Leon Cristiani, *St. John of the Cross: Prince of Mystical Theology*, transl. by Leon Cristiani (Garden City, N.Y.: Doubleday & Co., 1962), 159; and St. John of the Cross, *The Dark Night*, in *The Collected Works of John of the Cross*, transl. by Kieran Kavanaugh and Otilio Rodriguez (Garden City, N.Y.: Doubleday & Co., 1964), 371-81.

is a kind of darkness, John explains, in that it reaches out into that which cannot be seen.[304] Consistently with the nature of faith as darkness, we're told to distrust visions and supposed revelations, which are presented to the senses.[305] We are also advised that mystical understanding, God's manifestation of Himself to the soul, is not achieved by the application of reason, but rather through faith.[306] As with other mysticisms, initial and ongoing purification of the mind is needed.[307]

The soul in John's scheme goes through ten degree steps, which are (1) love of self-abasement (detachment from created things) and longing for God, (2) love of the soul's search for God, (3) love of work (i.e., willingness to be spent in the cause of the Lord), (4) love that suffers (with and for Jesus), (5) love of conquest (accompanied by impatience to be with the Lord), (6) love of the high-flown and rapid journey that leads to God, (7) love marked by bold confidence in God (and that there is no danger of being separated from Him), (8) love of the Divine Espousal (of the soul to the Divine Bridegroom), (9) love marked by blandness (in being now "perfect" in God's sight, which cannot be described in words), and (10) the state of Spiritual Marriage (in which perfection has finally been achieved).[308]

By the time of Teresa of Avila and John of the Cross, mysticism as a way of approaching the Christian God had been around for at least thirteen hundred years. This does not mean that a scriptural approach to contacting God was followed during any part of this period. Even in the time of early Christian theologian Origen (185?-254?), one of the Alexandrian "desert fathers," the prevalent mysticism rejected the idea, now a predominant feature of postmodernist thinking, that a text could have but one meaning,

---

[304] See Cristiani, *St. John*, 159; and St. John, *The Dark Night*, 365-66.

[305] See Cristiani, *St. John*, 162-63, 166-69; and St. John, *The Dark Night*, 370-71, 382.

[306] See Cristiani, *St. John*, 180; and St. John, *The Ascent of Mount Carmel*, in *The Collected Works*, 129.

[307] See Cristiani, *St. John*, 170-75; and St. John, *The Ascent of Mount Carmel*, 129.

[308] Based on portions of *The Ascent of Mount Carmel*, *The Dark Night of the Soul*, and *The Spiritual Canticle*, all contained in John of the Cross, *The Collected Works of St. John of the Cross*, cited supra.

and the allegorization of Scripture, encouraged by Origen, was common.[309] While mystic practice, as it does today, recognized the common elements of purgation, illumination, and union with God, and while there was even an openness to extrabiblical visions, revelations, and the like, the core of mystical experience, then as now, was purported contact with the Almighty that could not be rationally described, could only be experienced.[310] Also absent, it seems, was any acknowledgment that nature (the creation and the laws by which it operates) can play a role in leading human beings to a more complete understanding of God's Word. To a greater love for the Son and the Father, perhaps, but what else?

Phenomena reported by British clergyman E. W. Trueman Dicken in *The Crucible of Love*,[311] his study of the mysticism of Teresa of Avila and John of the Cross, include visions,[312] raptures,[313] locutions,[314] trances,[315] and ecstacies,[316] However, as vicar Dicken notes and both Sr. Teresa and Br. John take pains to make clear, the greatest holiness, the highest perfection as it were, is not in the enjoyment of spiritual delights such as those mentioned above, but rather in being in complete, voluntary conformity to the will of God.[317] The mystic will then have what Prof. Huston Smith has referred to as "altered traits rather than altered states."[318]

Being "conformed" to the will of God in the case of John of the Cross would have meant being "in union with God," as that phrase

---

[309] See Gary E. Gilley, "Roots of the Spiritual Formation Movement," August 2014 *Think on These Things*, at http://www.svchapel.org/resources/articles/133-spiritual-formation-m.,2.

[310] Ibid., 3-4.

[311] E. W. Trueman Dicken, *The Crucible of Love: A Study of the Mysticism of St. Teresa of Jesus and St. John of the Cross* (New York: Sheed and Ward, 1963).

[312] Ibid., 35, 38-39, 118, 232-33, 317, 343-49, 375-406, 410-11, 428, 478.

[313] Ibid., 39, 91, 118, 202, 365, 375, 395-403, 478.

[314] Ibid., 343, 347, 375, 380, 382-84, 391-95.

[315] Ibid., 38, 118, 200-202, 210, 212, 258-59, 270, 317, 375, 395-405, 407-18, 459, 478.

[316] Ibid., 38, 204-05, 270, 281, 396.

[317] Ibid., 38-39.

[318] Downing, *Awe*, 156, 190.

was defined in the lexicon of John's day.[319] The connection spoken of by John is a "union of love," which implies a likeness between the mystic and God, one that is of supernatural origin.[320] Since God is ineffable (not describable in words), so too is the nature of the union between the human soul and God, according to John.[321]

The so-called unitive way, Vic. Dicken indicates, is less clearly defined in the case of Teresa of Avila, for whom union with God often takes on the appearance of a trance.[322] There is, however, another form of union, "the true union," which consists in habitual conformity to the will of God.[323] This type of union dispenses with all mediation on the part of created things; therefore, it cannot arise from the practice of nature mysticism.[324] The test for this union, as suggested above, lies not in whether there is a "mystical experience" but rather in the results that are produced in the subject soul.[325]

One searches in vain, notwithstanding, for any indication in Dicken's exhaustive summary that the explication of texts—as opposed to the ability to live in accordance with their mandate—is a gift one may expect to be given through mystic experiences such as those reported by Teresa of Avila and John of the Cross.

*Soul Searching*

As I've tried to convey in various ways, I would be hoping for more than personal sanctification or "thrills, spills, chills, and excitement" were I to set sail on the mystic sea using any particular medium of transport. The questions most important to me at this point involve not the course of the voyage itself but rather what port I will land at. May I expect that "the mystic way" will lead to understanding in the normal sense—i.e., that it will give me an

---

[319] See Dicken, *Crucible*, 352-55.

[320] Ibid., 355-62.

[321] Ibid., 361-65.

[322] See ibid., 407-17.

[323] See ibid., 417-21.

[324] See ibid., 421-28.

[325] Ibid., 429.

intellectual experience touching on divine reality? If not, will I be left with anything beyond pleasant feelings about God? If my mystical experience does produce bread for the intellect, can I expect that the understanding thus gained by me will extend to hermeneutics? That it will help me deal with the inconsistency that the Bible at first glance seems to display (such as, for example, with respect to human free will)? That it will in any way assist with or bolster doctrine? Or will it be in derogation of orthodoxy (as has been the case with the Emergent Church's brand of mysticism)?

Biographer David Downing may have provided an answer to some of the above questions. Downing remarks that C. S. Lewis "must certainly have been one of the most mystical-minded of those who never formally embarked on the mystical way."[326] This is shown particularly, Downing says, in Lewis's "vivid sense of the natural order as an image of the spiritual order,"[327] which Downing comments more fully on later in his biographical *Into the Region of Awe*.[328] Nonetheless, for all the things he saw in nature that made him think of God's attributes, Lewis did not consider himself a mystic.[329] In *Letters to Malcolm*, written by him just months before his death, an ailing Lewis recalled how he loved walking in the hills and mountains in his younger days, but confessed that even during the early period of his life he was "no climber."[330] Similarly, he tells the fictional Malcolm, he does not now "attempt the precipices of mysticism."[331] Concerning the occult "higher level—the crags up which the mystics vanish out of (his) sight," Lewis states, "I don't think we are all 'called' to that ascent."[332] Was Lewis therefore someone who was not able to establish "direct relations" with God? Does this help my own case at all? It appears that if we have

---

[326] Downing, *Awe*, 33.

[327] Ibid.

[328] See, in particular, ibid., 45-46.

[329] Ibid.

[330] C. S. Lewis, *Letters to Malcolm: Chiefly on Prayer* (New York: Harcourt, Brace & World, 1963, 1984), 63.

[331] Ibid.

[332] Ibid.

nothing else in common, C. S. Lewis and I are both "people of the foothills."[333]

Lewis also tells his friend Malcolm that there is a growing view that what the mystic finds when he or she "puts to sea" will likely be the same regardless of the sailor's religious or occupational background.[334] "(T)he land sinking below the horizon, the gulls dropping behind, the salty breeze," is the way Lewis describes it.[335] This seascape will have little to do with doctrine; indeed, it seems, it will align nicely with Evelyn Underhill's open-mindedness, which allows for mysticism "without an opinion as to the exclusive truth of any one system or revelation."[336] The identical experience of all mystics, Lewis says, "vouches for nothing about the utility or lawfulness or final event of their voyages."[337]

Reading the above statements, I find myself once again wondering if mysticism will be of much use in my present pursuit of metaphysical truth. If we're all going to see the same things in our voyages of discovery, then it will almost certainly be the case that mysticism will not validate the religion in whose context it happens to occur (in this case, orthodox Christianity).[338] The intellectual depths of my religion will thus remain unplumbed; only feeling, although most likely pleasurable, will be achieved. This is not a satisfactory "bottom line" for me. I'm not willing at this point to give up on the enterprise of seeking insights that will help me to better understand and fully believe the Bible. I'm here to learn rather than luxuriate.

The above discussion touches on things about mysticism that should now be giving me pause. I've been aware of these negatives, at least in a general way, for some time now. Nevertheless, I've spent days circling The Islands automotively, poring over the best maps of the area that I've been able to find, sitting out under God's

---

[333] Ibid.

[334] Ibid., 63-64.

[335] Ibid., 64.

[336] Underhill, *Mysticism*, 104.

[337] Lewis, *Letters to Malcolm*, 64.

[338] Ibid., 65.

sky, walking paths that have promised helpful vistas, and reading or listening to available writing and lectures on mysticism and cosmology. Why? In the hope, as I said earlier in different words, of creating a spiritual chart, a theological or philosophical counterpart of the NOAA map that hangs on The Cottage's dining room wall.

To date, I've only gotten as far as the Trinity in my search for correspondences. It should be apparent, to anyone who is even vaguely familiar with Christian doctrine, that The Islands, the three land masses that are within binocular range from The Cottage and also are not camper-trampled — I'm referring to Six Town Point, Gull, and Bass — can be seen as corresponding, in an undifferentiated way, to the Trinity. There's a lot more, I admit, that needs to be filled in as far as my understanding of what I see is concerned. The Islands, as part of a vast inland sea, have a context that must also be considered for correspondence, for one thing.

As physical topography, The Islands and the capes, points, and bays that surround them have been mapped with great thoroughness. The task that remains, as I'm picturing it in my mind, is for me to see if I can notice things about the land and water configuration before me, which I see whenever I look out from Campbell's Point toward the open water (our horizon), that will help me sort out the being-related questions that I posed at the beginning of this book. I'm not sure a metaphysical mapping would have to involve mystic encounter, but some kind of divinely-furnished intuition would be helpful. If my path still has to be through mysticism, all well and good. If I have to travel by some other route, I'm open to that, also.

Why should the Trinity be important to me? It's only theology, after all. Why not just concentrate on what C. S. Lewis calls "practical religion"? Why not simply go around doing good or at least doing what you're told and be cheerful about it? Answer: because then you have only a works religion. The works are your acts of obedience to Christ's commands. Although commendable, they won't save you. What you need for salvation is faith, defined as trust in who Jesus is (God the Son) and what He's done (paid the penalty for your sins and thus satisfied God the Father's justice principle). Otherwise, you make Christ's sacrifice worthless. It wasn't necessary; it was just a vain, quixotic gesture.

It was some time after I began thinking of The Islands as the Trinity that I saw the Trinity and my salvation coming together in my mind as a matter of nature-based realization. It had to do with the fact that Father, Son, and Holy Spirit all partake of the same water system, which ultimately sustains the entire world. It's a unity principle, really. You can't have the needed trust in what Jesus did on the cross unless you understand fully that the Son and the Father are one (1 John 5:7). Only then is Christ's sacrifice sufficient to allow forgiveness of your sins. But you can't believe that way— believe unto salvation—without the help of the Holy Spirit, the third person of the Trinity, who comes to you because God the Son and God the Father have come to you (John 14:16, 26; 15:26; 16:7). Salvation arrives as a package. You can't have it unless there's a Trinity—all three Persons— at work in your situation.

The above is all basic information, which needs to be given to non- or new believers as soon as they're introduced to the Gospel. I had hoped that my sojourn around The Islands would give me more than this, additional insights. I'll just have to be content, for the present, with what the Lord has provided me. Anyway, could the relationship of the first two persons of the Trinity be better explained than as the relation of father and son? I don't see how. The apostle John says that Christ was *begotten* of the Father (John 1:14, 18; 1 John 4:9)—and thus united bodily with the Father— rather than *made*, as would be the case with a being merely created by the Father and for that reason separate from Him. The relationship of all three members of the Trinity is explained by the apostle Paul in Ephesians 2:18, where it's stated that "through [Christ] we have access by one Spirit unto the Father." The Spirit admits us to the presence of Father God; we only have to knock on the door (Matthew 7:7-8). Perfect. An approach without reproach.

Is there anything else that needs to be said concerning the relationship between the born-again believer and the members of the Trinity, particularly the Holy Spirit? Yes, much that's relevant if we're to understand what the believer's relationship to God the Father through the Holy Spirit consists of. Jesus assures his disciples prior to Calvary that the Spirit will "*teach* (them) all things" (John 14:26), will "*guide* (them) into all truth" (John 16:18). This is to be expected because God's Spirit "searcheth all things, yea, the deep things of God," as Paul tells the Corinthian church (1 Corinthians 12:2). The apostle John in his gospel quotes Jesus as saying that the Father will *send* the Holy Spirit to the disciples (John 14:26), that the Spirit "*proceedeth from* the Father" (John 5:26), and that He (Jesus) must depart from their midst before He can *send* the Spirit to them (John 16:7). (emphasis supplied in all instances). As thus framed, the believer's *receipt of the Spirit* is a different thing from a believer's being *absorbed into God*, which is the mystic's goal. In the above statement of the matter, the believer to whom God's truth is sent remains separate from the Father, Son, and Holy Ghost, mutual love of them all notwithstanding. We are close to but not united with God (Acts 17:27-28), crucified with but not part of Christ (Galatians 2:20).

In view of the above, I must then ask: Is there, indeed, any sense in which mystical encounter may be viewed as the mystic's transcending the normal limits of self? Does it ever involve something like what happened with the young seeker of truth in C. S. Lewis's *That Hideous Strength*?

> A boundary had been crossed. She had come into a world, or into a Person, or into the presence of a Person. Something expectant, patient, inexorable, met her with no veil or protection between. In this height and depth and breadth the little idea of herself which she had hitherto called *me* dropped down and vanished, unfluttering, into bottomless distance, like a bird in a space without air.[339]

Doesn't an account such as the above presuppose that a boundary of the self can be established? Where, pray tell, does that boundary fall? Is it, as suggested earlier, defined by remembered experience? Is there something that's added holistically that makes "self" more

---

[339] C. S. Lewis, *That Hideous Strength: A Modern Fairy Tale for Grown-Ups* (New York: Macmillan Publishing Co., First Paperback Edition 1965), 318-19.

than the sum total of experience? Is self a character or nature that comes from elsewhere, that we can't help having?

The above questions need to be asked, certainly, but there's another inquiry that must come first: is a supposed mystical encounter anything more than an experience that you have? If not, then you as a would-be mystic will remain, comfortably or uncomfortably, within the cocoon of self but with an expanded inventory of impressions. If, on the other hand, you've escaped from the ego prison, actually merged your being with God's, then what has happened to you should be nothing less than transformational, to use a cliché of the day, but is this really how God works? He transforms, through the wondrous working of his Spirit (Romans 12:2; Ephesians 3:16; Titus 3:5), but does He do this from within or from outside the believer? If from within, how much of Him is within? These are questions that need to be answered, along with others, before you and I can settle into any kind of metaphysical comfort, gentle reader.

As I write the above, the thought engine is still dieseling (refusing to die after the ignition has been switched off). If a believer's insight concerning the nature of things comes mostly from "out there" rather than from within the believer, can we still call it mystic intuition? If we can't, does it matter?

# 4
# PARADIGM PARALYSIS

*The Canal and Other Wonders*

We're back in Rochester for the winter, having closed up The Cottage a week or so ago. I'm out walking on this partly cloudy late October day, on the Erie Canal path between Lock 33 and Lock 32. There's enough breeze to make ripples on the Canal—feather-like (pinnate) highlights of gray, yellow, and blue that are traveling from east to west, an unruly migration in roughly the direction the wind is blowing. I parked the car at Lock 33, which is where Edgewood Avenue spans the Canal. I used to leave my vehicle a mile or two west of the Edgewood lock, at a pull-off where Clinton Avenue goes over the Canal on its way south to its terminus at Brighton-Henrietta Town Line Road, but began to notice glass on the ground at the Clinton pull-off and decided that it would be better to leave my relatively new SUV where it would be within sight of a lock operator.

I'm headed east as I walk, but probably won't make it as far as the village of Pittsford. I'll be satisfied with my usual turn-around a mile or so past Lock 32 (Clover Street). Going that far will take me past the wide part of the Canal where the Pittsford Rowing Club holds its practices, which are always of interest to me. There's a flag on a homeowner's dock along the Rowing Club stretch that summarizes perfectly my attitude since I retired from the County Law Department: IT'S ALWAYS 5 O'CLOCK SOMEWHERE.

The Canal wasn't always rowing sculls, pleasure boats, and crazies on bicycles and roller blades, as it is now. When it was conceived in the early nineteenth century, the idea was to provide an efficient means by which settlers could be moved west and wheat and other commodities from the Midwest could reach the Port of New York and be shipped to other parts of the world. Shipping on the Canal fell off, however, with the opening in 1957 of the Saint Lawrence Seaway, which provided a route ocean-going freighters could take from the Great Lakes to the Atlantic Ocean.

The final nail in the lid of the coffin of Canal commerce was hammered with an accidental stroke in 1973, when a drainage tunnel that was being dug under the canal bed at Bushnells Basin collapsed, flooding a large area some three miles east of Pittsford. Since a great amount of earth was displaced, it took more than a year to restore the Canal to commercial usability. During that time, shippers found other ways of transporting their cargoes. Thereafter, with the need to share water space with tugs and barges now reduced, recreational boaters took to the Canal in record numbers.

As suggested in my description of the Bushnells Basin disaster, the Canal at various locations has walls and a water level that are higher than the surrounding terrain. It's been that way from the start and remains so, even with the rerouting that was done at the beginning of the twentieth century to get the Canal out of city downtown areas. Construction challenges were not limited to the need to build up the canal bed in order to cross swamps and other low areas or to cut the canal bed into hillsides. The vessels navigating any such waterway, at intervals dictated by the terrain, need to be raised or lowered so that they can proceed into water of a higher or lower level than the water from which they've just come. This stepping up or stepping down, which is effected through a system of locks, is the fundamental feature of canals. The water level in the lock needs to be raised whenever ascent to a higher level is required, and lowered whenever descent to a lower level is sought. As long as the water depth at the higher of the two navigation levels enclosing the lock is sufficient, the lock can be operated by simply opening a filling valve or an emptying valve, depending on whether a step-up or step-down of the canal boat or other craft is needed.

The above assumes dependable sources of water to keep the canal filled, which may be connected rivers or lakes, lateral canals, or even reservoirs. Thus, construction of the Erie would have involved more than digging a long ditch and putting in locks at strategic locations. Feeders had to be built or provided for. Sometimes the topography dictated that there be multiple locks within a short distance of each other, as at Lockport (near Buffalo). The Lockport locks took the Canal over the Niagara Escarpment, a ridge of rock that required extensive use of black gunpowder (dynamite was not yet invented) to cut through it.

A hundred miles or so east of Lockport, the obverse was presented as an issue, as the Canal diggers had to wade in the waters of the mosquito-infested Montezuma Swamp. Building the Canal in this and other wet areas required the development of cement that hardened underwater. The nature of these and other obstacles that were overcome when the Canal was built should serve to illustrate that a high level of engineering and other technical skill is not something exclusive to the twentieth and twenty-first centuries.

The construction of the Erie Canal, it is agreed, is an example of science being applied for the benefit of mankind. Author/lecturer Dave Breese notes that in the past hundred years or so science has given us many things that are "of practical use in the world beyond the laboratory(citing as examples) the internal combustion engine, the flashlight, the rocket, the space shuttle, the satellite, radio transmission, and a thousand other dependable products that once would have been the astonishment of society (parenthetical added)."[340] The science that produces such things is called *operational* (or empirical, or observational) science.[341] It deals with testing and verifying ideas in the present.[342] What is termed *his-*

---

[340] Dave Breese, *Seven Men Who Rule the World from the Grave* (Chicago: Moody Press, 1990), 39.

[341] Cloud, *Non-Existent*, 33; see also Roger Patterson, *Evolution Exposed* (Hebron, Ky.: Answers in Genesis, 2006), 20.

[342] Ibid.; see also Jason Lisle, *The Ultimate Proof of Creation: Resolving the Origins Debate* (Green Forest, Ark.: Master Books, 2009), 109-10.

*torical* (origins) science attempts to interpret evidence pertaining to the past and thus includes the models of evolution and special creation.[343] Biblical creation apologist David Cloud offers the following comment on operational versus historical science:

> Scientists have accomplished wonderful things through empirical science, such as building technological devices and exploring the living cell, but when they try to look beyond the physical world and beyond the constraints of time, they enter into a sphere about which they are not qualified to speak. They leave the evidence and enter into speculation.[344]

Operational science, as envisioned in the above analysis, is not limited to toasters and televisions. British philosopher Bertrand Russell more than eighty years ago claimed not only that "(w)ith our present industrial technique we can, if we choose, provide a tolerable subsistence for everybody,"[345] but also that "hatred and fear can, with our present psychological knowledge and our present industrial technique, be eliminated altogether from human life."[346] Speculation is an element of operational as well as historical science, it must be noted. Ominously, Russell advocated a use of mind science that would have removed the enterprise of dealing with sin not only from the church but also from the home. He stated:

> It has become clear that, while the individual may have difficulty in deliberately altering his character, the scientific psychologist, *if allowed a free run with children*, can manipulate human nature as freely as Californians manipulate the desert. *It is no longer Satan who makes sin but bad glands and unwise conditioning*.[347]

---

[343] Cloud, *Non-Existent*, 33; Patterson, *Evolution Exposed*, 20.

[344] Ibid.

[345] Russell, *Why I Am Not a Christian*, 47.

[346] Ibid.

[347] Ibid., 159; emphasis mine.

## PARADIGM PARALYSIS

A communist or socialist society was apparently Russell's choice for an administrator of his child welfare agenda. What should be noted in this regard is that science's being of an operational character, even when it involves technology that is not applied to making weapons of war or environmental toxins, does not necessarily make it of a higher dignity than origins science, no matter how irresponsibly undertaken the origins science may be (as, for example, in attempting to prove evolution without producing fossils of transitional species).

*Normal and Abnormal Science*

Whether empirical or historical, science suffers from the presence and persistence of paradigms. A "paradigm," as the term has been used from the mid-20th century on, is a matrix of assumptions, shared by a particular professional community, which guides research and informs the interpretation of results.[348] The paradigm as a feature of scientific practice was described by Thomas Kuhn in his book *The Structure of Scientific Revolutions* (1962). Kuhn was a Harvard graduate student in theoretical physics at the time. Some 350 years earlier, English philosopher Sir Francis Bacon had identified and warned against certain "idols of the mind." At least two of Bacon's "idols"—those arising out of individual temperament or biases ("idols of the cave") and those stemming from undue value being given to certain ideas simply because they are in common currency ("idols of the marketplace")—would probably have been regarded as symptoms or precursors of the Kuhnian paradigm[349]

The paradigm warned of by Kuhn poses dangers that will likely be felt outside the scientific community. As he expresses it in *The Structure of Scientific Revolutions*, Kuhn's basic concern is that the public has been given a misleading picture of science. It's assumed

---

[348] See Thomas S. Kuhn, *The Structure of Scientific Revolutions* (Chicago: The University of Chicago Press, 1962, 1970, 1996), 176-77, 182-84; also Stokes, *Essential Thinkers*, 203; and Rohmann, *Ideas*, 295-96.

[349] See Rohmann, *Ideas*, 34; also Steven L. Goldman, *Science Wars: What Scientists Know and How They Know It* (DVD) (Chantilly, Va.: The Teaching Company, 2006), lecture 2.

that science is totally objective, he says, while in reality its practice is governed by paradigms. You gain entry to your particular scientific community by studying paradigms.[350] Research within that community is paradigm-directed.[351] Unwillingness to accept the community's paradigm will lead to professional isolation or ostracism,[352] since, as Kuhn puts it, "(a) paradigm is what the members of a scientific community share, *and*, conversely, a scientific community consists of men who share a paradigm."[353]

But they don't necessarily share that paradigm forever. Paradigms are continually being abandoned in response to newly-discovered anomalies that show that the existing paradigm-nature fit is no longer viable,[354] but there is no guarantee that the new paradigm will be any better a fit.[355] Indeed, Kuhn says, the process for determining which theory will be accepted as the new paradigm is more like natural selection than perfection, in that it "picks out the most viable among the actual alternatives in a particular historical situation."[356] Kuhn's answer to the question of whether science is progressing toward truth is therefore somewhat unsettling:

> In the sciences there need not be progress. We may, to be more precise, have to relinquish the notion, explicit or implicit, that changes of paradigms carry scientists and those who learn from them closer and closer to the truth.[357]

The above statement is made by Kuhn notwithstanding the supposed fact that science has what he terms a "progress" consciousness

---

[350] Kuhn, *Structure*, 10-11.
[351] Ibid., 18.
[352] Ibid., 18-19.
[353] Ibid., 176.
[354] Ibid., 52-53, 82, 90.
[355] See ibid., 145.
[356] Ibid., 146.
[357] Ibid., 170.

embedded in it, a characteristic said by Kuhn to be not shared with art, philosophy, or even political theory.[358]

Sandwiched between paradigm arrivals and departures are periods of what Kuhn calls "normal science." "Normal science" is research or investigation acknowledged by the practitioner's professional community, the "thought collective" to use immunologist Ludwig Fleck's label.[359] "Normal science" involves "mopping up"—clarifying ideas that are already there rather than identifying new phenomena that need attention.[360] Solving existing puzzles rather than mapping out new ones.[361] Extraordinary problems emerge only on special occasions prepared for by the advance of normal science to the point where anomalies become crises and must be resolved.[362] Only then do you have a chance to be part of a scientific revolution, Kuhn says.[363] Only then are you given the opportunity to upset existing assumptions such as geocentricity of the universe or gravity as action at a distance rather than a curvature of space-time.[364]

When you automatically defer to the present dominant paradigms, when you value ideas for no other reason than that they're the currency of the day (Bacon's "idols of the marketplace"), you're practicing what I call popular science. You're therefore vulnerable to the siege of Truth. One of the most striking ways in which popular science has shown such vulnerability is by positing a universe that's billions of years old. It's theorized that the cosmos began with a gigantic explosion, a "big bang," which propelled outward in all directions the matter that became the various heavenly bodies. If the most distant galaxies that we can now observe are, say, twelve

---

[358] Ibid., 160.

[359] See Goldman, *Science Wars*, lecture 16.

[360] Kuhn, *Structure*, 24.

[361] Ibid., 35.

[362] Ibid., 34.

[363] Ibid.

[364] Copernicus's and Einstein's revolutions, respectively. More on Einstein's theory of gravity in the "On the Trampoline" section of my Relativity chapter, infra.

billion light years away,[365] then the alleged big bang must have happened at least twelve billion years ago, the reasoning goes. The earth with its inhabitants and the sun, moon, and stars thus could not have been created in six days as the Genesis account says they were.[366] As physicist Russell Humphreys has pointed out, however, the length of a day, like the length of a year, may vary greatly from location to location under Einstein's general relativity theory. The critical factor is the presence or absence of substantial gravitational influence.[367]

In pursuit of the above point, Dr. Humphreys begins his book *Starlight and Time* with the statement that he takes as givens various standard *observational* assumptions (a universe that has expanded, most distant galaxies at least twelve billion light years away, speed of light around 186,000 miles per second, et cetera) and adds a non-party-line supposition that the universe is bounded and therefore has a center.[368] Applying these assumptions in combination with Einstein's equations, he arrives at a solution to general relativity that has the universe appearing out of a "white hole"—the opposite of the familiar "black hole" but still a valid concept theoretically, according to Humphreys. Instead of constantly expanding its event horizon (the distance from its center at which escape from its gravitational grasp is no longer possible), a white hole *spews out* its contents and therefore has an event horizon that shrinks, eventually down to nothing.

The spewing out of the white hole's contents has the effect of making the universe's mass less concentrated, and at the outer edge of the expanding cosmos it becomes many times less dense than the mass that remains at or near the location of the white hole, in

---

[365] A light year is the distance that light travels in one year. However, under Einstein's special and general theories of relativity, your one year and my one year may not be the same. More on this momentarily.

[366] See Gen. 1:1-31; 2:1-3; and *Henry Morris Study Bible* nn. thereto.

[367] See D. Russell Humphreys, Ph.D., *Starlight and Time: Solving the Puzzle of Distant Starlight in a Young Universe* (Green Forest, Ark.: Master Books, 1994, 2006), 9-29, and the summary that follows herein.

[368] The prevailing "no boundary" view is "an arbitrary assumption," he says. Humphreys, *Starlight and Time*, 18. See further, in this regard, Mark W. Cadwallader, *Creation Spelled Out for Us All* (Conroe, Tex.: CTS Publications, 2007), 48-49.

which area is located the matter that becomes the earth. Dilution of matter is dilution of gravity, and dilution of gravity allows time to pass at a faster rate than where gravity is stronger. Thus, under Humphreys' calculations, billions of years are able to pass out on the edge of the universe while only a few days have passed on Earth, where gravity has slowed time to a comparative standstill. This means that by the fourth day of creation, when God set the sun, moon, and stars in place (Genesis 1:16), there would have been ample time for the light from galaxies billions of light years away to reach the earth. The creation, including stars both visible and invisible, could properly have been said to be complete within six Earth days, as the Bible says it was (Genesis 1:31; 2:1-3).[369]

The above is an example of a physics theory having more than one solution as regards given particulars. A theory such as Einstein's general theory of relativity is normally expressed by a set of equations, and it's that set of equations that we're talking about when we say that the theory has multiple solutions. A "solution" to the equations would be a mathematical working out of the result(s) that the principles of the theory, when applied, can be expected to produce. How do you choose the correct solution for your situation when more than one solution is possible? You might simply opt for the one that seems to work best with the facts presented.

But what if those facts are ambivalent or confusing? Then you could wind up applying a subjective measure such as elegance. Personal bias might come into play. Yes, it bears repeating, multiple solutions to the equations of physics place theoretical physicists in the position of being able to choose solutions based on prejudice. The popular scientific paradigms re the beginning of the universe and human origins (big bang, evolution over eons) are therefore no more entitled to truth status than a history of the cosmos and mankind that is in keeping with the Bible, assuming (as is indeed the case) that the Bible version of the story finds support in available evidence and logic.

Even the speed of light, when considered without regard to the influence of gravity, cannot be counted on by popular science to help make its case that we're living in a universe that's billions of

---

[369] For a fuller development of Humphreys' thesis, see *Starlight and Time*, 9-29, 37.

years old. By way of background, it should be noted that in 1676 Danish astronomer Olaf Roemer proposed that light did not make its presence known throughout the universe all at once; it traveled, he said, at a speed that was finite. Controversy regarding this basic proposition continued until 1729, when another researcher confirmed a finite speed of light. After that, the bone of contention was whether the speed of light was constant.

The discussion as to light speed constancy, which has continued to the present and in the past thirty-five years or so has been spearheaded by Australian astronomer Barry Setterfield, arises from the fact that during the almost three hundred years since 1729, measurements have shown a steadily declining speed of light.[370] The farther back in time the mathematical curves are projected, the more rapidly the speed of light seems to have been decreasing.[371] As might be expected, this has thrown the scientific establishment into consternation.

An obvious reason for the tumult, during the hundred and fifty or so years since Charles Darwin's appearance on the scene, has been the reality that slowing light, like Russell Humphreys' fast-outer-universe-time formulation, threatens an idea that is essential to maintaining the tattered evolution hypothesis. Evolution, with its hypothesized baby steps from lower to higher forms of life, requires billions of years. Positing fast light in the past in comparison to now takes away "billions of years." By the calculations of Russian scientist V. S. Troitskii, the speed of light was some ten billion times faster at the supposed beginning of the universe than it was in 1987, when Troitskii's paper on the subject was released.[372] This, like the Humphreys formulation, could mean that the cosmos is only a few thousand years old as years are presently reckoned, since under Einstein's special theory of relativity, estimates of rates of time passage during earlier periods of the history

---

[370] Walter T. Brown, Jr., *In the Beginning: Compelling Evidence for Creation and the Flood* (Pheonix: Center for Scientific Creation, 2008), 322.

[371] Ibid.

[372] Ibid.; see also Helen Setterfield, "History of the Light-Speed Debate Part I," excerpting July 2002 *Personal Update NewsJournal*, cited in *http//www.khouse.org.news article 2013/2045*.

of the cosmos would have to be adjusted upward to accommodate the faster speeds of light during such earlier periods.[373]

Attacks on the slowing-speed-of-light idea have been based *inter alia* on the alleged need for stability in science. This was supposedly the reason for the scientific community's declaration, in October of 1983, that the speed of light was officially established, as a constant of nature, to be 299,792.458 kilometers per second, or roughly 186,000 miles per second. In the wake of Einstein's special and general relativity theories, it was said, the speed of light needed to be given the status of an atomic constant—i.e., established with reference to the decay of radioactive isotopes.[374]

It now turns out, also, that radiometric dating, a traditional bugaboo of "young earth" theorists, is also sensitive to changes in the speed of light. Higher light speeds in times past will produce a calculated present age of the matter being tested that is younger than would have been the case if light speed had been constant over the life of the material.[375]

The idea that light is variable-speed is not wholly unscriptural. The verses stating that God "stretcheth (out) (forth) the heavens"[376] imply an elasticity in the fabric of space-time. Elasticity implies inconstancy of dimension; thus, it's easy to imagine that light, which is a primary element of the created universe (Genesis 1:3), is something that's flexible as far as its speed is concerned.

Another kind of uniformitarian assumption—specifically, that the ratio of carbon-14 to carbon-12 in the earth's atmosphere has remained roughly the same throughout history—has likewise been a deceiver in the past, it appears, by leading to grossly inflated estimates of the age of organic material. The theory of carbon-14 dating is similar to that of radiometric dating: the age of the material being tested can be established by finding out how much decay

---

[373] Helen Setterfield, "History of the Light-Speed Debate Part I," July 2002 *Personal Update NewsJournal*, cited in *http//www.khouse.org/news article 2013/2044*.

[374] Ibid.

[375] Helen Setterfield, "History of the Light-Speed Debate Part II," July 2002 *Personal Update NewsJournal*, cited in *http//www.khoouse.org/news article 2013/2045*; Brown, *In the Beginning*, 322-23.

[376] Isaiah 40:22; 44:24; Zechariah 12:1.

has occurred. For radiocarbon testing, this means measuring how much carbon-14 is left in the sample, whereas for radiometric testing, which is used on inorganic material, the question is one of which isotope the radioactive material has become.[377]

It is now known that a worldwide flood, which would have uprooted and buried pre-flood forests and would have been fueled by escaping subterranean waters (see Genesis 7:11, also the *Henry Morris Study Bible* note thereto), could have substantially reduced the amount of carbon-12 in the atmosphere that would have been available to dilute the carbon-14.[378] Whereas carbon-14 is produced by cosmic radiation acting on nitrogen in the upper atmosphere, carbon-12 forms from decaying vegetation and is found in subterranean waters, which are shielded from the sun's rays and other radiation.[379]

But suppose that for the moment we lay aside the questions surrounding time speed, light speed, and radiometric and radiocarbon dating, any of which may affect in a profound way what we can say about the age of the cosmos. We still need to examine critically another axiom, which bears on the questions both of how and when the universe came into being, and ask: has popular science at least shown that the universe began with an explosion of the totality of all matter? This humble student of such matters submits that it has not.

*The Big Bomb*

British philosopher of science William Whewell (1794–1866) is quoted as having said, "Man is the interpreter of nature, science the right interpretation."[380] Contrariwise, a study of developments with respect to the "big bang" theory since its promulgation by American physicist George Gamow in the late 1940s discloses at best an awkward fit between science and nature. Attempts to

---

[377] Brown, *In the Beginning*, 342-43.

[378] Ibid.

[379] Ibid.

[380] *Philosophy of Inductive Sciences*, aphorism 17 (1840), quoted in *Bartlett's Familiar Quotations*, 432:19.

resolve the disparities between theory and observation have relied in a major way on the "inflation" hypothesis formulated in 1979 by another American, Alan Guth.[381]

Theoretical physicist Michio Kaku defines "inflation" as "(t)he theory that the universe underwent an incredible amount of superluminal expansion at the instant of its birth."[382] Theorist/lecturer Brian Greene defines "inflation" and "inflationary cosmology" as a "(m)odification to the earliest moments of the standard big bang cosmology in which [the] universe undergoes a brief burst of enormous expansion."[383] The universe could expand at a rate faster than the speed of light, Kaku explains, because it was empty space that was expanding rather than the universe expanding into empty space;[384] thus, the materiality of the universe was unavailable as a factor to slow the initial expansion of the universe down. That there was never anything outside the universe is pure, unprovable theory, of course.

What Guth's theory modified, specifically, was the Standard Model of Cosmology, which was based on Russian meteorologist Alexander Friedmann's solution to Albert Einstein's general relativity equations (big bang, no inflation).[385] Guth found a solution to Einstein's equations that was different from Friedmann's, in other words (here we go again!). For me at least, Guth's "inflation" has the appearance of an idea jury-rigged in desperation, of a last-ditch, *ad hoc* attempt to rescue the big bang theory from the consequences of its logical inconsistencies.

One of the big bang issues supposedly addressed by "inflation" is the "horizon" problem. This is a conundrum that arises because under the present universal constant governing its speed, light can travel through space at no more than the leisurely pace of 186,000 miles per second. If, as popular science assumes, light can travel no faster than that, then neither can information transmitted

---

[381] See Kaku, *Parallel Worlds*, 13-16; also Hawking, *A Brief History*, 131-34.

[382] Kaku, *Parallel Worlds*, 391.

[383] Brian Greene, *The Elegant Universe: Superstrings, Hidden Dimensions, and the Quest for the Ultimate Theory* (New York: W. W. Norton & Co., 1999), 417.

[384] Kaku, *Parallel Worlds*, 13.

[385] Greene, *The Elegant Universe*, 346.

by gravitation, radio waves, or any other medium. Thus, if our sun should die, we wouldn't know about it until some eight minutes later.

Combining the sluggishness of information transmission through space with popular science's estimate that the universe is 13 or so billion years old, one is led to the conclusion that there is no reason why either the configuration of the stars and galaxies or the cosmic microwave background radiation should be as remarkably uniform as they are.[386] And yet they are (remarkably uniform). No matter which direction you turn in, the universe looks the same and the microwave radiation is the same.[387] This seems to signify that stars, galaxies, and microwave radiation everywhere in the universe all have some kind of "knowledge" of what stars, galaxies, and microwave radiation elsewhere in the cosmos are doing. Celestial bodies and phenomena at the far reaches of the universe seem to have been influenced by—i.e., conformed to—information that would not yet have had time to arrive at where they are.

Conformity assumes contact, to put it another way. As Stephen Hawking has aptly stated, taking the microwave background radiation into consideration is "a bit like asking a number of students an exam question. If they all give exactly the same answer, you can be pretty sure they have communicated with each other."[388] And yet communication of any kind between regions at opposite ends of the observable universe would be impossible under the unmodified Standard Model, since such information would have had to travel twice as far as the estimated age of the cosmos would have allowed.[389]

Alan Guth's solution to the distance problem was to say that it was not a problem inasmuch as not only were the particles in

---

[386] See Kaku, *Parallel Worlds*, 88-89; Hawking, *A Brief History*, 125; and Greene, *The Elegant Universe*, 352-54. The cosmic microwave background radiation is microwave radiation that is said to be suffusing the universe and to be responsible for a uniform temperature of 2.7 degrees above absolute zero. The popular science view of the CMBR is that it is the residue of the supposed big bang fireball. Bible-believing Christians would see it as something else, as will be explained shortly.

[387] Kaku, *Parallel Worlds*, 88; Hawking, *A Brief History*, 125.

[388] Hawking, *A Brief History*, 125.

[389] See, in this regard, Kaku, *Parallel Worlds*, 88-89.

question at first packed tightly together and thus in thermal contact; they were then driven apart in an exponential expansion of the universe that far exceeded the speed limit for the passage of light and other transmissions through the cosmos. Under the Standard Model, expansion outruns communication and therefore, contrary to observation, distant regions of the universe are in theory free to do their own thing temperature-wise. In Guth's inflationary universe model, communication is not outrun by expansion and thus all regions remain in thermal contact, allowing the microwave background to be uniform.[390]

Even without specific reference to the horizon problem, the inflation scenario described above raises more questions than it provides answers. Where did the energy for the fabled exponential expansion come from? Was the mysterious antigravity force the source?[391] What set off the inflation?[392] How was the inflation turned off?[393] What is the exact numerical value of the hypothesized antigravity force (also known as the cosmological constant or Lambda)?[394]

Since communication is the central issue of the horizon problem, it's not necessary to invoke "inflation" and then trust in the "dark energy" of antigravity to deal with the problem. Russell Humphreys' explanation of how distant starlight has been able to reach Earth on a timetable that matches the Bible's would seem to work equally well with respect to communication between the various parts of the universe. Actually better, since it presents fewer unanswered questions than the present big bang model. Because

---

[390] Greene, *The Elegant Universe*, 355-56.

[391] The antigravity force is exactly what the name implies. See Kaku, *Parallel Worlds*, 381. Einstein invoked this concept, prior to Edwin Hubble's telescopic observations, in defense of a static universe, but then threw in the towel when Hubble's work showed an expanding universe. Ibid., 37, 104. The antigravity force was later revived to explain inflation. Ibid., 381. As will be seen in subsequent discussion, Lambda has also been a matter of interest as far as the continuing expansion of the universe is concerned.

[392] See ibid., 86, 91-92.

[393] See ibid., 92.

[394] Compare Hawking, *A Brief History*, at 160, with Greene, *The Elegant Universe*, at 225.

of the passage of billions of years "out there" (but not here on Earth), all regions of the universe would presumably now be in causal contact with all other regions of the universe—i.e., would have had time to by electromagnetic wave transmissions make themselves "known" everywhere. Earth time, cosmological "snail mail," would no longer be a factor limiting the amount of light (information) that could by now have traveled from anywhere to anywhere in the cosmos.

I would be remiss, gentle reader, if I did not at this point give you a Bible believer's thoughts regarding what the so-called cosmic microwave background radiation might be. The book of Genesis begins with the earth in darkness.[395] God then decrees light, which is distinct from darkness so that there can be day and night, and thus allows His creation to have a normal Earth day, its first day.[396] It is not until the fourth day of the creation week that God crafts the sun, moon, and stars.[397]

For help in understanding where the light for the first three days is coming from, then, we need to look at some other key passages in the Bible. In Exodus 34, we see that when Moses comes down from Mount Sinai with the tablets of stone on which the Ten Commandments are written, the skin of his face shines, so that the other Israelites are afraid to come near him unless he covers his features with a veil.[398] In Matthew 17, Jesus takes Peter, James, and John up to the top of an unnamed mountain and there is transfigured before them, so that His face "shine(s) as the sun"[399] In the second-to-last chapter of The Book of Revelation, we find it stated that in the apostle John's vision, the new Jerusalem "had no need of the sun, neither of the moon, to shine in it, for the glory of God did lighten it, and the Lamb (Jesus) is the light thereof."[400]

---

[395] Genesis 1:3.

[396] Genesis 1:3-5.

[397] Genesis 1:16-19.

[398] Exodus 34:29-35.

[399] Matthew 17:2.

[400] Revelation 21:23.

It's not surprising to learn that both Father and Son are light sources, since the Bible tells us that they share personhood. At John 14:9, Jesus declares, "(He) that hath seen me hath seen the Father." At John 10:30, Jesus makes it even clearer: "I and my Father are one." The light that lit the creation during the first three days was thus very likely from Christ's face, since in the beginning He "was with God," in fact "was God," and "without him was not made anything that was made."[401] The light of God's countenance has a residual effect, as Moses discovered at Sinai and as the face of the incarnate Son of God demonstrated at the Transfiguration.[402] Therefore, if we believe the Bible, we need not indulge the fantasy that the cosmic microwave background radiation is the after-glow of a naturally-generated explosion, in particular one that features "inflation" as described above. The light that Genesis speaks of was all directly God-generated.

A second area in which there has been an attempt to press "inflation" into service to rescue the big bang and its underlying "billions of years" hypothesis is with respect to the bang's "flatness" problem. As mentioned earlier, an example of revolutionary science would be Einstein's general theory of relativity, which says that what we call gravity is not a magnetic-like clinging together of things that have mass (eggs and Earth, for example, or think of it as on some enchanted evening being attracted to a stranger by a force that pulls you across a crowded room).

Instead, as Einstein showed, gravity is an effect of the curvature of space-time, which varies in proportion to the matter or energy in a given region of the universe. Therefore, one of the factors considered in attempting to determine whether the universe will expand forever or instead collapse in a "big crunch" is the average density of its matter, known since the early twentieth century as *Omega*.[403] As the universe expands, assuming the total amount of matter remains the same, the Omega value necessarily decreases, or, to

---

[401] John 1:1, 3.

[402] Exodus 34:29-35; Matthew 17: 2.

[403] Alexander Friedmann's solutions to Einstein's equations depended on this and two other parameters: $H$, the rate of expansion of the universe; and *Lambda*, the energy associated with empty space, "dark energy." Kaku, *Parallel Worlds*, 40-44, 251.

put it another way, space-time becomes on average less curved, more "flat," because there's less gravity per cubic centimeter This apparently hasn't happened to anywhere near the extent it should have by now if the universe is the billions of years old that popular science says it is. Omega should be almost zero by now; instead, we see a cosmos with an Omega value of close to one. Michio Kaku finds this quite remarkable. He states:

> For any reasonable value of Omega at the beginning of time, Einstein's equations show that it should be almost zero today. For Omega to be so close to 1 so many billion years after the big bang would require a miracle. This is what is called in cosmology the fine-tuning problem. God, or some creator, had to "choose" the value of Omega to within fantastic accuracy for Omega to be 0.1 today.[404]

The above comment, then, suggests one of the possible "flatness" explanations: the hand of God. Another is inflation. In the latter regard, Kaku paraphrases Alan Guth as follows:

> This could be explained if the universe, like a balloon that is rapidly being inflated, was flattened out during the inflation period. We, like ants walking on the surface of a balloon, are simply too small to observe the tiny curvature of the balloon. Inflation has stretched space-time so much that it appears flat.[405]

Elegance does not necessarily explain. My earlier questions regarding the actual mechanics of inflation are still unanswered.

How could there have been an initial explosion sufficient to produce inflation? The principal explanation popular science offers for how the universe's matter could have been thrown outward with enormous force is the "false vacuum" theory, under which energy

---

[404] Ibid., 87.

[405] Ibid., 78.

confined behind a dam of sorts would have "broken through" and caused the universe to grow explosively.[406] However, the Grand Unified Theory (GUT), which is the ostensible basis for the "false vacuum" idea,[407] unites only the strong nuclear force, the weak nuclear force, and electromagnetism.[408] Gravity, the fourth fundamental force of nature, is excluded.[409] This would seem to make reliance on the GUT in support of the big bang theory questionable, especially in view of the inclusion of the negative energy of gravity in theoretical constructs that attempt to show that positive energy and matter could have arisen in a universe having zero net total energy.[410] Such constructs, it might be said, come tantalizingly close to acknowledging *ex nihilo* creation.[411]

It happens, anyway, that gravity is unfit for inclusion in any unified field theory because it would introduce quantum fluctuations that are infinite. Quantum fluctuations are variations from what would be expected under the classical theories of Newton and Einstein, due to the uncertainty principle of quantum physics.[412] The quantum uncertainty principle, put forth by German scientist Werner Heisenberg in 1926, is that one can never know with certainty both the position and velocity of a particle. The more accurately you try to establish the position of the particle, the less accurately you will be able to measure its speed, and vice versa.[413]

Thus, for example, if you attempt to ascertain the path of an electron by bouncing a high-frequency laser beam off it, your

---

[406] See ibid., 85-86.

[407] Ibid., 85.

[408] Hawking, *A Brief History*, 201; Greene, *The Elegant Universe*, 175-78; Kaku, *Parallel Worlds*, 389-90.

[409] Ibid.

[410] See Hawking, *A Brief History*, 133-34; Davies, *God and the New Physics*, 26-32; and Kaku, *Parallel Worlds*, 93-96.

[411] See Davies, *God and the New Physics*, 31.

[412] Kaku, *Parallel Worlds*, 396. The "Newton" reference is to English philosopher/mathematician Isaac Newton (1642-1727).

[413] Hawking, *A Brief History*, 56-57, 203; Kaku, *Parallel Worlds*, 101, 400; Paul Davies, *About Time: Einstein's Unfinished Revolution* (New York: Simon & Schuster, 1995), 90-91.

attempt at measurement will result in a transference of energy from the beam to the electron and thus a disturbance of the particle's path.[414] If you seek to reduce the disturbance by making the photon energy lower, thereby lowering the beam's frequency, the result will be a less localized photon and therefore less precision in measuring the electron's position.[415] In the ordinary macroscopic world, the correction to Newton's laws of motion may be minor, according to Michio Kaku, but when gravity is turned into a quantum theory (i.e., when the quantum of energy is a graviton), the fluctuations are infinite, an untenable situation.[416] This is why GUT-based big bang models cannot take gravity into account in a meaningful way.

In any model, there must be initial energy, whether positive in nature and thus from motion or mass, or negative in nature and thus from gravitational or electromagnetic fields.[417] Energy converts to matter under the first law of thermodynamics.[418] Under the Standard Model of Cosmology, additionally, there has to have been a big bang that involved ultra-high temperatures in order to foster a nonsymmetrical excess of matter over antimatter, so that stars and planets could form.[419] If matter and antimatter were present in equal amounts, their particles would simply have annihilated each other completely, leaving nothing in the universe but gamma rays.[420]

I have no reason to question the existence of antimatter and its annihilation with matter—these phenomena are well-established in the annals of physics.[421] However, aren't we showing a lack of good faith in using the big bang to prove the big bang? That's

---

[414] See Richard Wolfson, *Einstein's Relativity and the Quantum Revolution: Modern Physics for Non-Scientists*, Pt.II (VCR) ( Chantilly, Va.: The Teaching Company, 2000), lecture 19.

[415] Ibid.

[416] See Kaku, *Parallel Worlds*, 194-95, 396.

[417] See Davies, *God and the New Physics*, 31.

[418] Kaku, *Parallel Worlds*, 289, 384.

[419] See Davies, *God and the New Physics*, 29-30.

[420] Kaku, *Parallel Worlds*, 95-96. Stephen Hawking describes gamma rays as electromagnetic rays of very short wavelengths, produced in radioactive decay or by collisions of elementary particles. Hawking, *A Brief History*, 201.

[421] See Davies, *God and the New Physics*, 27-28.

what we're doing, it seems to me, when we account for the presence of matter by saying that the heat of the big bang gave matter the edge over antimatter. How did whatever was there before the alleged big bang, if there *was* a "there" and a "before," get there? Doesn't the prevailing scientific lore say that the big bang was the beginning of *everything* (in the words of Bible scholar and lecturer Chuck Missler, "First there was nothing and then it exploded")? I'm confused.

Alan Guth's resolution of a third big bang puzzle, the monopole problem, was similar to his explanation regarding how inflation solved the flatness problem: deletion by dilution. A monopole is a theoretical entity that has a single magnetic north or south pole.[422] In nature, contrastingly, poles are always found in oppositely-charged pairs.[423] You may recall from high school science that it's impossible to destroy this duality: if you cut a magnet in two, no matter what its shape is, you'll simply create two magnets that have both a north and a south pole. Not surprisingly, then, no one has ever seen a monopole, and there is no conclusive evidence of their existence.[424] The sticky wicket is that applications of the Grand Unified Theory (GUT) have often predicted the monopole's existence.[425]

Guth's solution? Inflation, once again. The exponential expansion of the universe immediately after the big bang took the monopoles that were present at that time and spread them over so vast an area that the odds are heavily against our ever seeing any of them, he said.[426] The monopole problem isn't directly relevant on the question of the age of the cosmos as are the thermal communication and flatness problems, but it does in my opinion serve to underscore the sketchiness of the arguments by which popular science has attempted to legitimize inflation as a prop for the big bang theory.

---

[422] Kaku, *Parallel Worlds*, 86, 393.

[423] Ibid., 86.

[424] Ibid., 86, 393-94.

[425] Ibid., 86.

[426] Ibid.

Back to our first issue, the "horizon" problem, for some additional thoughts. If "inflation" fails as an explanation for the uniformity of the cosmic microwave background radiation (CMBR), then the "big bang after-glow" characterization of the CMBR also fails. Relevant in this regard is what science writer Martin Gardner has offered:

> The remarkable thing about this microwave radiation is its "isotropy"—that is, its uniformity in all directions. It is this that rules out the possibility that the black-body radiation is *coming from* some single, unknown source (emphasis supplied). If such were the case, it could not be so isotropic.[427]

The big bang theory nonetheless presupposes a single source, which would have had to cast its radiation over an area hundreds of millions of light years across. The excluded "possibility" mentioned by Gardner appears to involve radiation being generated at a single location on an ongoing basis (since it "is coming"). Assume, for a moment, that (1) "inflation" is not available as an explanation for CMBR uniformity, and that (2) the radiation in question is deposited in the universe on *only one occasion* in the past. The conclusion under these givens would be the same as would be reached in an *ongoing* one-location radiation discharge situation: the CMBR could not possibly be as isotropic as it is observed to be. "Inflation" was an admission that the Standard Model version of the big bang theory needed rescuing in view of the CMBR's isotropy. It still does.

Perhaps I should have begun this section by relating how all the speculation re a "big bang" origin of the universe got started in the first place. For the beginning of the story, we must go back to the early 1920s, when, as mentioned earlier, Russian meteorologist Alexander Friedmann found what is now known as the "big bang" solution to the equations of Albert Einstein's general theory of relativity, of which theory we will speak more later. This

---

[427] Martin Gardner, *Relativity Simply Explained* (Mineola, N.Y.: Dover Publications, 1962, 1976, 1997), 158.

solution declared that the universe had its beginning when it burst forth violently out of a state of infinite compression, and that the cosmos was still expanding as result of that explosion.[428] Friedmann used Einstein's equations to show, as cosmologist Brian Greene puts it, that the galaxies would be "carried along on (a) substrate of stretching spatial fabric" in moving away from each other.[429] Expansion *of* space rather than *into* space, as was often later noted. Interestingly, Einstein's equations were consistent as well with a cosmos that was shrinking in size.[430]

The idea that the universe was not static so upset Einstein that he revisited his equations and inserted a *cosmological constant*, an antigravity term that he hoped would allow him to continue in the comfort of a universe that stayed the same size.[431] This was before Hubble and his telescope. When American astronomer Edwin Hubble a few years later made observations that suggested that the universe as a whole was expanding, Einstein revised his equations back to their original form and famously stated that the "cosmological constant" was "the biggest blunder of (his) life."[432]

As implied above and stated elsewhere herein, mathematics can mislead. The assumption, based on Einstein's general relativity theory, that the universe began in a state of infinite compression — central to the big bang hypothesis — is a case in point. Brian Greene explains the problem posed by the Standard Model in this way:

> By blindly applying the equations of general relativity, physicists have found that the universe continues to get ever smaller, ever hotter, and ever denser, as we move backward in time toward the bang. At time zero, as the size of the universe vanishes, the temperature and density soar to infinity, giving us the most extreme signal that this theoretical model of the universe, firmly rooted in the

---

[428] Greene, *The Elegant Universe*, 346.

[429] Ibid., 82.

[430] Ibid., 81-82.

[431] Ibid., 82.

[432] Ibid.

> classical gravitational framework of general relativity, has completely broken down.[433]

Greene then uses the above conclusions as a springboard for string theory advocacy, stating:

> Nature is telling us emphatically that under such conditions we must merge general relativity and quantum mechanics—in other words, we must make use of string theory.[434]

Unfortunately, string theory at present has no experimental support; therefore, no discussion of it can be included herein.

To understand how Hubble arrived at his conclusions—specifically, that the galaxies were getting farther away from Earth and from each other, and also that the most distant galaxies were moving away the fastest—one needs a rudimentary understanding of the Doppler effect. The Doppler principle, named after an Austrian physicist of the nineteenth century, is that the movement of an object toward or away from an observer will affect the frequency of the electromagnetic waves by which that object is made known to the observer. Frequency is defined as the number of complete wave cycles per second.[435] Thus, if a star is moving toward you the observer, the frequency of its light will decrease, causing the light to look slightly bluish. If, on the other hand, the star is moving away from you, the frequency of its light will increase and the light will be shifted toward the red end of the spectrum.[436]

One needs to know, also, that astronomer Hubble had spent considerable time cataloging the distances to galaxies outside the Milky Way, which he did by comparing the intensities and spectra of similar stars in the various galaxies.[437] To his surprise, Hubble found not only that most of the galaxies he observed were red-shifted

---

[433] Ibid., 357.

[434] Ibid.

[435] Hawking, *A Brief History*, 201.

[436] Kaku, *Parallel Worlds*, 386.

[437] Hawking, *A Brief History*, 38, 41; Kaku, *Parallel Worlds*, 47.

in the light that was coming from their stars, but also that the size of a galaxy's red shift on the spectrum was directly proportional to the galaxy's distance from Earth. The farther away a galaxy was, the faster it was moving away, it seemed to Hubble.[438]

Was Hubble correct in his understanding of the galactic red shifts? It all depends on how you state the question. If the question is framed as one of whether Hubble correctly saw on the basis of the red shifts that the galaxies were getting farther apart and farther away, without regard to whether they were accelerating in their recession, the answer is almost certainly yes. There have been various other attempts to explain the reddening of starlight—as, for example, its being caused by its passage through cosmic dust, or as reflecting "tired light" in the sense that the longer light travels, the slower it vibrates[439]—but these efforts have met with little encouragement in the scientific community.[440]

Albert Einstein brought some light to the cosmic expansion question (no play on words intended) by adding to Hubble's interpretation the qualifier that the red shift of a galaxy is not, technically speaking, caused by the galaxy's speeding away from Earth; instead, he said, it's a result of expansion of space itself between the galaxy and the earth.[441] The origin of the red shift, he explained, is that light coming from a distant galaxy is stretched—and thereby given a longer wavelength (distance between adjacent troughs or crests) and thus a lower frequency—by the expansion of space.[442] Distant galaxies don't produce a red shift by accelerating through space; instead, space itself by its expansion causes red shifts, which make the galaxies appear to be moving away faster and faster when in reality they're merely stretching space more and more. Ph.D. physicist Russell Humphreys, a prominent creationist, agrees with

---

[438] See Hawking, *A Brief History*, 40-41; Kaku, *Parallel Worlds*, 48-50.

[439] Cf., in the latter regard, the "light slowing down" discussion supra this chapter.

[440] See Gardner, *Relativity*, 132.

[441] Kaku, *Parallel Worlds*, 49-50.

[442] Ibid., 50.

this interpretation.[443] Such a view calls into question any idea of "runaway expansion."

There is now another, seemingly larger fly in the red-shift ointment. Since 1976, University of Arizona astronomer William Tifft has been finding that the red shifts of distant stars and galaxies typically differ from each other by only certain fixed amounts.[444] There is no smoothly-graduated spectrum of red shifts. Australian astronomer Barry Setterfield has characterized this as a "quantization" of red shifts, so called by him because of its similarity to the phenomenon in quantum physics in which atoms are seen to give off tiny bundles of energy *(quanta)* that are only of certain amounts, with no energy levels in between.[445] Scientist/educator Walt Brown finds red shift quantization "very strange if stars are actually moving away from us."[446] He explains:

> It would be as if galaxies could travel only at specific speeds, jumping abruptly from one speed to another, without passing through intermediate speeds. If stars are not *moving* away from us at high speeds, the big bang theory is wrong, along with many other beliefs in the field of cosmology.[447]

Other astronomers have obtained results similar to Tifft's, Walt Brown notes.[448] The observed "quantization," Barry Setterfield believes, is an atomic effect rather than a recessional-velocity effect.[449] It has to do with space *absorbing energy* in fixed increments, which like stellar and galactic recession will produce a red-shift effect as far as starlight is concerned, with the farthest

---

[443] See Humphreys, *Starlight and Time*, 37- 38, 79, 98, 121.

[444] See Brown, *In the Beginning*, 324.

[445] Ibid., 324-25.

[446] Ibid., 324.

[447] Ibid (emphasis supplied).

[448] Ibid.

[449] Ibid., 324-25.

stars' light being red-shifted the most.[450] As might be expected, Setterfield, a leading proponent of the idea that light is slowing down, is working on a theory to tie the above quantization and the decay in the speed of light together.[451]

As noted earlier, the Bible speaks of God "stretch(ing) out" the heavens. The verses using this or similar language are numerous. See, e.g., Job 9:8 ("spreadeth out the heavens,"), Psalm 104:2 ("stretchest out the heavens like a curtain:"), Isaiah 42:5 ("created the heavens, and stretched them out;"), Isaiah 45:12 ("stretched out the heavens,"), Jeremiah 10:12 ("stretched out the heavens by his discretion"), Jeremiah 51:15 ("stretched out the heaven by his understanding"), and Zechariah 12:1 ("stretcheth forth the heavens,"). Thus, Scripture allows for expansion of some kind, be it past or continuing, and in fact, by repeatedly using forms of the word "stretch," seems to accommodate the Einstein/Humphreys interpretation of the red shift.

The idea that it's space rather than the universe in space that continues to expand allows us to harmonize the perceived outward movement of the galaxies with the statements of Scripture that "(t)hus the heavens and the earth were finished, and all the host of them" (Genesis 2:1), and that on the seventh day God "ended his work which he had made; and rested from all his work which he had made" (Genesis 2:2). The above verses leave no room for universe growth taking place after the sixth Earth day of creation in the sense of material being added to the cosmos, but do allow for later expansion of *space*—specifically, of the interstitial areas between the galaxies created by the Lord during the six days of creation.

My quarrel is not with the general idea of a red-shifted expanding cosmos, in any event, as long as matter is not said to be coming into being except by conversion from energy, which would be a change of form but not a creation of substance. The issue for me is the way in which secular science has made use of the discovery that the universe may have expanded in the past and may still be expanding. Based on these findings, taken in combination with Friedmann's calculations, science reasoned that if the universe

---

[450] See ibid., 325.

[451] See ibid.

had gotten bigger, it must have been smaller in the past—so small that under the equations of general relativity theory it began as an infinitely-compressed bit of matter from which was released almost boundless energy.

One now wonders whether popular science has allowed us to be bullied by mathematics into reaching an unwarranted conclusion regarding the original condition of the cosmos. Was any solution to Einstein's equations other than Friedmann's seriously considered? Was there any observational data directly supporting the conclusion that the perceived expansion of the universe was attributable to an enormous explosion of all of the universe's matter? Remember, we shouldn't be counting the cosmic microwave background radiation as support for the bang given the inexplicable uniformity of the CMBR. What's left?

Nucleosynthesis, some will say, the formation of new atomic nuclei by successively adding particles to hydrogen, the smallest atom. George Gamow and the graduate students working with him in the late 1940s thought that the entire periodic table of the elements could be accounted for as the product of the fusion of subatomic particles in the supposed intense heat of the big bang.[452] Gamow liked to refer to the hypothesized early nuclear reactions as the elements being "cooked" in "the prehistoric kitchen of the universe."[453] You begin, he said, with hydrogen atoms colliding and being fused together to produce helium.[454] Subsequent collisions between hydrogen and helium atoms would, he theorized, produce a set of elements that included lithium and beryllium, and so on.[455]

Only trouble was, when Gamow got to elements in the progression that had 5 and 8 neutrons and protons, he encountered the problem that these elements are extremely unstable and hence cannot act as a "bridge" to the heavier elements in the universe.[456] Like the Otisco causeway, Gamow's continuum was not continuous.

---

[452] See, in this regard, Kaku, *Parallel Worlds*, 55-56; also, Hawking, *A Brief History*, 122.

[453] Kaku, *Parallel Worlds*, 55.

[454] Ibid.

[455] Ibid.

[456] Ibid., 56.

The scientific community was thus being forced to conclude that the heavy elements of the periodic must have had some origin other than the alleged big bang. It was left to British astronomer Fred Hoyle to provide a solution for the problem. Sir Fred had the idea that the hundred or so elements of the universe that were known of at that time were cooked not in the big bang, but in the stars. In a series of papers published in the 1940s and 1950s, Hoyle and his colleagues detailed how the nuclear reactions in the core of a star could add protons and neutrons to create all the heavier elements up to iron.

Hoyle got past the "orphan heavy elements" problem that had stumped Gamow by employing an unstable form of carbon, which he was able to create, that lasted just long enough to be the needed "bridge" to the heavier elements.[457] For the elements above iron in the periodic table, such as copper, nickel, zinc, and uranium, even more intense heat than what the stars could provide was needed. The needed super-oven, according to Hoyle, was supernovae—the explosion of giant stars in their death throes.[458] Dependence on the big bang to explain the configuration of the elements was avoided, but so was reliance on God as an explanation. How much easier it would have been if the great minds involved in the above scenario had been willing to accept the fact of divine in-place creation!

Other difficulties for the big bang theory were outlined in plasma physicist Eric Lerner's 1991 book, *The Big Bang Never Happened*.[459] Not the least of these was the existence of clusters of galaxies many times too large to have been able to form in the maximum of twenty billion years allowed by the most generous calculations as the time since the big bang.[460] These clusters, Lerner said, additionally violated the big bang theory's assumption that

---

[457] Ibid., 63.

[458] Ibid.

[459] Eric J. Lerner, *The Big Bang Never Happened: A Startling Refutation of the Dominant Theory of the Origin of the Universe* (New York: Random House, 1991). Plasma physicists are a group of scientists who hold that the fundamental stuff of the universe is hot, electrically conductive gases, and that the cosmos had no beginning and will have no end, since there is no observational support for either a beginning or an end.

[460] See Lerner, *Big Bang*, 23-25.

at the largest scales the universe is smooth and homogeneous, not clumpy.[461] The observation that the cosmic microwave background radiation is smooth, Lerner said, "makes it impossible to explain how today's clumpy universe could have come to be."[462] Lerner found it strange, also, that the matter needed to allow gravity to do its work in galaxy and supercluster formation was not observable and had to be accounted for as "dark matter"—i.e., as matter that no one had ever seen and the existence of which was not shown by any proof, direct or indirect.[463] Lerner thus gave the following summary of the big bang theory's shortcomings:

> The test of scientific theory is the correspondence of predictions and observation, and the Big Bang has flunked. It predicts that there should be no objects in the universe older than twenty billion years and larger than 150 million light-years across. There are. It predicts that the universe, on such a large scale, should be smooth and homogeneous. The universe isn't. The theory predicts that, to produce the galaxies we see around us from the tiny fluctuations evident in the microwave background, there must be a hundred times as much dark matter as visible matter. There's no evidence that there's *any* dark matter at all. And if there's no dark matter, the theory predicts, no galaxies will form. Yet there they are, scattered across the sky. We live in one.[464]

In view of the above cumulation of problems, it's not hard to see why conventional cosmologists would have found it necessary to resort to strange concepts such as "inflation."

Lerner's book was published a few months before completion of the initial analysis of data gathered by the Cosmic Background

---

[461] Ibid., 15.

[462] Ibid., 31.

[463] See ibid., 32-39.

[464] Ibid., 39-40.

Explorer (COBE) satellite.[465] For two years prior to that, COBE had been searching the cosmic microwave background radiation for any irregularities such as could be relied on as evidence that the universe even in its earliest stages had a certain clumpiness to it.[466] In the absence of such evidence, astronomers and physicists, not willing to believe in a divine special creation, felt obliged to include in their cosmologies the tens of billions of years required for the universe to get from total unclumpiness to its present condition, which had the effect of requiring a finding that the cosmos was far older than the prevailing big bang model said it was.[467]

By 1992, examination of the COBE data had shown that there were tiny "ripples" in the microwave background radiation, hot and cold patches superimposed on what was otherwise a complete uniformity.[468] These ripples were interpreted as evidence of early cosmos clumpiness, which if proven could foreshorten the time period needed for the formation of celestial bodies. Still missing, however, was the "dark matter" needed to allow the gases of the universe to clump together and form stars and galaxies.[469]

An early 2014 experiment using a telescope mounted on the Antarctic Plateau similarly showed "ripples" that were said to be evidence of the big bang and inflation, according to a *USA Today* article published in March of 2014.[470] The "ripples" that Harvard-Smithsonian researchers claimed to have found were "faint, swirly patterns" that were said to have been imprinted on the cosmic microwave background (CMB) as a result of "inflation;" however, the new data had not (and apparently still have not) been confirmed, and in any event do not overcome the isotropy (uniformity) in the

---

[465] Davies, *About Time*, 146-47.

[466] Ibid.

[467] Ibid., 147-48.

[468] See ibid., 146-47.

[469] Ibid., 148-49.

[470] Traci Watson, "'Smoking gun' rocks the universe," *USA Today* for *Rochester Democrat & Chronicle* (March 18, 2014), 3B.

CMB that was previously noted as a problem for the big bang/inflation model.[471]

A sidebar to the abovementioned *USA Today* article admitted, additionally, that the dark matter mentioned by Eric Lerner had still not been found.[472] The more dark matter, the more gravity there is, and the more gravity, the longer it would have taken for the universe to reach its present size.[473] It should also be noted that no explanation for the space-time ripples observed in 2014 other than the now shopworn inflation thesis appears to have been seriously pursued, even though secular cosmologists concede that these "signals" could be the result of factors other than inflation.[474]

*Science and Religion Generally*

The reason I'm contending with science on its terms should be obvious: As I indicated in my Preface, part of my intended readership is people who consider themselves educated. I want to give such persons the ability to believe, or to at least open themselves to the only true source of the ability to believe, the Word of God as written and interpreted by the Holy Spirit of God. We're talking about individuals who think they know what Thoreau meant when he said, in *A Week on the Concord and Merrimack Rivers* (1849), that there is "more religion in men's science than there is science in their religion."[475] Thoreau later owned that there is "a chasm between knowledge and ignorance which the arches of science can never span."[476] The second quote might be taken as a rejection

---

[471] Jake Hebert, PhD., "'Smoking Gun' Evidence of Inflation?" *Institute for Creation Research* March 21, 2014, cited in *http://www.icr.org/article/8031//utm_source=dlvr.it&utm_medium=fa.*, 1-2.

[472] Doyle Rice, "Is anyone out there? - and 2 other big questions," *USA Today* for *Rochester Democrat & Chronicle* (March 18, 2014), 3B. See also, in this regard, Kaku, *Parallel Worlds*, 11-12, 70-74; Greene, *The Elegant Universe*, 225, 235; and Davies, *About Time*, 148.

[473] Davies, *About Time*, 153-54.

[474] Hebert, "Smoking Gun," 2.

[475] Thoreau, *A Week*, 91.

[476] Ibid., 116.

of the idea that science is "the final arbiter of truth;"[477] the first clearly is not.

A sharper line of demarcation between science and religion is drawn by Bertrand Russell. Although he appears to concede that the church is no longer a flat earth society (it never was, that I'm aware of), Russell nonetheless sees the interests of the church and science as unalterably opposed. In this regard, he states:

> The church no longer contends that knowledge is in itself sinful, though it did in its palmy (halcyon) days; but the acquisition of knowledge, even though not sinful, is dangerous, since it may lead to pride of intellect, and hence to a questioning of Christian dogma.[478]

Is the great English philosopher here punishing the children for the sins of the fathers (see Deuteronomy 24:16)? We are a few years removed from pre-Copernican geocentrism, after all. How, in the past two centuries, has the church attempted to restrict the acquisition of scientific knowledge? Do honest challenges of the big bang theory and the evolution hypothesis, both of which are attackable without recourse to Scripture, qualify as attempts to muffle science?

Rather than seethe over this most recently discovered unfair statement of Bertrand Russell, I can pause for a moment to enjoy a pleasant scene. The Parnassus Imprints edition of *A Week on the Concord and Merrimack Rivers* that I borrowed from our local public library has a cover illustration that depicts a slender river cutting through field and forest. The only signs of civilization are an individual in a skiff and, farther in the distance, a stone fence and a monument. The river, while exhibiting ripples, appears to be dilatory in its flow. The narrative in the book describes an unhurried boat tour taken by Henry David Thoreau and his brother John through the New England town-and-countryside in the summer

---

[477] The view and words of American philosopher W. V. O. Quine. Stokes, *Essential Thinkers*, 207.

[478] Russell, *Why I Am Not a Christian*, 45; parenthetical mine.

of 1839, which gives occasion for later philosophical musings on Thoreau's part.

Both cover and contents of the Parnassus edition make me think of an annual canoe day trip that my son Jeremy and I had begun taking before my arthritis became a major factor. From a canoe and kayak rental shop on the southern end of Irondequoit Bay, we were bussed to a launch site on Irondequoit Creek in Ellison Park, a County of Monroe picnic ground, from whence we would paddle down the creek, passing through the park and an extensive wetland area and ending our journey where the creek flows into Irondequoit Bay, although we didn't rule out venturing out onto the Bay, wave height permitting.

For me, a fascinating part of the above excursion, a bonus added to the wildlife that we would see in the swamp and the fresh air and the father-and-son time, was the experience of passing through a picnic area and not being able to see or be seen by the picnickers. The banks of the creek, as it flows through an unflooded Ellison Park, are just high enough that from a sitting position in your canoe you can see barbecue smoke and hear laughter and horseshoes clinking but not see anyone. You move from one sphere of happy noise to another. Eventually you may hear the shouts from a softball game in progress, or a volleyball game, or some other kind of weekend-warrior team sport. You're involved in the lives of people but also detached from them. The boundaries of self are blurred, obscured metaphysically.

Back to the topic I was pursuing before I got sidetracked by happy memories, belief versus scientific skepticism. C. S. Lewis, our archetypal Christian man of letters, separates science and faith while giving due credence to science. After stating that the role of science is to experiment and observe, Lewis continues:

> But why anything comes to be there at all, and whether there is anything behind the things science observes—something of a different kind—this is not a scientific question. If there is 'Something Behind,' then either it will have to remain altogether unknown to men or else make itself known

> in a different way. The statement that there is any such thing, and the statement that there is no such thing, are neither of them statements that science can make.[479]

Are you listening, Richard Dawkins? Oxford University professor Dawkins, a scientist, ventures into the realm of theology—that is, tries to make a statement about spiritual reality—in *The God Delusion*.[480] His principal arguments concerning God's existence are contained in a chapter entitled, "Why There Almost Certainly Is No God."[481] In this chapter, Dawkins equates creationism and the "intelligent design" movement, stating that the creationist often "chooses to masquerade in the politically expedient fancy dress of 'intelligent design'."[482]

Dawkins' explanation for the world as we know it is Darwinian natural selection. This is not the same as "chance," he claims, even though both depend on random mutations.[483] His explication of how natural selection, a "bottom-up" process, might have worked to produce what we now see is unconvincing. For one thing, Dawkins fails to address the familiar problem that mutations do not add new information to the genome.[484] The ability to add DNA information, which is at the heart of the debate over "natural selection," is commented on by Dr. Jason Lisle, Ph.D. researcher with the organization Answers in Genesis. Dr. Lisle explains that whenever or wherever we find any kind of information, certain rules or "theorems" apply. Two such theorems, he says, are as follows:

> (1) There is no known law of nature, no known process, and no known sequence of events that can cause information to originate by itself in matter.

---

[479] Lewis, *Mere Christianity*, 23.

[480] Richard Dawkins, *The God Delusion* (New York: Houghton Mifflin Co., 2008).

[481] Ibid., 137-89.

[482] Ibid., 138.

[483] Ibid., 146-47.

[484] See Cloud, *Non-Existent*, 137.

> (2) When its progress along the chain of transmission events is traced backward, every piece of information leads to a mental source, the mind of a sender.[485]

Genetic information added to a species in a supposed "bottom-up" process such as natural selection would essentially consist of data that are transmitted by a lower being to a higher being. Such a process cannot happen under the rules outlined by researcher Lisle, since Lisle's theorems presuppose an information source that has more of the data needed for species improvement than does the information recipient. Thus, the upward modification of species by natural selection is an impossibility.

Also, Dawkins seemingly refuses to assume that a God fitting the Judeo-Christian definition of "God" would be omnipotent. Accordingly, he questions whether God could do the "fine tuning" needed to create an intelligence-friendly universe.[486] A God capable of creating the universe would be complex and therefore improbable, he argues.[487] Such a God would need "an even bigger explanation than the one he is supposed to provide."[488] Not surprisingly, Dawkins reduces God to the status of needing the odds to be on His side in order to exist.

Faith has no significance, Dawkins indicates; objective likelihood rather than God-provided belief is what's needed. However, an attitude of philosophical positivism with respect to anything about God ("give me proof") is directly contrary to the apostle Paul's counsel at Ephesians 2:8 ("For by grace are ye saved through *faith* [emphasis supplied]; and that not of yourselves: it is the gift of God"). You need faith rather than proof, Paul says, and God will give you the faith.[489]

Some men of science have been accepting of the idea that God exists but have stripped Him of His immanence. As I indicated in my Region of Awe chapter, saying that God is *immanent* is saying

---

[485] Lisle, *Ultimate Proof*, 18-19.

[486] See Dawkins, *God Delusion*, 171-72.

[487] Ibid., 176.

[488] Ibid.

[489] See Morris, *Study Bible*, n. to Eph. 2:8.

that He is active in the universe. He hears prayer, counsels, assists, saves. The opposite of an immanent deity is the *deist* God, who allegedly set the world in motion and then walked away from it, leaving it and its inhabitants to their own devices or physical principles.

Scientists who have spoken on the question of God have shown varying degrees of proximity to the deist position. Thus, you have Don Page of the University of Alberta, a former student of Stephen Hawking, saying that there's "definitely a purpose" to the universe, that the universe might have been brought into existence so that God could "create man to have fellowship with (Him)," and that "a bigger purpose maybe was that God's creation would glorify God."[490] This all sounds positively evangelical.

More toward the deist side, it seems, is Charles Misner of the University of Maryland, who expresses a "feeling that in religion there are very serious things, like the existence of God and the brotherhood of man, that are serious truths that we will one day learn to appreciate in perhaps a different language on a different scale," and who concludes that "the majesty of the earth is meaningful, and we do owe honor and awe to its Creator."[491]

Misner was an early analyzer of the general relativity theory of Albert Einstein, who provides one of the best examples of the deist point of view. Einstein was of Jewish ancestry; indeed, this appears to have been one of the reasons why he did not return to Germany from America after Hitler came to power in 1933. Nonetheless, he rejected the biblical idea of God, whom he did not see as someone who answers prayers and performs miracles or who is in any way concerned with the actions and fates of individual human beings. Consistently with his view of God as non-immanent, Einstein wrote, "I cannot imagine a God who rewards and punishes the objects of his creation...Neither can I believe that the individual

---

[490] Don Page, as quoted by Michio Kaku, *Parallel Worlds* at 356, quoting Alan Lightman and Roberta Brawer, *Origins: The Lives and Worlds of Modern Cosmologists* (Cambridge,Mass.: Harvard UniversityPress, 1990), 409.

[491] Charles Misner, as quoted by Kaku, *Parallel Worlds* at 356, quoting Lightman and Brawer, *Origins*, 248.

survives the death of his body."[492] Einstein made mention in his writing of "the Old One," nevertheless, and spoke of wanting to "read the mind of God."[493] He further stated, "I want to know how God created this world. I am not interested in this phenomenon or that. I want to know God's thoughts. The rest are details."[494]

Parenthetically, Einstein would add, "Science without religion is lame. But religion without science is blind."[495] Thus, Einstein appears to have recognized intelligent design of some kind at work in the universe. It strikes me as odd, however, that he would allow for intelligent design and then neglect to ask the logical next question: to what end? Would an intelligence of any description create human beings only to destroy them? Or is the organizing activity contemplated by Einstein so undirected as to disqualify it altogether from being thought of as the purposive work of an intelligence? In that case, the moral and logical constraints implicit in these questions would seem not to apply.

Also noteworthy is the fact that during his search for a unified field theory a "theory of everything" that would unite all of the forces of nature, Einstein reflected on whether God "could have made the Universe in a different way; that is, whether the necessity of logical simplicity leaves any freedom at all."[496] Brian Greene comments that with this query, Einstein articulated a view that is now beginning to be shared by many physicists: if there is a final theory of nature, one of the most convincing arguments in support of the form in which it's presented would be that the theory

---

[492] Albert Einstein, as quoted by Kaku, *Parallel Worlds* at 357, quoting Gary Kowalski, *Science and the Search for God* (New York: Lantern Books, 2003), 24. For the same quote, see Alice Calaprice, ed., *The Ultimate Quotable Einstein* (Princeton, N.J.: Princeton University Press, 2011), 330.

[493] Einstein, as quoted by Kaku, *Parallel Worlds* at 344, quoting Alice Calaprice, ed., *The Expanded Quotable Einstein* (Princeton: Princeton University Press, 2000), 202. See also Calaprice, ed., *Ultimate Quotable Eintein*, 380.

[494] Kaku, *Parallel Worlds*, 344; Calaprice, ed., *Ultimate Quotable Einstein*, 324.

[495] Kaku, *Parallel Worlds*, 344; Calaprice, ed., *Ultimate Quotable Einstein*, 335.

[496] Einstein, as quoted by Greene, *The Elegant Universe* at 283, quoting John D. Barrow, *Theories of Everything: The Quest for Ultimate Explanation* (Oxford, U.K.: Oxford University Press, 1992), 20. Similar language is quoted by Alice Calaprice (Calaprice, ed.,*Ultimate Quotable Einstein*, at 344).

couldn't be otherwise.[497] Such a theory would hold that things are the way they are as a matter of logical necessity—i.e., because they *have to* be that way.[498] Where does that leave God? one may ask. Isn't it logic rather than God ruling the universe under these circumstances? Wouldn't a God who had no choice in how he created the universe be less than omnipotent? Lacking the power to choose, would the entity that we call God be nothing more than an organizing principle (not a person, in other words)?

Paul Davies says that while the idea of an "inevitable" universe makes redundant the idea of God-the-Creator (why would you need a Creator in the scheme of things if the scheme of things couldn't be otherwise?), it does not rule out a universal mind existing as part of that universe: a natural as opposed to a supernatural God.[499] Some troubling questions would be raised, however, if we had this kind of deity. It is not just that pantheism, which erases any distinction between the created and the Creator, is unscriptural. If the universe is suffocated by its own entropy under the second law of thermodynamics, God, also, would seem to be in danger of dying a slow death.

The entropy could coincide with the universe's continuing to expand forever, resulting in a "big freeze" ending to all life. If, on the other hand, gravity causes the cosmos to collapse back on itself in a "big crunch," then we're talking about a final *singularity* consisting of a state of infinite gravity, where general relativity theory would no longer apply and where all scientific predictive power would presumably vanish.[500] In that event, one may ask, is it not just the universe that's extinguished, but also the logic that enables the cosmos to exist? Does this mean that nothing can ever exist again?

Returning to the question of God's role in creation, I prefer to think that God would not be in any way diminished by a logically necessary universe. He is, the Bible tells us, eminently logical. "Come now, and let us reason together, saith the Lord," the

---

[497] Greene, *The Elegant Universe*, 283.
[498] Ibid.
[499] Davies, *God and the New Physics*, 223.
[500] Kaku, *Parallel Worlds*, 398; Hawking, *A Brief History*, 49, 189, 199.

prophet Isaiah beckons.[501] Concerning this saying, Henry Morris assures us that faith in the God of creation and redemption "is not gullibility but fully consistent with all true spiritual reason—a reasonable faith."[502]

We could not reason with God were He not logical—i.e., reasonable. That which is reasonable is pleasing to the mind; it does not jar the sensibilities. There is a beauty and a symmetry to it. Paul Davies notes that when physicists talk of beauty and symmetry, the language through which these concepts are expressed is mathematics.[503] It's more than a matter of aesthetics, however. Science *depends on* mathematics. The universe, created by God, has a *scientific foundation* that may be explained mathematically.

Why would God have chosen to make His universe explainable by means of mathematics? Answer: because mathematics is based on logic, and, as we just noted, God is logical. A potential problem, notwithstanding, is that logic itself is a paradigm—allowing, perhaps, for a greater latitude in thinking than the paradigms described by Thomas Kuhn, but a paradigm nonetheless. Although it may seem strange to you to hear it, logic straitjackets God's thoughts and actions. To be logical in the sense that the term is here applied to God means to be unable to think or act illogically.

We may safely conclude, therefore, that God does not act arbitrarily. He doesn't just say, "This is how it is." Everything is the way it is for a reason. God has set everything up logically, and that includes His gospel. Let's say—because it's true—that a holy God wants to draw man, created in His image (Genesis 1:27), to Himself. Let's also say—because it's true—that individual man by his own efforts is unable to become righteous enough to participate in God's kingdom. What might a loving and merciful God do to overcome the problem of human unrighteousness under these circumstances? Answer: He might allow *His* righteousness to be the currency used by believers—in essence, impute His goodness to individual man.

---

[501] Isaiah 1:18.

[502] Morris, *Study Bible*, n. to Isa. 1:18.

[503] Davies, *God and the New Physics*, 221.

How would God go about conferring His blamelessness on believing humanity? Answer: through a device understandable by Old Testament man: blood sacrifice. But this time around—in 33 AD, with a new covenant between God and man going into effect—the most perfect sacrifice imaginable, that of Himself made flesh. This would be the sacrifice that would most assuredly be sufficient to atone for sin. Told of this plan of redemption, a thinking person should be unable to miss the fact that God's style is logic rather than fiat, the latter being the exercise of authority arbitrarily, without regard to whether the action taken makes sense. God's plan of salvation flows forth naturally from the circumstance (sin) it seeks to overcome; it doesn't have to be coaxed out of hiding by dint of imagination.

A *fiat*, it should be noted, is issued by someone having absolute authority. God by definition is such a person. He can do whatever He wishes; thus, He can operate by fiat but He also can choose *not* to exercise His authority arbitrarily. The subject of belief unto salvation is one of the infinitely many areas in which God has chosen the latter course. And the utter logicality of God's plan of redemption assures that man will be accorded what in the law we would call procedural due process, which is notice of the Law Giver's thoughts and the ability to please the Law Giver. Because He is logical, God can be comprehended in His statements to man. Rather than leaving His subject clueless as to the why and how of his spiritual status, God can give man an understandable reason to believe, a rationale available even in advance of any ministration of the Holy Spirit. And that is exactly what has happened. God has not just said to man, "Believe or die!" That would have been arbitrariness on His part. Instead, He has said to man, "Here is *why* you should believe. It has to do with the fact that you are not me but Jesus is. It has to do with the fact that I created you and love you enough to die for you. Does it all make sense to you now?"

The circumstance that the universe has a foundation that may be explained mathematically, taken in conjunction with the basis of mathematics being logic, should tell you that science and religion have much in common, as already suggested. It's easy to see in the laws of nature the creative hand of a supreme yet logical intelligence. Science and faith are differing world-views, certainly; nonetheless, they should be able to coexist because of their common basis in reason. The circumstance that one is attuned to the head while the other operates primarily out of the heart is not necessarily fatal to the cooperation of the two disciplines. As Human Genome Project director Francis Collins states in his book *The Language of God*, there yet remains the possibility of "a richly satisfying harmony" between the two world-views.[504] It is my fervent hope that in the remainder of this book I will be able to integrate science and faith in a way that buttresses both.

---

[504] Francis S. Collins, *The Language of God: A Scientist Presents Evidence for Belief* (New York: Free Press, 2006), 5-6.

# 5
# PROVING GOD

*Induction and Deduction*

It's February, and Sue and I are back at the Point. We're up here to let the bank's appraiser in, for the purpose of yet another mortgage refinance. Interest rates have been dropping, and getting our monthly expense down is crucial if we're to keep The Cottage. The municipality just hit us with a nineteen-hundred dollar increase of our property tax, and we're at the point of being maxed out as to how many weeks per summer we can rent and how much more we can raise the rent. Localities must regard cottagers as the best possible situation from a government-interest perspective: they can tax our vacation homes based on a high fair market value and they don't have to provide schools for our children, since most of us that have school-age kids live at least several school districts away during the off-season.

Though there's a lot of snow, getting into the Point wasn't all that difficult. The Town of Hounsfield plows the access road, which needs to be negotiable by fire trucks, and our neighbor Tom (the guy with the imposing cottage next door, one of the Campbell's Pointers who stays here all winter) lets me use his driveway.[505]

---

[505] It's not really his driveway, anyway. Campbell's Pointers, under the standard deed, own only to the overhang of their eaves.

Getting from the end of Tom's driveway to the back door of our cottage is somewhat more challenging, since the drifts are three feet high in places.

Before we enter The Cottage, we fight our way through a hundred feet or so of knee-high snow (I'd rather it was corn on the Fourth of July)[506] to get around to the "front" of our house, to where we can see the Lake and the condition of our sand fence. The fence—basically a low plywood wall—is something we put up every year when we leave in October, its purpose being to keep the beach from migrating into our front yard when the November winds howl. All that's showing of this fence today is the tops of the metal stakes to which we fasten two-by-eight-foot sheets of plywood; the drifts have otherwise covered up our wall (see Fig. 9). The Lake is frozen as far as the eye can see, an expanse of white that ends at some indistinguishable point out beyond The Islands, whose skeletal winter vegetation is little more than gray shadows as it appears from this distance.

As I stand looking out at the Lake, the wind shaking my pant legs and blowing granules of snow against my face, I ask myself if I would be here today if I didn't have to be. Assume for the moment that The Cottage has sidewall and rafter insulation, an efficient woodstove, and a drilled well (all of which Tom has). Leave out of our givens the circumstance of our having children and grandchildren in Rochester, a hundred and ten miles away. But keep our Rochester house in the picture as one of our homes, as it is now; let's make The Cottage, in our imaginations, just a place for winter getaways, not a winter home. Sue's feelings aside, would I personally think that our being here was a good idea? And by what thought process would I reach my conclusion on this question?

Inductive reasoning would involve my starting with known (existent) facts, such as those recited or supposed above, and working to a generalization as to what they show regarding whether we should or should not be here in winter. It would have me building a case, stone on stone, for one or the other conclusion. On the positive side, if we were to winterize The Cottage, then

---

[506] Upstate New York farm people are fond of saying that corn should be "knee high by the Fourth of July."

our family—Sue and me, and the kids and grandkids whenever they could come for a weekend—would be able to spend cozy winter evenings in front of the woodstove, playing board games or reading, free from the interference of television (we don't have cable at The Cottage, and the only non-CATV programming you can get here is a local public station and a French-speaking channel out of Montreal). I'm assuming that such idylls would actually happen. We would take hikes (well, walks) in the snow. We would dust off our cross-country skis.

Reveries come with reservations. It could turn out that Sue and I would be the almost exclusive enjoyers of The Cottage in winter, since it might be impractical for our sons and their families to spend much time here during the school year—too many school activities going on, for one thing. Also, there's the snowiness of the roads. Getting stuck in the drifts, especially with young children in the car, is not an ideal situation. I recall an instance of our visiting Sue's parents at their Sackets Harbor home in the month of February, after they had stopped wintering in Fort Meyers, when Sue and I had arrived for our visit and Sue's brother called the house and said that he and his girlfriend had reached a point north of Oswego where a white-out forced them to turn around and head back to Rochester. It was the middle of the day and the sun was shining at Sackets Harbor when we received this call.

My Rochester son and his wife, fortunately with no kids aboard, had experienced near turn-arounds on the same route, in particular one in which they had to wait for a snow plow to catch up to and get ahead of them before they could proceed farther. If Sue and I were staying in The Cottage for prolonged periods during the winter, I would be worried that people would feel that they had to subject themselves to such conditions in order to visit us.

What we have, then, is a mixed bag of existent facts; therefore, no inductive generalization is possible. I could have saved myself the trouble of the above analysis, of course, if I had simply ruled out *ab initio* the possibility of our winterizing The Cottage. As it is, we can barely afford maintenance that keeps the place suitable for spring, summer, and fall use. But fully contemplating the possibilities of life can be a salutary thing. If you want access to everything that life has to offer, rule out nothing.

The deductive approach to the winter-use question would have me starting with assumptions and proceeding syllogistically to a conclusion. Vacation or get-away homes are a beneficial thing as far as family togetherness is concerned, I might assume. The more use of the get-away, the more benefit, it follows as a corollary. We have a get-away and we desire family togetherness. Therefore, we should use The Cottage as much as possible. Should we be spending time here during the cold months of the year? Here, I encounter the reality that no assumptions can be made about upstate New York winter weather, certainly none that could be characterized as sanguine; therefore my deductive process breaks down at this point. The cost of cottage winterization would probably have killed the discussion in any event.

What about induction, deduction, and the truth of the Bible? Again, we hear Kierkegaard declaring to us that true faith has nothing to do with rationalistic proofs of the existence of God;[507] In so stating, Kierkegaard showed mystical tendencies that were to later influence the development of the Emergent Church.[508] The compulsion that I feel at the moment to rationalize God is irresistible, nonetheless, even though I'll admit that my eagerness to proceed with such a project may be a reaction to Kierkegaard's anti-intellectualism. But before we can re-enter the domain of reason—a land that may well be populated by monsters waiting to devour us—I believe we need to consider two questions.

The initial threshold inquiry for us has to do with God's role as a default explanation. Dietrich Bonhoeffer, a Dutch scholar imprisoned and finally executed by the Nazis during World War II, was able through a favored status during the time of his confinement to have access to writing materials and to send and receive letters. The product of his pen was a collection that was later entitled *Letters*

---

[507] See Hong and Hong, eds., *Essential Kierkegaard*, 204; also Stokes, *Essential Thinkers*, 145.

[508] See, in this regard, Roger Oakland, *Faith undone: the emerging church—a new reformation or an end-time deception?* (Eureka, Mont.: Lighthouse Trails Publishing, 2007), 24-25.

*and Papers from Prison.*[509] In a letter to his friend Eberhard Bethge dated 29 May 1944, Bonhoeffer urged that it is "wrong to use God as a stop-gap for the incompleteness of our knowledge."[510] As the frontiers of knowledge were pushed farther and farther back, he warned, God would be "continually in retreat."[511] The number of questions for which only God had the answer would constantly shrink. This was true not only with respect to scientific knowledge, Bonhoeffer said, but also of "the wider human problems of death, suffering, and guilt."[512]

Bonhoeffer's statement regarding society's ills evokes Bertrand Russell's pronouncement in 1930 that "hatred and fear can, with our present psychological knowledge and our present industrial technique, be eliminated altogether from human life."[513] The "wider human problems" identified by Bonhoeffer have not yet been solved as far as I know, and the hatred and fear targeted by Bertrand Russell also seem to be still with us; thus, "the God of the gaps" can seemingly be waved off only when the question is one of pure science, and perhaps not even then, since it still cannot be said with certainty that the critical scientific observations will be free of theory bias.

The second major consideration is the "falsification" theory of Viennese philosopher of science Karl Popper (1904-94). Popper held that the mark of a scientific theory is that it makes predictions that could possibly be proven false. Since deductive reasoning proceeds from unchallenged assumptions, such as that all frogs can swim, then by definition deductive reasoning cannot lead to a conclusion that can be shown to be false. Only inductive statements qualify as falsifiable; in fact they all do, according to Popper. Thus, Popper agrees with David Hume's statement that any generalization goes beyond the possible evidence for it.[514] This means, ulti-

---

[509] See Dietrich Bonhoeffer, *Letters and Papers from Prison*, Enlarged Ed., ed. Eberhard Bethge, transl. by Reginald Fuller and others (New York: Simon & Schuster, 1971).

[510] Ibid., 311.

[511] Ibid.

[512] Ibid.

[513] Russell, *Why I Am Not a Christian*, 45.

[514] See Stokes, *Essential Thinkers*, 195.

mately, that any conclusion arrived at by induction has only the logical status of a conjecture.[515]

How, exactly, would we apply induction or deduction or both in assessing the claims of the Bible? As far as induction is concerned, we would presumably begin with statements that are as close to deductive premises as they can be and yet still be falsifiable. From there we would reason to a generalization that we could justifiably *choose to treat* as something more than conjecture. Take, for example, the observation that the natural world is a beautiful place, as I could see as a youth when Mother and Dad and I picnicked at the Wheeler farm on Otisco Lake (the property of another of Dad's handyman connections), under willow trees that were like maidens leaning over and sweeping their tresses across the water. The logical response when presented with what I've just described would be to conclude that such beauty must reflect a God or overarching principle of beauty.

Immediately, a skeptic would falsify our premise by noting that there is much in the world that is physically repulsive—decomposing road kill, for example. The skeptic might also characterize our observation of beauty as "theory-laden" or subjective: you see beauty because you want to see beauty. Finally, we would have to deal with the complaint that our conclusion in favor of God is nothing more than thinking borrowed from Plato, who held that the imperfect is evidence that the Perfect exists. Nonetheless, it would seemingly be an inescapable conclusion that there is aesthetic harmony in the universe, suggesting at a minimum the presence of a harmonizing principle, the antithesis of discord and almost certainly something (Someone) of non-accidental origin.

The question of how we happen to have life requires even more resourceful argument on the skeptic's part if the assertion that God is a reality is to be successfully resisted by him. Our inductive premise in this instance, accepted without question throughout the scientific world, is that a huge number of physical factors must have coalesced to allow our universe to include an environment suitable for life. The Bible believer sees this premise as requiring the conclusion that the universe was divinely created. To think that

---
[515] Ibid.

the universe and particularly the earth could have come into being through some self-organizing principle of nature, one that orchestrated a near-infinity of factors in favor of life, requires a larger step of faith by far than to believe the Bible.

Our skeptic might, I suppose, cry "falsifiable" on the ground that a life-sustaining environment is possible—indeed, is inevitable under the law of averages—if there is a large enough number of universes. Of course, the supposition that there are other universes "out there" is a conclusion arrived at neither inductively nor deductively, since there is no evidence whatsoever of multiple universes and no logic requiring their existence other than the argument that the odds are against life existing if there is not a multiplicity of universes.

As a related inductive premise, there is the fact that no human agent has ever been able to produce life or give any explanation for it. The law of *biogenesis*, that life can only come from life, has been recognized since the time of Louis Pasteur (1822-95), and no exception to this law has ever been found. The 1950 Miller-Urey experiment, which attempted to produce life in a laboratory setting, was a failure, as already noted. For life, as for beauty and the universe, Bible believers posit a supernatural source, while the atheist clings tenaciously to the hope that *abiogenesis*, the theory that living things can arise from inorganic matter, will someday be resurrected from its well-deserved oblivion. Like the many-worlds conjecture, abiogenesis has no evidentiary support. The best that the person arguing for a non-supernatural explanation of life can offer is that it's not yet time to invoke "the God of the gaps." When *will* it be time? I ask.

Also worth considering are the reports of healings for which there is no physical explanation. These accounts may certainly be dismissed with the response that there *is* an explanation; we just don't know what it is. Such an answer is the equivalent of saying that we would be unwarranted in relying on naturally unexplainable recoveries from sickness as proof of God because the state of our medical knowledge is bound to improve and thus take the mystery out of healings that baffle us at present. Therefore, as Bonhoeffer said we would be, we're positing a God who's subject to shrinkage *whenever* we invoke the "gaps" God to explain something that for now is medically inexplicable. Fair enough.

Before we dismiss the healing God, however, we need to consider the small matter of Jesus Christ. Our Lord and Savior performed many physical healings, well-documented by the Gospel writers (Matthew, Mark, Luke, and John), during His time on Earth. These were not of the leg-lengthening variety; they were what you would have to call real miracles. Jesus restored Lazarus to life when he had been dead four days and his body had begun to putrefy.[516] He raised from the dead, in the presence of many witnesses, the only son of the widow of Nain.[517] Before another crowd of mourners, he raised the ruler Jairus's daughter.[518] Without even seeing him personally, Christ restored to health the Roman centurion's servant, who was suffering from palsy and said to be "grievously tormented."[519] He cleansed lepers,[520] healed other sick,[521] cured the blind and dumb.[522] There would seem to be no explanation for these things other than the supernatural power residing in Jesus, God the Son. The physical changes undergone by the recipients of the healings described above were just too drastic to be explicable in any other way.

As might be expected, nonbelievers will simply protest that the miracles attributed to Jesus never happened. The Gospel writers made up stories to validate their claim that their crucified leader was in fact God in the flesh, it will be said. But to what end? The early Christians were already on political thin ice by reason of their rejection of the pagan Roman gods and their refusal to recite "Caesar is Lord," the latter of which caused them to be seen as disloyal to the Roman Empire and by itself was enough to subject

---

[516] John 11:1-44.

[517] Luke 7:11-15.

[518] Matthew 9:18-25; Mark 5:35-43; Luke 8:41-55.

[519] Matthew 8:5-13; Luke 7:1-10.

[520] Matthew 8:1-4; Mark 1:40-42; Luke 5:12-13.

[521] Matthew 4:23-24; 8:14-17; Mark 1:29-31, 34; Luke 4:40; 5:16-26; 13:10-13; John 5:2-9.

[522] Matthew 9:27-34; 20:29-34; Mark 7:31-35; 8:22-25; Luke 18:35-43; John 9:1-41.

them to persecution.[523] Why would they have further jeopardized their position by making outlandish claims of miracles attributed to their god? The obvious answer: because their accounts of miracles were true.

The biggest fib of all, supposedly, was the story of Christ's resurrection, but this pivotal event of history is hardly described in such a way as to make prevarication seem likely. The consistency, detail, and logic of the Gospel writers' accounts of the event and of the crucifixion that preceded it, written at various times over a period of some thirty years, are amazing. All Gospel accounts record that Jesus, on the cross, "gave up the ghost."[524] The Roman soldiers who were there confirmed that Jesus was dead; therefore, they did not need to break His legs to hasten His death.[525] This circumstance notwithstanding, one of the soldiers thrust a spear into Christ's side.[526] Thus was fulfilled the psalmist's prediction that none of Jesus' bones would be broken,[527] and thus was laid the groundwork for fulfillment of Zechariah's prophecy that Israel at Christ's second coming will "look upon [Him] whom they have pierced."[528]

Joseph of Arimathaea then came and got Pontius Pilate to release Jesus' body to him, whereupon he took the body and laid it in a tomb that he had recently purchased for himself.[529] At the requests of the chief priests and Pharisees, who feared a bogus resurrection staged by the apostles, Pilate secured the tomb in which Christ was buried and assigned a guard to it.[530]

At dawn on the first day of the week, women that were part of Jesus' party came to see the sepulcher where Jesus' body had been placed and found the tomb empty, with the stone that blocked

---

[523] Shelley, *Church History*, 38-45; E, H. Broadbent, *The Pilgrim Church* (Grand Rapids, Mich.: Gospel Folio Press, 1931, 1999), 40-41.

[524] Matthew 27:50; Mark 15:37; Luke 23:46; John 19:30.

[525] John 19:31-33.

[526] John 19:34.

[527] Psalm 34:19-20.

[528] Zechariah 12:10.

[529] Matthew 27:57-60; Mark 15:43-45; Luke 23:50-53; John 19:38-42.

[530] Matthew 27: 62-66.

its entrance rolled away.[531] The women would presumably have not been able by themselves to move the stone, which was "very great."[532] The soldiers guarding the tomb would not have moved the stone and allowed Christ's body to be taken away because they would thereby have violated their orders. The only other possible stone movers were Jesus himself and one or more agents (angels) of the Lord, who would have moved the stone not to let Jesus out (seemingly unnecessary now that Jesus was in His supernaturally-endowed resurrection body and able to go anywhere without physical constraint)[533], but rather to let Christ's followers see in.[534] All four gospels (Matthew, Mark, Luke, and John) note an angelic presence at the tomb on the morning when the tomb was found empty.[535] The intervention of the angel of the Lord and the accompanying circumstances were reported to the chief priests by some of the tomb guard, whereupon the Jewish leaders bribed the soldiers to say that Jesus' disciples came and took his body while they slept.[536] Following the discovery of the empty tomb, Jesus appeared repeatedly to the disciples, giving them proof of His bodily resurrection.[537] Finally, Jesus was physically taken up into heaven, with a promise that He would return "in like manner."[538]

Do the above facts, unrefuted, make an inductively proven case for Jesus' resurrection? The unbeliever will interject that in the absence of timely medical resuscitation, the dead always stay dead. This is so invariably true, it's said, that we may conclude as a matter of *deduction* that Christ could not have risen from the dead. Even if we view this case as a proper subject of inductive analysis, our skeptic further urges, accounts of Jesus raising Lazarus and the son of the widow of Nain are arguably fabrications and therefore can't

---

[531] Matthew 28:1-6; Mark 16:1-6; Luke 24:1-7; John 20:1-2.

[532] Mark 16:4.

[533] See Morris, *Study Bible*, n. to John 20:26.

[534] See Morris, *Study Bible*, n. to Matt. 28:2.

[535] See Matthew 28:1-6; Mark 16:5-6; Luke 24:4-6; John 20:11-13.

[536] Matthew 28:11-15.

[537] Matthew 28:16-17; Mark 16:9-14; Luke 24:13-43; John 20:11-29; 21:1-23.

[538] Mark 16:19; Luke 24:50-51; Acts 1:9-11.

be relied on to falsify the conclusion that no one has ever climbed out of the grave. But in the case of Jesus, it should be kept in mind, the individual who rose from the dead is someone who previously had done many other miracles and who claimed to be God in the flesh. May we not, therefore, assume that there's a falsifiable inductive premise because the person said to have been resurrected in our case is unique in the ways indicated above? This person, it must be remembered, was recognized and held himself out not only as fully God, but also as fully man.[539] The premise falsified, and thereby made scientific and inductively provable, is that no *man* has ever risen from the dead.

The circumstances surrounding the birth, death, and resurrection of Jesus Christ were predicted in detail by the Old Testament prophets. He would be born of a virgin, according to Isaiah.[540] He would be born in Bethlehem, the prophet Micah predicted.[541] Joseph and Mary, to protect the baby Jesus, would find it necessary to flee to Egypt and remain there until the death of Herod, the prophet Hosea indicated.[542] The voice of one crying in the wilderness (John the Baptist) would prepare the way for the coming Lord, Isaiah said.[543] Isaiah further foretold that Jesus would be rejected by His own people, the Jews.[544] Prior to His rejection, according to the prophet Zechariah, Jesus would triumphantly enter Jerusalem riding on a donkey.[545] He would be betrayed by a friend, the psalmist David predicted.[546] His betrayer would receive thirty pieces of silver, according to Zechariah.[547] His garments would be gambled for, David said, as happened at the crucifixion site.[548] None of Christ's

---

[539] Matthew 1:1, 23; 18:11; 21: 9; Luke 24:41-43; John 1:14; Galatians 4:4; 1 Timothy 2:5; and see Matthew 26:36-46.

[540] Isaiah 7:14; cf. Luke 1:26-27, 30-31.

[541] Micah 5:2; cf. Luke 2:4-5, 7.

[542] Hosea 11:1; cf. Matthew 2:14-15.

[543] Isaiah 40:3-5; cf. Luke 3:3-6.

[544] Isaiah 53:3; cf. John 1:11; Luke 23:18.

[545] Zechariah 9:9; cf. Mark 11:7, 9, 11.

[546] Psalm 41:9; cf. Luke 22:47-48.

[547] Zechariah 11:12; cf. Matthew 26:14-15.

[548] Psalm 22:17-18; cf. Matthew 27:35-36.

bones would be broken, David predicted.[549] But His side would be pierced, according to Zechariah.[550] He would then receive the burial of a rich man (i.e., be interred in the tomb of a wealthy individual), Isaiah foretold.[551] His body would not be allowed to molder in the grave, according to David and an unidentified psalmist.[552] He would ascend in returning to the Father, David indicated.[553]

The number of facts in the narrative of Christ that are pointed to by the above and other prophecies is so great as to suggest knowledge that could only be possessed by God, speaking through His prophets.[554] Moreover, there is no credible support for the contention that Old Testament books of prophecy were written "after the fact." Nor can "lucky guessing" on the part of the prophets be plausibly urged as a theory, since no wrong guesses are in evidence. Thus, no conclusion can reasonably be reached other than that the Old Testament prophets were divinely inspired when they spoke of Jesus, and that the birth, death, and resurrection of God the Son was part of a plan laid out by God the Father long before the events referred to by the prophets.[555]

Yet another proof of the truth of Scripture is the universe operating in accordance with physical laws. The existence of laws implies a Law-Giver, you may have heard. If there were no laws of physics, there would be undifferentiated, uninhabited chaos. We would not be here to ask Martin Heidegger's preliminary question, "Why is there something, rather than nothing?"[556] The "argument from something"—the idea that existence can't come out of that

---

[549] Psalm 34:20; cf. John 19:32-33, 36.

[550] Zechariah 12:10; cf. John 19:34.

[551] Isaiah 53:9; cf. Matthew 27:57-60.

[552] Psalm 16:10; 49:15; cf. Mark 16:6-7.

[553] Psalm 68:18; cf. Mark 16:19; 1 Corinthians 15:4; Ephesians 4:8.

[554] The above are less than a third of the Old Testament prophecies that have been identified as pertaining to Jesus. See the "Prophecies of the Messiah" appendix in *The King James Study Bible* (Nashville: Thomas Nelson Publishers, 1988), 2026-32.

[555] See again Revelation 13:8, which states that Christ was "slain from the foundation of the world." God's gracious plan of salvation has thus been in effect at least since the sin of Adam if not longer.

[556] Stokes, *Essential Thinkers*, 151.

which is nonexistent— is, of course, a line of reasoning that can be used not only to prove the existence of the God of the Bible; it can also be applied to attribute the presence of the universe to an organizing principle, whatever that is. It's therefore not a convincing apology for God, some will say.[557] Nonetheless, I'm not about to withdraw it.

*Where Thought Experiments Lead*

The quality of proof and attached arguments notwithstanding, things are bound to get out of control when you begin talking about religious faith and either induction or deduction in the same breath. You're attempting to bring the infinite (the Almighty, the author of creation) within the bounds of the finite (language and logic), which creates an inherently precarious situation, God's logicality notwithstanding.[558] Can there ever, in actuality, be a trustworthy basis for faith other than faith? Is it not the Kierkegaardian "leap" that makes belief effective and worth holding?[559] Does the Word not say that without faith it is impossible to please God?[560] Yes, but after you believe, reason steps in, if Augustine was right, and helps your faith.[561] Therefore, God should not be offended, it seems to me, when you subject your faith to logical testing. I readily acknowledge that the ultimate source of wisdom is God,[562] and I'm still counting on Heaven-sent intuition (call it Holy Spirit guidance) concerning the things that continue to puzzle me. But can't I in the meantime use my logic to piece things together when I need to, proceeding analogously to what we hope will happen when we instruct children to "use your words"?

---

[557] I use the word "apology" in the sense of a defense. Nobody's saying they're sorry.

[558] See the "Science and Religion Generally" section of my Paradigm Paralysis chapter, supra.

[559] See the discussion in the "Existentialism" section of my Essence chapter, supra.

[560] Hebrews 11:6.

[561] See, in this regard, Stokes, *Essential Thinkers*, 45; also Richard Price, *Augustine*, Great Christian Thinkers Series (Liguori, Mo.: Liguori Publications, 1996), 10 (referring to Isaiah 7:9, Latin Version).

[562] James 1:5.

# WATER

The answer to the question of whether reason may be called on to assist faith is still a qualified yes nearly sixteen hundred years after Augustine. But care must be taken by the twenty-first century believer to assure that "reason" and "faith" have not been redefined for that believer in the meantime. The test is the centrality of the traditional Bible. Stray from Scripture as your ultimate guide and you run the risk of becoming a rudderless ship, as has happened with New Age mysticism, which, as I stated before, has consistently devalued and marginalized the Bible.[563]

As an example of how God might work with us through deduction, a mental process based principally on reason rather than empirically-developed evidence, I'm thinking it might be worthwhile for us to consider the "thought experiment," a staple of some of science's supposed greatest thinkers. In *About Time*, his exploration of the consequences of Albert Einstein's special and general theories of relativity, physicist Paul Davies makes the following statements concerning thought experiments:

> Science is based on the assumption that the world is rational, and that human reasoning reflects, albeit in a somewhat shaky way, an underlying order in nature. Logical consistency requires that the various laws and principles which govern the natural world must fit together consistently. It is sometimes possible, by tenaciously following a logical thread, to make discoveries about the real world without ever conducting an experiment, simply by imagining a particular physical state of affairs. In practice, it is essential to confirm such theoretical predictions experimentally, as there are many examples in history of apparently rational thought producing absurd conclusions.[564]

---

[563] See, again, the discussion and authorities in the "Intuition and Exposition" section of my Region of Awe chapter, supra, especially Smith, *"Wonderful" Deception*, 100-101, 104-06, 125-26, 187.

[564] Davies, *About Time*, 92.

Einstein, Davies continues, was "a master of the thought experiment."[565] Indeed, Davies observes, Einstein "believed more in the power of thought than the power of experimentation to help us unravel nature."[566]

An example of an Einstein thought experiment would be the famous "twins" paradox, mentioned earlier.[567] In that hypothetical, a twin traveling through space at close to light speed returns to Earth from a long voyage to find that his brother, who stayed home, has aged more than he has, because the traveler's velocity hurtling through space has made his body clock run more slowly than his earthbound sibling's.[568] This imagined state of affairs follows solely from the equations of Einstein's special theory of relativity, which dictate that events that are simultaneous (i.e., that occur at the same time) in one frame of reference are not simultaneous in another frame that's moving relative to the first.[569]

Our continuing lack of a space vehicle capable of both attaining the needed speed and sustaining human life over a prolonged period of time has kept the "twins" proposition from being verified in the form presented by Einstein. Nonetheless, a similar time dilation effect may now be experimentally demonstrated by comparing the times of a rapidly moving atomic clock and a companion clock kept at rest.[570]

As Paul Davies has noted, the history of thought experiments relating to gravity stretches at least as far back as Galileo.[571] In reaching the conclusion that bodies that are of various weights will fall at the same speed when air resistance is not a significant factor, Galileo asked the thinking public to imagine that a heavy body ($h$)

---

[565] Ibid., 93.

[566] Ibid., 101.

[567] See the "More Openness" section of my Fluid Reality chapter, supra.

[568] See Davies, *God and the New Physics*, 120-21.

[569] See Richard Wolfson, *Einstein's Relativity and the Quantum Revolution: Modern Physics for Non-Scientists*, Pt. I (VCR) (Chantilly, Va.: The Teaching Company, 2000), lecture 10.

[570] Ibid., lecture 9.

[571] Davies, *About Time*, 93. The full name of Italian physicist/astronomer Galileo (1564-1642) was Galileo Galilei.

was connected by a thin string to a light body ($l$) and that the two bodies were dropped from a tower. If Aristotle's theory that heavy objects fall faster than light objects was correct, Galileo said, then $l$ should lag behind $h$ at the end of a taut string and have the effect of *slowing* the fall of $h$.[572]

But, Galileo then asked, shouldn't the fact that the combined weight of $l$ and $h$ was greater than the weight of $h$ alone have the effect of making both objects fall faster and thus of *speeding* the fall of $h$?[573] This would leave us with $l$ both slowing and speeding the fall of $h$, would it not? Thus, Galileo argued, Aristotle's theory that heavy and light objects fall at different speeds had led to contradictory conclusions.[574] The only way out of this inconsistency, he said, was to hold that $l$ and $h$ must fall at the same rate.[575] Galileo was therefore able to demolish Aristotle's theory without having to scale the leaning tower of Pisa, Davies notes.[576]

Granting the usefulness of the thought experiment in science, should it also be allowed in the realms of metaphysics and theology? If so, should it be given the empirical credence that's accorded to scientific thought experiments? It may be argued that Christianity has already had *observational* validation in the supernatural events that are prophesied in the Old Testament, depicted in the New Testament, and confirmed by the testimony of witnesses, but you, gentle reader, may be unwilling to accept this. The inductive reasoning exercise at the beginning of this chapter and the logic exercise at the end of the preceding chapter, both of which are premised on Bible facts, were certainly satisfying for me, but they may not have been so for you. Should we also, therefore, inquire as to whether there are thought experiments that support the Gospel?

More fundamentally, should we even be concerned about modes (methods) of reason at this point? There are, after all, still-pending exegetical questions that I have regarding the Bible, which were posed in my Introduction and elsewhere herein and are of an

---

[572] Ibid., 93-94.

[573] Ibid., 94.

[574] Ibid.

[575] Ibid.

[576] Ibid.

elementary nature. On the sensibility of the answers that I provide to these questions may rest the cohesiveness of the Gospel that I entrust to you the reader. It's hard to know where to start, however. I worry that I'm giving you milk, which you've already had, when I should be starting you on solid food (1 Corinthians 3: 2). But first, like mothers of small children in underdeveloped countries, I need to chew the food myself before I place it in your mouth. Until I break the meat down to something that's easily swallowable, I feel embarrassed offering it to you. Nonetheless, we can in the meantime still take a stab at constructing a thought experiment or two that establishes God's existence.

Let's start with a set of conditions that we should be able easily to agree on: (1) I'm here (in existence). (2) There was a time during which I don't remember being here (in existence). (3) There was a time during which nobody—no sentient creature who may have inhabited the universe—remembers my being here (in existence). (4) If I wasn't always here (in existence), then I'm not God as people in the Western religions (Christianity, Judaism, and Islam) have traditionally thought of Him, mystical considerations of oneness with Him aside.

From the above conditions, I think we can reason logically to our sought-after ultimate conclusion. If I wasn't always here, I must have had a beginning. There was something that triggered my existence, even if it was nothing more than the operation of natural law. A beginning implies a cause. No cause, no beginning. But does *existence* imply a cause? Only if you're a being or entity that needs a beginning in order to exist. Only if you're like me in that respect. Assuming the existence of the traditional western God, a factor that distinguishes me from Him is that He's always been here whereas I haven't. His existence is beginningless, in other words. He purely and simply *is*. When He speaks to Moses on Mount Horeb from a bush that burns but is not consumed, God identifies himself as "I AM," the self-existent One.[577]

But before we can consider or put any stock in the abovementioned theophany, we need to return in our thinking to my personal need for a cause. Tracing my ancestry back to the inception

---

[577] Exodus 3:14; and see Morris, *Study Bible*, n. to Ex. 3:14.

of the human race, we must conclude that I had a beginningless cause—i.e., that whoever or whatever my progenitors sprang forth from always was. That would be the only way ontologically to secure my existence, since a cause that needed a beginning would itself require a cause and therefore would be contingent in nature and not an ultimate cause at all. Such a cause could not inhabit eternity as God does.[578]

What about, as suggested above, a natural (as opposed to supernatural) cause, perhaps of the kind Paul Davies was inquiring about?[579] The problem with this idea, as Prof. Davies himself has noted, is that a cause of this description, being dependent on the operation of physical law, would exist only as long as the universe exists.[580] It would not have a foundation in eternal reality and thus would lack ultimacy. My existence, if it is to have any real being, therefore has to proceed from a supernatural cause, and we might as well call that cause God.

The above, you may recognize, is a variation on the "first cause" argument propounded by Thomas Aquinas in the thirteenth century, which was given by him as his second of five "ways" of proving the existence of God.[581] Aquinas's "first cause" proof is tied to related arguments to the effect that neither change nor somethingness can be its own cause, which are his first and third "ways." Everything has to have a cause, Aquinas said, until the chain of causation recedes back to what Aristotle called the "Prime Mover." Personally, I'm not particularly concerned that Aquinas allows God to be causeless while not granting the same privilege to anyone or anything else. God is, after all, *sui generis*.

Let's experiment in our thinking with another hopefully agreeable set of conditions: (1) Things living and inanimate exist. (2) Things living and inanimate that now exist did not always exist. (3) Things living and inanimate cease to exist. From these premises,

---

[578] Isaiah 57:15.

[579] See the "Science and Religion Generally" section of my Paradigm Paralysis chapter, supra.

[580] Ibid.

[581] St. Thomas Aquinas, *Summa Theologica*, transl. by Fathers of English Dominican Province, 5 vols. (Westminster, Md.: Christian Classics, 1981), vol. 1, pp. 13-14.

also, we can reach the conclusion that God exists. Our thought experiment this time bears an even more striking resemblance to one of Aquinas's "ways."[582] Thomas reasoned that not everything can be in the category of things that come to be and pass away, since this would mean that there was a time when nothing existed (not even time). Because there is now something, he said, there was never nothing, since something cannot come from nothing. The something that must always have existed has to be God, he concluded.

Aquinas's fifth "way" of proving God's existence, which is based on Aristotle's concept of *telos*, or purpose, bears a close resemblance to the "intelligent design" argument touched on earlier.[583] My version of the teleological argument is based on conditions that everyone *should*, but not necessarily will, be able to agree on: (1) For there to be anything, the elements of the universe must be directed so as to allow that something to exist. Nothing can exist in the absence of at least an organizing principle. (2) What operates to produce the something of the universe cannot do so accidentally, since randomness is inconsistent with structure, which is required for existence. (3) Nonaccidental activity implies a mind, or something that works like a mind, guiding the activity.

The required mind or mind-like entity has to belong to God. Looking at the universe, we must conclude that it could not work the way it does by accident.[584] Instead, what we see in what we observe is the manifestation of a purpose.[585] This argument was made by English theologian William Paley in his influential 1802 work, *Natural Theology*.[586] Paley there compared God to a master watchmaker, and argued that if considering the design of a fine

---

[582] See Aquinas, *Summa Theologica*, vol. 1, p. 13; also Stokes, *Essential Thinkers*, 51; and Stephens, *Philosopher's Notebook*, 28, 29.

[583] See the "Whence Cometh Life" and "Animal Kingdom" sections within my Essence chapter, supra.

[584] Rohmann, *Ideas*, 163.

[585] Ibid.

[586] See, generally, William Paley, *Natural Theology* (England: 1802; reprint, Houston: St. Thomas Press, 1972).

watch makes us appreciate the work of a skilled craftsman, so much more should we recognize intelligent design in something like the human eye or in the universe as a whole.[587]

No attempt to develop or analyze arguments for the existence of God would be complete without mentioning the "ontological argument" offered by Anselm, who was a predecessor of Aquinas in the Middle Ages' "scholastic" movement and has been referred to as "the father of the Scholastic tradition."[588] I've already summarized his argument, in the "Other Philosophers Weigh In" section of my Fluid Reality chapter, but for your convenience I'll repeat my summary here. Anselm begins by defining "God," in Philip Stokes' paraphrase, as "something than which nothing greater can be thought of."[589] Since even a nonbeliever accepts that this is what the concept of God entails, Anselm argues, the existence of God follows from the definition of God.[590] If you think of something as perfect, he says, you can't think of it not existing.[591]

Philosophers have generally been of the opinion that something is wrong with Anselm's "ontological" argument, but have been unable to agree on what it is.[592] Personally, I'm concerned that in the form in which it's framed above, the "ontological argument" allows individual man to limit God by reason of the poverty of his imagination.[593] God is the greatest thing that can be humanly imagined, but how much can you or I imagine! Wouldn't it be better simply to say that there's an ideal Platonic form of everything somewhere out there and let it go at that?

Aquinas later offered a version of the "ontological argument," his fourth "way," that was based on the observation that things exhibit varying degrees of any particular quality. Some things are hotter than others, for example, or more good, or more noble,

---

[587] Ibid.; also, Rohmann, *Ideas*, 163-64, 274.
[588] Stokes, *Essential Thinkers*, 48.
[589] Ibid., 49.
[590] Ibid.
[591] Ibid.
[592] Ibid.
[593] I made a similar argument in the "Other Philosophers" section of my Fluid Reality chapter, supra.

and thus more closely resemble the highest degree of that quality.[594] The maximum of any genus is the cause of all in that genus, Aquinas said, and with respect to goodness, as an example, that ultimate degree and first cause is God.[595] This "way" seems to suffer from the same basic defect as Anselm's argument, dependency on human imagination: How much heat, goodness, or nobility can you imagine?

*The Mother of All Thought Experiments*

In his classic work *Christian Apologetics*, author/educator Norman Geisler draws together much of the thinking outlined above. In chapter 5 of his book, Dr. Geisler provides "positive reasons for believing that theism is true."[596] These "reasons" may fairly be characterized as a smooth progression of thoughts, moving from undeniable to undeniable, that culminates in the conclusions that God exists and that the God who exists is the God of the Bible.[597] "Theism" is defined elsewhere as "the belief that there is a God both *beyond* and *within* the world, a Creator and Sustainer who sovereignly controls the world and supernaturally intervenes in it" (emphasis supplied).[598] With apologies to Dr. Geisler for giving you a condensation of his very thoroughly stated argument, I offer the following paraphrase/summary:[599]

> (1) *Something exists.* It is undeniable that something exists. No one can deny his or her own existence without affirming it; at the least, the denier must exist.

---

[594] Aquinas, *Summa Theologica*, vol. 1, p. 14; Stokes, *Essential Thinkers*, 51.

[595] Ibid.

[596] See Norman Geisler, *Christian Apologetics*, 2nd Ed. (Grand Rapids, Mich.: Baker Academic, 2013), 265-87.

[597] Ibid., 278-79.

[598] Ibid., 137.

[599] My summary covers pages 269-79 of the Second Edition of *Christian Apologetics*. Book page citations for specific points of Geisler's proof will not be included herein for the most part.

(2) *My nonexistence is possible.* My existence must fit into one of three categories: impossible, possible, or necessary. It is clearly not impossible since I undeniably exist. My actuality proves instead that my existence is possible. My existence is not necessary, however, since it is possible for me to not exist. A necessary existence would be one that cannot not exist. It would be pure *actuality* with no *potentiality* inasmuch as there would be no chance that it could be anything other than existent or anything other than what it is. If it were impossible for it to not exist or to be something different, it would be changeless. This would make the necessary being *nontemporal* (eternal) and *nonspatial*, since space and time by their respective natures involve going from position to position and from moment to moment.

The lack of potentiality to be anything other than what it is also means that there can be only one necessary existence, since having more than one such being would imply a potential for the two or more to differ, whereas the potential to differ from another being is inconsistent with the idea that a being is necessary. Additionally, the necessary existence would have to be infinite in the attributes it possesses, since pure actuality—i.e., complete necessariness—cannot be limited in any way. A being's having a potential to be something other than what it is—i.e., its being less than infinite in its attributes—would mean that it is not necessary for it to be what it is. If you can be something other than what you are, your existence is not necessary. A necessary being would also be uncaused, since whatever is caused passes from potentiality to actuality and having potentiality as part of its history prevents an existence from being necessary.

(3) *Whatever has the possibility of nonexistence is currently caused by another being.* Here, again, we must examine three choices: a potential existent is self-caused, caused by another, or uncaused. Self-causation is impossible for anything that has the potential to not exist, since that person or

object would have to have existed prior to what it caused—itself—and have been simultaneously in states of actuality and potentiality as far as its being was concerned, which is logically possible. A cause of being must first exist in order to cause something. If some being is uncaused, mere possibility would be offered as the basis for that being's actuality, which would entail the problem that nothing cannot present something. The mechanism, then, by which some things that could possibly not exist do exist is that the something that could be nothing is caused to exist by something that cannot be nothing.[600] All contingent beings are caused by a necessary being, to paraphrase. But what causes me to be when I need not be or continue to be? This is " the real metaphysical question that only theism can answer adequately," Geisler says.[601]

(4) *There cannot be an infinite regress of current causes of existence.* All efficient (working) causality of existence is *current*, in that it supports not my becoming but rather my continued be-ing (current existence), which is required if I am to continue to be. But it cannot also be the case that *every* cause is simultaneously both actual and potential, by reason of its causing existence and having its own existence caused at the same moment; infinite regress of current causes of contingent beings is impossible. Either the series is grounded ultimately in a noncontingent being or it cannot exist. Adding up an infinite number of dependent (i.e., contingent) beings within a series does not provide an adequate ground for them. What does not account for its own existence cannot possibly ground the existence of someone or something else, which is another way of saying that whatever is in a state of potentiality regarding its own existence cannot possibly be in a state of actuality for the

---

[600] This would seem to answer Heidegger's question, "Why is there something rather than nothing?"

[601] Geisler, *Christian Apologetics*, 273.

existence of another. Only what is actual can actualize. No contingent being can cause being.

(5) *Therefore, a first, uncaused cause of my current existence exists.* Since I exist but my nonexistence is possible, I must have a cause that actualizes my existence. The ultimate cause of a contingent being cannot itself be contingent, however; it must be a necessary first cause of my existence. There is, in fact, an uncaused cause of all that exists, including myself.

(6) *This uncaused cause must be infinite, unchanging, all-powerful, all-knowing, and all-perfect.* We have already seen that the necessary being on which my existence depends must be pure actuality, changeless, nonspatial, nontemporal, one, undivided, infinite, and uncaused. As far as this being's power is concerned, we need only note that this involves the ability to effect change in another—i.e., to cause something or someone else to be or not be in some way. This is precisely what we see in the uncaused cause described above, which causes the being of all else that exists.

As far as all-knowingness is concerned, we need only note that the world is filled with knowing beings, the source of whose knowledge is the uncaused cause. Whereas a human knows finitely and imperfectly, the cause of all knowledge knows infinitely and perfectly. He knows the totality of what all those dependent on him know and more.

Finally, God *is* goodness. We humans only *have* goodness. We are only the cause of the becoming of good acts, through our free choice, while the Creator is the cause of the be-ing of all goodness. The Creator's goodness must be infinite, since all actualities actualized in an effect must preexist in the cause of the effect. Since the infinite Cause is simple and has no parts, He cannot be partly anything, including partly good.

(7) *This infinitely perfect being is appropriately called "God."* Anything less than what is ultimately and intrinsically worthy of our admiration is not really "God." An ultimate commitment to what is less than ultimate is idolatry. Nothing has more intrinsic value than the ultimate ground and source of all value, the infinitely perfect uncaused cause of all that exists.

(8) *Therefore, God exists.* We may conclude, then, that a theistic God exists. We're talking not just about an abstract, unmoved Prime Mover, but a personal Object that we can love with all our heart, soul, mind, and strength (Mark 12:30). As Geisler puts it, "The God the heart needs is a God the head has good reason to believe really exists."[602]

(9) *This God who exists is identical to the God described in the Christian scriptures.* The God described in the Bible is said to be eternal (Colossians 1:16-17; Hebrews 1:2), changeless (Malachi 3:16; Hebrews 6:18), infinite (1 Kings 8:27; Isaiah 66:1), all-loving (John 3:16; 1 John 4:16), and all-powerful (Matthew 19:26; Hebrews 1:3); thus, he matches the God who is made evident through the above proofs. For reasons already set forth above, there could not be more than one infinitely perfect, changeless, eternal being that matches this description. The existence of more than one being that is said to match the description of Christianity's God would imply deficiency on the part of each such being and would therefore negate the above description.

(10) *Therefore, the God described in the Bible exists.* If there is only one God and the God described in the Bible is identical in characteristics to that God, then it follows logically that the God described in the Bible exists. There cannot be

---

[602] Geisler, *Christian Apologetics*, 278.

two (or more) infinitely perfect beings or ultimates or absolutes. God is all in all.[603]

*A Final Note:* You may have found the above discussion somewhat off-putting, because it's head rather than heart theology. I urge you nonetheless to read through it again, in fact to do so as many times as are needed for you to understand the arguments therein, and to visit the unabridged version of this apology in the Second Edition of Norman Geisler's classic *Christian Apologetics* if you have the opportunity.[604] For an unsaved person or a believer at any level of spiritual maturity, it is essential to grasp that the Gospel intellectually has both content and coherence. The above is, of course, stated without prejudice to Soren Kierkegaard's position, repeatedly noted supra, that true faith has nothing to do with rationalistic proofs of the existence of God.[605]

*Where We Stand Now*

I'm feeling at this point, gentle reader, that I've done all that I can to give you arguments that show the pure reasonableness of believing in God. If, nonetheless, a person has made it a matter of steadfast resolve that he or she will refuse to believe in God under all circumstances, then it won't matter how many thought experiments we dream up, adapt, or adopt from Aquinas or Anselm or Norman Geisler; "reason" will never get us past the threshold question of whether God exists.

You'll recall that God's existence was not identified as an issue in my Introduction.[606] I didn't bring it up at that point because by

---

[603] Dr. Geisler's tenth argument point goes on to indicate that the above conclusion should not be read as a claim that God did or said everything attributed to Him in the Bible, only that it is possible that He did or said anything that the Bible says He did that would not be inconsistent with His nature as previously described (see Geisler, *Christian Apologetics*, 279). I wish to make it clear to my readers that I am not willing to accept such a qualifier, because of my views on biblical inerrancy.

[604] Geisler, *Christian Apologetics*, 265-87.

[605] See, again, the discussion in the "Existentialism" section of my Essence chapter, supra. See also the discussion that follows.

[606] See, in particular, the "Cottage" section of the Prologue.

the time I was into the writing of this book I had moved on to other questions. God's existence, once I began to feel his Holy Spirit indwelling me, was not something I felt I needed to spend much time convincing anyone else of. In adopting this attitude, I now realize, I was being selfish. Instead of using the help of the Holy Spirit and what I knew at that time to immediately start bringing others to the saving knowledge of Jesus Christ, I hung back and focused primarily on how I should arrange the furniture in a house to which I had already been admitted.

I wanted to see everything fit together perfectly so that I could satisfy my lawyerly instincts, which, I thought, required a cohesive presentation of the case for Jesus. I wanted to be able to talk convincingly about matters such as free will and immortality, to have a house that was fully together, all theological loose ends tied up, before I invited anyone else to come in. I wanted to give a good account of myself as a lawyer. Motivated by pride more than passion, is how someone could have seen me.

Eventually the scales did fall from my eyes, so that I could see clearly what I was doing and not doing (Acts 9:18), but it took a long time and this book is the proof of it.

Perhaps it would be better, then, if you and I were to hearken to Soren Kierkegaard's urging that Christians eschew metaphysical schemes and instead just believe (and present) the Gospel simply and passionately.[607] I apply this especially to myself. I've sometimes wondered if I'm not a party to what Kierkegaard calls "pencil-pushing modern speculative thought (that) takes a dim view of passion," perhaps ignoring or relegating to secondary importance Kierkegaard's statement that passion is "existence at its very highest."[608] Pencil pushing I will admit to, but I could not have continued in the present endeavor for as long as I have without some passion. Passion propels but does not predominate, is the way I would describe my involvement in this writing project to date.

In any event, we would be premature in giving reason her walking papers. As may be seen from the above discourse, more

---

[607] See Hong and Hong, eds., "Concluding Unscientific Postscript," *Essential Kierkegaard*, 207.

[608] Ibid., 204.

than metaphysics is now in play as far as our understanding of God is concerned, and also more than mysticism, the latter of which I'm still not ready to embrace other than by trying to commune with God through nature. We have Gospel-related statements (as, e.g., with respect to the resurrection) that have become falsifiable and therefore may be taken as scientific. In fact, we have copious inductive proof of Christ's divinity, since we can look to His miracles in addition to His resurrection from the dead. We also have a large body of Old Testament prophecy that is fulfilled by Christ's life and ministry. In short, we bring things to the table that are more tangible than the proofs offered by Aquinas and Anselm and Geisler, although the arguments of these men for the existence of God have not been refuted.

Religion became the enemy of science, or at least a relative that should not be allowed in the house, during the so-called Age of Reason (roughly the eighteenth century), when, in the words of historian Bruce Shelley, "(r)espect for science and human reason replaced the Christian faith as the cornerstone of Western culture."[609] Now, as evidence referred to above suggests, some members of the thinking class are buying into the idea of a new rapprochement between science and faith. Whether that happens in the world at large remains to be seen, since Western society is still in a postmodern (we don't believe anything is real) condition and since evangelical faith is increasingly taking the form of mysticism. As I indicated at the end of the preceding chapter, I'll do what I can to assist with the reconciliation of Christianity and secular science. Other than the Bible and naked reason, there's nothing but science that's left for us to turn to at the present time, it seems.

---

[609] Shelley, *Church History*, 309.

# 6
# RELATIVITY

*The Initial Epiphany*

It's June and Sue and I are back at the Point. A warm night, at least for June, and therefore I'm sitting out under the stars. Sue is inside; she doesn't like the bugs. Since it's past ten o'clock, the beach lights, which are on a timer, have been turned off, so there isn't much to interfere with my view of things celestial. The moon, which will be first quarter, has yet to arise from the woods behind The Cottage. The stars, as is typically the case when I'm sitting out at this location this time of night, seem infinite in number. The night sky has a feeling of depth to it. There's depth because there are layers. You have the brightest stars, then myriad fainter points of light, then a grainy background that suggests stars close enough to contribute some small quantum of light but too distant to be distinguishable with the naked eye.

There's universe beyond what's visible above us, stars that we can't see even with the most powerful telescopes, astronomers tell us. The reason we can't see them, we're told, is that their light hasn't had time to reach us. A German philosopher named Heinrich Olbers in 1823 wrote a paper concerning the puzzle of the dark evening sky, a condition that was inconsistent with the then-popular model of a static and infinitely old universe.[610] Only by theorizing that the cosmos was of a finite age—i.e., that it had a beginning— could astronomers explain why the light of every star in the universe was not presently reaching Earth, causing the night sky to be bright as day.[611] If the universe had a beginning of some kind, they reasoned, then there might not have been time for the light of all or even a majority of the stars to reach Earth.

Will the light of the stars that are still beyond the communication horizon ever make its way to Earth? Revelation 21:23, mentioned earlier, speaks of the New Jerusalem being lit by the Father and the Son rather than the sun and the moon.[612] To interpret that verse as referring to the light of all the stars finally reaching the earth would be contrary to its plain language, however, and would tend toward pantheism (God as nature). The New Jerusalem is situated in "a new heaven and a new earth,"[613] and thus it may be assumed that we no longer have all of the physical limitations associated with the first heaven and first earth, which are "passed away."[614] Superseded realities might include an expanding universe, finite light speed, or anything else that would be inconsistent with the glories described in Revelation 21. Therefore, I don't think that we should be concerning ourselves with Olbers' paradox. It will be a whole new ball game physically when we reach "the other side." But I'm getting ahead of myself.

I awake from stargazing and dreaming dreams of eternity. My mind is now in the same place as my body—back at Campbell's Point, where I've situated my folding chair on a cement stoop that projects toward the lake from the front of The Cottage and is some twenty-five feet from where the beach begins. I'm sitting where I'm sitting for a very practical reason: the sand fleas or gnats or

---

[610] Hawking, *A Brief History*, 6-7; Paul Davies, *The Mind of God: The Scientific Basis for a Rational World* (New York, Simon & Schuster, 1992), 46.

[611] Ibid.

[612] See the "Big Bomb" section of my Paradigm Paralysis chapter.

[613] Revelation 21:1.

[614] Ibid.

# RELATIVITY

whatever they are that we get this time of year. They're thicker out on the beach. These bugs were swirling around my son Jeremy and me, bouncing off our bare legs, when a few years ago we stood on the beach on a similar warm night and my brainy son pointed out the various constellations to me. The insects didn't bother me then; it was just such a joy having one of my children share the universe with me.

When the same son was a teenager, some years earlier, Campbell's Point employed a live-in caretaker who had a boy about Jeremy's age. The two boys were fast friends until Danny's father found other employment. On one occasion, these two hardy youths tried spending the night in sleeping bags on the beach, but the bugs made the experience a challenging one. Less in the forefront than insect life, indeed nonexistent, was the issue of social boundaries (The caretaker's son? Really?), which simply didn't matter to our family. I was not in a law firm serving the upper crust. I didn't know anyone on the Kodak board. So my kids could associate with anyone they wanted to, except for the dope smokers, children of privilege as it happened.

During election season a few years later, after Danny's family had moved on, we (Mom and Dad Case) drove past a sign on Route 3 that had Danny's name on it as someone running for the Hounsfield town board. Good for him.

The thought that light is coming to us from stars billions of light years away brings to mind what I know about Albert Einstein and his special and general theories of relativity. Special and general relativity have a direct bearing on the question of what we're supposed to think concerning the light that reaches us from distant stars, a question explored by physicist Russell Humphreys as indicated above,[615] and I'm also beginning to see other areas in which Father Einstein may be able to help us.

---

[615] See the "Normal and Abnormal Science" section of my Paradigm Paralysis chapter, supra.

It's hard to write about Einstein's work without including biographical material. Almost as hard as writing about science without bringing in the Bible. Biographies of Einstein have him as a teenager asking what a light beam would look like if you could catch up to it.[616] Even at that point it was known that light traveled at a finite speed, finally established for the scientific community' benefit as 186,000 miles per second.[617] The adolescent Einstein's question was not framed as simply as I've stated it above. He would later speak of

> a paradox upon which I had already hit at the age of sixteen: If I pursue a beam of light with the velocity $c$ (velocity of light in a vacuum), I should observe such a beam of light as a spatially oscillatory electromagnetic field at rest. However, there seems to be no such thing, whether on the basis of experience or according to (James Clerk) Maxwell's equations (second parenthetical added).[618]

As a pre-adolescent child, Einstein had thought that if one could race alongside a light beam, the beam would be found to have the appearance of a motionless wave.[619] In both instances, it appears, light was envisioned by Einstein as something that could be in a state of rest relative to other moving things.

The above reveries bring to mind God's question to Job, "Where is the way where light dwelleth?"[620] God knows that while a space traveler can in theory reach the sun or any of the other stars in the universe, it's impossible to arrive at the location of light itself, to

---

[616] See, e.g., Brian, *Einstein*, 12.

[617] See, again, the "Normal and Abnormal Science" section of my Paradigm Paralysis chapter, supra; see also Richard Wolfson, *Einstein's Relativity and the Quantum Revolution: Modern Physics for Non-Scientists*, Pt. I (VCR) (Chantilly, Va.: The Teaching Company, 2000), lecture 4.

[618] Paul Arthur Schilpp, ed., *Albert Einstein: Philosopher-Scientist* (New York: Tudor Publishing Co., 1951).

[619] See Kaku, *Parallel Worlds*, 31-32.

[620] Job 38: 19.

pull up alongside it as it were, because light moves at a speed of 186,000 miles per second relative to *any* observer (see discussion following). One can therefore never arrive at a point where it's possible to look over at a beam of light and have the impression that it's standing still. Thus, God is asking Job a question that He knows Job will be unable to answer. From the context, it's clear that God is trying to show Job how little he (Job) knows. In the pre-scientific era in which the book of Job was written,[621] only God could have known that this question was unanswerable by a mere mortal. The answers to this and related questions would come only after age-old paradigms had fallen by the wayside.

To know anything about light's position relative to a human observer, it seemed to pre-Einstein thinkers, it would be necessary to find a way of measuring or establishing the motion of the earth in space. This brought in the concept of the luminiferous ether. In the nineteenth century, physicists thought of space as filled with an invisible substance, the ether, which was said to be the means by which light waves were conducted throughout the universe. This substance did not itself move, it was thought; instead, objects such as planets and stars moved through it.[622] Showing the earth's movement through the ether would therefore be a way of establishing the earth's motion through space, it was supposed. The nature of the experimentation by which the earth's movement through this medium could be demonstrated remained unclear, however.

Enter, in the 1880s, American physicists Albert Michelson and Edward Morley. Michelson, the first of the two to appear on the scene, seized on a suggestion made in 1875 by Scottish physicist James Clerk Maxwell: measure the speed of light as it travels in different directions on the earth's surface.[623] Because of the "ether wind" generated by the earth's movement through space, it was thought, light would travel faster in one direction than in another.[624] An 1881 experiment in which Michelson applied the above assumption was

---

[621] Job is believed to be one of the oldest books of the Bible. See the *Henry Morris Study Bible*'s introduction to Job.

[622] See Gardner, *Relativity*, 14.

[623] Ibid., 19.

[624] Ibid.

unsuccessful, however, in that light seemed to travel at the same speed in all directions.[625] Perhaps a more sensitive measurement technique was needed, Michelson conjectured.

By 1887, Michelson had joined forces with Morley, and the two had designed an apparatus which consisted of a large stone slab that floated on a bed of liquid mercury and had an arrangement of mirrors by which light from a single source was split into beams that ran perpendicular to each other and thereafter were recombined.[626] The idea was that the different orientations of the light beams with respect to the hypothesized ether wind would result in different light speeds that would be reflected in an interference pattern when the beams were recombined.[627] The experiment using this apparatus was carried out repeatedly and with variations, but no unusual interference was seen in the recombined light.[628]

In the wake of the failure of Michelson and Morley to turn up any evidence of an ether wind, other scientists made strenuous efforts to "save the ether." These efforts smacked of ad hocery even more than would later be the case with the "inflation" hypothesis that was appended to the big bang theory. Holding onto the ether idea, it was thought, was necessary in order to be able to establish exactly what it was that light and other electromagnetic waves traveled at a fixed constant speed relative to ("speed $c$ relative to what?" as Prof. Richard Wolfson succinctly phrases the question[629]). Otherwise you had physics roommates who were not at all comfortable with each other—on the one hand a Newtonian principle of relativity which implied that the speed of light should vary according to the motion of an observer in relation to a light pulse, and on the other hand James Clerk Maxwell's finding that light always travels at the same speed regardless of the motion of the observer.[630] If motion through the ether couldn't be proven,

---

[625] Ibid., 20.

[626] Ibid., 21.

[627] Ibid.

[628] Ibid., 23.

[629] Wolfson, *Einstein's Relativity*, Pt. 1, lecture 5.

[630] Ibid.; see also Gardner, *Relativity*, 14-17.

then there was no way of knowing what an observer's absolute motion was.[631]

As luck would have it, the answer to the ether wind conundrum and other scientific puzzles was provided in 1905 by a twenty-six year old Albert Einstein, who had not been particularly impressive as either a physics student or a teacher and was then working as an examiner in the Swiss patent office.[632] In that year, Einstein published a paper that contained two "fundamental postulates," which I'm paraphrasing as follows:

> (1) Only relative motion matters. There is no way to tell as a certainty whether an object is at rest or in uniform motion (i.e., moving at a constant speed in a straight line). This conclusion seemed to be unavoidable given the failure of Michelson and Morley to prove that bodies in the universe move through a fixed ether.[633] A corollary of this postulate is that there is no such thing as absolute simultaneity of events. Observers in relative motion will not agree on which events occur at the same time.[634] Events that are simultaneous in one frame of reference are not simultaneous in another.[635]
>
> (2) Regardless of the relative motion of its source, light always moves through empty space (i.e., a vacuum) at the same constant speed. This means that the speed of the light that's coming at you will always be the same when measured from your vantage point as an observer no matter how fast the light source is moving toward or away from you in uniform motion.[636]

Since both of Einstein's "fundamental postulates" applied only to "uniform" motion, the provisos of Einstein's 1905 paper became

---

[631] Gardner, *Relativity*, 17.

[632] Ibid., 29-30.

[633] Gardner, *Relativity*, 31, 34; Wolfson, *Einstein's Relativity*, Pt. I, lecture 7.

[634] Greene, *The Elegant Universe*, 36; and see Gardner, *Relativity*, 35-39.

[635] Wolfson, *Einstein's Relativity*, Pt. I, lecture 10.

[636] Gardner, *Relativity*, 31-32, 34-35; Wolfson, *Einstein's Relativity*, Pt. I, lecture 7; Brown, *In the Beginning*, 323.

known as his *special* theory of relativity. Some ten years later, with the publication of his *general* theory of relativity, Einstein removed the "uniform motion" restriction, thus making his theory applicable to all motion.

Because light speed in any frame of reference (place that shares your motion) is a constant, other things have to give way when we increase the energy of the relative motion of an object or body traveling through space. These things are dimension (length in the direction of motion decreases), time (which runs slower, in the view of outside observers, as speed increases), and mass (faster means heavier).[637] Time dilation (slowing, in an outside observer's frame of reference) would produce effects such as the twins paradox.[638]

Keep in mind that the only assumption that we have to make for purposes of our current discussion is as to the *present* speed of light. As mentioned earlier, researchers have concluded that the speed of light has indeed undergone dramatic decay over time.[639] Various measurement techniques have been used in arriving at this conclusion.[640] This is not a concern that affects the present discussion.

*Going Farther*

As you wade out from the shallows at the edge of the water at Campbell's Point, you find that the lake doesn't get deep right away. In fact, you'll go a considerable distance before the cold water—which is a shock when you first get in—gets above your waist. When I stand in the water a hundred feet or so out from shore, with the waves slapping against my trunks and the currents pushing past my legs—first one way, then the opposite way, in a schizophrenic attempt to alternately attain and escape the beach—I

---

[637] See Gardner, *Relativity*, 35-43, 45-49; also Wolfson, *Einstein's Relativity*, Pt. I, lectures 8-10.

[638] See the "Where Thought Experiments Lead" section of my Proving God chapter, supra.

[639] See the discussion and references in the "Normal and Abnormal Science" section of my Paradigm Paralysis chapter, supra.

[640] See Brown, *In the Beginning*, 322-27.

can see our portion of shoreline in a wider perspective than when I sit in a chair on the front stoop of The Cottage. When I'm out in the water, I have the feeling that the arc of cottages that surrounds the Campbell's Point swimming area is wrapping itself around me. It's around me rather than just *over there* someplace. So it is with relativity theory at the moment: I'm past the initial plunge and now I'm standing up and looking around, getting a view of things that's more big-picture in nature, I like to think.

The responsiveness of both time and dimension to increased energy in the relative motion of an object is one reason for concluding that space and time are inseparable under relativity theory. Another is that in the relativistic way of looking at things, space defines the future and the past. Stephen Hawking, in his classic *A Brief History of Time*, uses drawings of light cones—spatial representations—to help the reader understand how this works.[641] The cones are actually a way of showing *spheres* of light or information, which expand or spread out as time passes.[642] Time passes between *events*. An event, as Hawking defines the term in his glossary, is "(a) point in space-time, specified by its time and place."[643] In text, Hawking explains that an event is "something that happens at a particular point in space and at a particular time."[644]

Events that can be reached (i.e., given information) from a present event P by a particle or wave traveling at or below the speed of light are in the future of P, Hawking further explains.[645] These events will be within or on the sphere of light (information) that has thus far, by any given time, been generated by event P.[646] Similarly, Hawking says, the past of P can be defined as the set of all events from which it is possible for information to reach event P traveling

---

[641] See Hawking, *A Brief History*, ch. 2, Figs. 2.3, 2.4, 2.5, 2.6, 2.7.

[642] See discussion at ibid., 25-26; also Fig. 2.3.

[643] Ibid., 200.

[644] Ibid., 24.

[645] Ibid., 27.

[646] Ibid.

at or below the speed of light.[647] As time passes, the sphere of the past, like the sphere of the future, will continue to expand.

Events that at the present time are neither in the future nor in the past of P lie within what Hawking calls the "elsewhere" of P.[648] These are events that at the moment can neither affect nor be affected by what happens at P.[649] For example, if our sun were to stop shining, the news that this had happened (and that we faced imminent extinction) would be in our "elsewhere" for some eight minutes (the amount of time it takes light to travel ninety-two million miles). If we were to send a radio message to the Mars Rover, as another example, the sending of that signal would be an "elsewhere" event for some eleven minutes as far as the Rover was concerned.

Since the past and the future can be spatially represented by Stephen Hawking's cones (spheres), we're led to conclude that time is as much a dimension as height, width, and depth.[650] Such was the thesis offered by Polish mathematician Hermann Minkowski (1864-1909) not long after the publication of Einstein's special theory of relativity.[651] We should not resist this conclusion, I contend, since the first verse of the Bible states, "*In the beginning* God created the heaven and the earth."[652] There is no interpretation that can reasonably be given to Genesis 1:1 other than that time, space, and matter came into being simultaneously, at what was the beginning of all things that can be touched or measured. Coexistence after co-creation implies cohabitation of reality. So we arrive at our four-dimensional space-time under both relativity theory and Scripture.

If we posit a four-dimensional space-time as our fundamental reality, then we need to think through the issue of determinism. What Minkowski's view requires of us is that we cease viewing

---

[647] Ibid.

[648] Ibid.

[649] Ibid.

[650] See Davies, *About Time*, 72.

[651] Ibid.; see also Gardner, *Relativity*, 77.

[652] Genesis 1:1; and see Morris, *Study Bible*, nn. to Gen. 1:1.

# RELATIVITY

time as a sequence or process and regard it instead as landscape, as something that's simply "there."[653] If space is simply "there," then time, also, is just "there." This is the concept that philosophers refer to as "block time,"[654] a term that might be seen as implying that history is already complete rather than working itself toward completion. "Block time" under this view would appear to include the idea that all events for all time are featured in a giant picture that has already been painted. "Past," "present," and "future" no longer exist in any meaningful way.[655] This isn't necessarily so, Paul Davies insists. Addressing the question of whether the uncertainty principle of quantum physics undercuts relativity's "block time" concept, Davies states:

> Any conflict is, in fact, illusory. Determinism concerns the question of whether every event is completely determined by a prior cause. It says nothing about whether the event is *there*. After all, the future will be what it will be regardless of whether it is determined by prior events or not. The four-dimensional perspective of relativity simply forbids us to slice up spacetime, in any absolute way, into universal instants of time. The notion of two events in different places being 'simultaneous' is relative to one's state of motion. They may be judged to occur at one moment by one observer, but one after the other by another observer...So in spite of the fact that past, present and future seem to have no objective meaning, the theory of relativity still allows a human being to decide later events by his earlier actions.[656]

Science writer Martin Gardner agrees. Gardner voices no objection to the proposition that each moment of any particle's trip through

---

[653] Davies, *About Time*, 72.

[654] Ibid.

[655] Davies, *God and the New Physics*, 127-34; Davies, *About Time*, 76-77.

[656] Davies, *God and the New Physics*, 137.

space will be an event by reason of which such moment is frozen in a timeless "block universe," as part of the particle's *world line*, its history in space-time, which is created by its passage through a succession of moments and locations.[657] A particle's world line will reflect not only where the particle has been, but also how long its movement in any direction has taken, subject to the limitation that a graph of the movement of a particle through a four-dimensional universe cannot fully depict such movement in a two- or even a three-dimensional drawing.

Gardner maintains, however, that the inclusion in the Minkowski graph of an event that has occurred "has no bearing on the question of whether the event had to happen the way it did."[658] I have some difficulty with this statement, I confess, but then I'm hearing it in the context of what I suppose to be a universe ruled by an omniscient and omnipotent God. If God knows of the event frozen in relativity's "block universe" and has in fact allowed it to be "there," has He not in effect decreed it? And if we say that He has decreed it, does this not posit a cause (the decree) and an effect (the event being "there")? Under these circumstances, cause and effect being treated as an arrow of time, how can we not say that the event is "determined by a prior cause?"[659]

To the above question, of why we may not assume a prior cause for an event that is "there," we might receive from one quarter the answer, "quantum physics." The fact that sub-atomic quantum fluctuations continually happen and appear to be outside the normal law of cause and effect is often offered as an antidote to deterministic thinking. While such an approach may be of some use in helping us understand what happens on a macro- level, by *suggesting* how God runs His universe (for example, that He allows things to happen that are outside His foreknowledge and direction), does it take us any farther than that? Can we make good our escape from the frozen landscape of "block time" riding on the back of the uncertainty principle?

---

[657] Davies, *About Time*, 75; Gardner, *Relativity*, 82-84.

[658] Gardner, *Relativity*, 84.

[659] Davies, *God and the New Physics*, 137.

A related question is that of whether we can manipulate the past so as to affect the present. This would involve actually being in the past, necessitating time travel. Physicists have traditionally dismissed the idea of journeying back to an earlier period because of time paradoxes: actions freely taken in the past could be inconsistent with present reality. A familiar example of such a paradox is presented in the movie *Back to the Future*,[660] in which Michael J. Fox's character, Marty McFly, goes back in time and meets his mother as a teenager, whereupon she falls in love with him. Incest considerations aside, there is the question of whether McFly will have an existence if his mother marries him rather than his father. But is it Marty McFly as we know him or nothing? A fundamental question of personal identity is presented in this situation. The choice for McFly seems to be between having no personhood at all or going back to the future in Christopher Lloyd's crazy automotive time machine, but does Marty McFly have to be defined biologically? Is there a Marty McFly that exists independently of his parents' physical union? Predictably, the film doesn't plumb this depth.

How would our protagonist have gotten into such a situation in the first place? you may ask. Michio Kaku opines that because a gravitational field implies that space-time is curved, a person's world line might loop back and intersect itself so as to physically visit its own past.[661] If the object (person) connects with itself to form a closed loop, the object might actually become its own past self.[662] Other time travel mechanisms have been proposed as well.[663] Quaere how the McFly idea fits with relativistic "block time." Can you see history at a glance and at the same time change it?[664]

---

[660] Steven Spielberg, *Back to the Future* (DVD) (Universal City, Calif.: Universal Studios, 2002).

[661] Davies, *About Time*, 236.

[662] Ibid., 241.

[663] See ibid., 236-51.

[664] For further discussion in this regard, see Kaku, *Parallel Worlds*, 143-44; also Davies, *God and the NewPhysics*, 126-27, 130-31.

## "Block Time" and Eternity

I may be supposing too much, but I'm guessing that the most urgent spiritual question for persons living their religious lives in our Western culture is whether one's final destination is heaven, hell, or nothingness. As stated earlier herein, one of the most prominent features of existentialism was dread of nothingness (see the "Existentialism" section of my Essence chapter, supra). Existentialist angst notwithstanding, I believe we should leave nothingness out of the present discussion. If the end for you is nothingness, you won't know it, is my simple-minded view. Thus, the options may be narrowed to heaven and hell, and relativity theory's view that the future of an object in space-time is simply "there" could be seen as implying that whichever it is, eternal bliss or eternal torment, the destination of your soul is a foregone conclusion. So say the determinists as they transpose relativity theory to the realm of theology.

Mssrs. Davies and Gardner respond to the determinist position with what is essentially the same argument expressed in different terms, the bottom line for each commentator being that the law of cause and effect still applies as far as human thought and action are concerned.[665] I agree. We're still able to pray the Sinner's Prayer and have it be effective to save us. As I hope I made clear above, I'm unwilling to let the question of whether an event is "determined by a prior cause" be decided negatively by reason of Einstein's principle of the relativity of simultaneity (two events may or may not be simultaneous depending on the states of motion of different observers)."[666] Past, present, and future may have no objective meaning, but that doesn't mean that human beings can't influence later events by their earlier actions.[667]

Having not yet dealt with quantum mechanics in any detail, we'll lay aside for now the argument that there is "free will" by analogy from quantum uncertainty. Even without considering quantum theory, there is much more than the above that can be

---

[665] See the preceding "Going Farther" section.

[666] See Davies, *God and the New Physics*, 137; and see again my "Going Farther" section.

[667] Davies, *God and the New Physics*, 137.

said at this point on the predestination issue. I humbly offer the following:

(1) As regards the question of whether heaven or hell is in your future, we're not operating under the laws of physics that apply to our present universe, which have relevance only as metaphor or as a comparative. As metaphor, physical phenomena can only *suggest* spiritual truth. When Jesus characterizes the operation of the Holy Spirit as the wind "blow(ing) where it listeth (John 3:8)," He is saying what the Spirit can be likened to without directly describing the Spirit. As a comparative, whatever happens within the laws of physics *tells us* what the spiritual reality is or is not. When Shakespeare asks his young patron, "Shall I compare thee to a summer's day?" and then states that the youth is "more lovely and more temperate"(Sonnet 18, lines 1 and 2), he is describing the subject by reference to a standard (he is "more") rather than an image.

(2) Instead of physical law, the foundation of our reasoning needs to be God. It is He who is the ultimate of all things, in the same way that light, to which the Bible compares Him metaphorically (see 1 John 1:5), is seen by popular science as an absolute of physics. Light has a finite value that is always the same, a supposedly unvarying velocity, $c$ when stated in equations. No speed limit applies to God, however. Therefore, what proceeds from God can travel with infinite speed—i.e., can instantly be at its destination (see, e.g., Acts 8:39-40). Physical laws other than the cosmic speed limit likewise do not constrain the Lord's movements (see Luke 24:30-31; John 20:19, 26).

(3) Salvation proceeds from God. Absent His grace (unmerited favor toward us), no one would be saved (Ephesians 2:8). But the "here and now" acceptance or nonacceptance of God's gracious offer of salvation is in man's control. It is not God's will that any one should perish (2 Peter 3:9);

thus, a saving decision on man's part unites man's will with God's will.

(4) Since God is the ultimate in all things and His will is not constrained by physical law, man's will is able to unite with God's in such a way that man's choice to be saved is projected instantly across all eternity, across what we will here call all everlastingness. The light-speed-based limitation on one's ability to affect the future is irrelevant. We're talking about something outside the laws of physics (John 4:24, "God is a Spirit:"). Thus we overcome the difficulty that would exist if acknowledging God's ability to save were an event that happens on or in a Hawking light sphere. If that were the case, then you could never have *eternal* life because the sphere would always have a limited radius—whereas everlastingness, eternal life, requires lack of limitation.

(5) Before you can accept God's free gift and be saved, you must deal with your sin. Sin separates us from God (Isaiah 59:2), and, if not repented of, will guarantee the sinner's destruction (Luke 13:3-5; 2 Peter 3:9). Every one of us approaches God with a past (Romans 3:10, "none righteous, no not one"). At this point we run into a bit of irony in applying a comparative: as stated above, Stephen Hawking's light spheres allow your past to include any event that can influence your "here and now" through information traveling at or below the speed of light. But the size to which the past has expanded under this rule as of the time of your death is irrelevant as far as your salvation is concerned, since the frame of reference after your death is eternity, which removes the universe's speed limit from consideration as a factor determining the length of God's memory. Therefore, there is no sin you could have committed that's too old to be remembered by God and thus to keep you out of heaven if not repented of.

## RELATIVITY

Your behaviors, if not repented of, remain "on the books" as far as God is concerned, and that's a fact of the greatest possible importance. Your past, if not erased from the ledger, will keep you out of God's house because God doesn't want sin there (1 Corinthians 6:9-10; Galatians 5:19-21; Ephesians 5:5; Revelation 21:27). He is a holy God, who wants His children to be holy (Leviticus 19:1-2; 1 Peter 1:16). Here again, God's ultimacy comes to your rescue. Through the atoning sacrifice of Jesus, your past—any trespass whenever it occurred—is wiped out the moment your will unites with God's through your acceptance of His offer of salvation. It is, indeed, just as if you never sinned (Acts 3:19; 2 Corinthians 5:17; 1 John 1:9). At Psalms 103:12, David states, "As far as the east is from the west, so far hath [God] removed our transgressions from us." A fitting description of temporal sin disappearing into the infinitude of God's mercy.

Stating the matter another way, there is no longer any "world line" for you as a saved person, since in opting for salvation you've stepped outside of space-time. No world line, no deterministic future that's just "there." Because you were within time when you chose salvation, your decision was a cause that had an effect. That effect was to rid you of any future (in this case, condemnation) or past (antecedent sin) that you otherwise would have had.[668] You were free to make this choice, as a metaphorical nod from God intimates, because you had world lines for every possible path that you might have taken in space-time. As will be explained in greater detail later (more delayed gratification!), American physicist Richard Feynman advanced the idea that all possible world lines for a given particle are equally real. But once you're outside of space-time—i.e., once you've physically died—you no longer have the alternative world lines. Since you're then outside of time, you're no longer able to avail yourself of God's marvelous cause-and-effect (repentance and reliance on Jesus in exchange for eternal life with Him). You're then looking at judgment without the protective cloak of Christ's righteousness (2 Corinthians 5:21; Hebrews 9:27).

---

[668] Romans 8:1 states, "There is therefore no condemnation to them which are in Christ Jesus, who walk not after the flesh, but after the Spirit."

There are other ways, also, in which the marriage of space and time under relativity theory is of value as a provider of metaphors or comparatives. One of the more useful ideas that arises in Minkowski's four-dimensional universe is the *space-time interval*. This conceptualization recognizes that although different paths may be taken in space and time to get from A to B, the total distance in space-time will always be the same. This is an example of a *relativistic invariant*, which does not depend on an observer's frame of reference. Different observers measure different times and amounts of space between two events, but they all agree on the "distance," or interval, in space-time.[669]

Another example of a relativistic invariant is the speed of light, which, without regard to the light speed decay debate, is the same for everyone regardless of the observer's motion.[670] This is a corollary of the larger principle, at the heart of the theory of relativity, that the laws of physics are the same for everyone.[671] There are, indeed, absolutes. One must therefore resist the temptation, often indulged by those who have a superficial understanding of relativity theory, to say that Einstein's work makes everything relative, including aesthetics, morality, and all human values. The Law Giver and His laws remain, unchanging and perfect (Malachi 3:6; James 1:17; Matthew 5:48; Psalm 19:7).

*On the Trampoline*

I've already let the cat out of the bag as far as a definition of gravity is concerned: it's a curvature of space-time. Now let's retrace physicist Albert Einstein's steps in arriving at that conclusion.

As indicated above, Einstein's special theory of relativity required that space and time be regarded as flexible concepts. Paul Davies notes that the first casualty of the special theory was the belief that time is absolute and universal; Einstein demonstrated that it is in fact elastic and can be stretched or shrunk by motion.[672]

---

[669] Wolfson, *Einstein's Relativity*, Pt. I, lecture 12.

[670] Ibid.

[671] Ibid.

[672] Davies, *God and the New Physics*, 120.

Space, also, is elastic, Einstein showed; when time is stretched, space is shrunk.[673] Thus, while time runs more slowly for a passenger on a train that is speeding through a railway station than it does for a person standing on the station platform, the same passenger in this situation will observe the platform to be slightly shorter.[674] This newfound flexibility violated assumptions that had been in place since at least the time of Sir Isaac Newton. Also now vulnerable was Newton's "action-at-a-distance" theory of gravity, under which gravity was seen as a force that reached instantaneously across space so as to, for example, keep the moon in orbit around the earth and the earth in orbit around the sun.[675] This picture of gravity, it turned out, was problematic for at least two reasons:

(1) The supposed gravitational attraction of the moon, earth, and sun under Newton's theory depended on each of these celestial bodies somehow "knowing" of the presence of the other body or bodies that were said to be acted on by its gravity. This implied a transmission of information from one body to another, but relativity theory forbids any information from traveling across space at a speed greater than the speed of light. Thus, Newton's view of gravity as instantaneous "action at a distance" violated the so-called cosmic speed limit.[676]

(2) The assumed gravitational attraction between bodies is stronger the closer the bodies are to each other, but special relativity theory holds that distance is relative, as is simultaneity, depending on the observer's frame of reference. Gravity such as that posited by Newton cannot operate on the basis of mixed signals concerning its fundamental premises.[677] Either body $a$ is a definite distance from body $b$ or there's no proper context for Newtonian gravity.

---

[673] Ibid., 121.

[674] See ibid.

[675] Wolfson, *Einstein's Relativity*, Pt. II, lecture 13.

[676] See ibid.

[677] See ibid.

Einstein was thus faced with the problem of how to explain gravity other than as an instant attraction of bodies regardless of the distance of the bodies apart. To successfully grapple with this question, it would be necessary to consider again the results of Galileo's thought experiment with the two balls that were never dropped off the leaning tower of Pisa.[678] Two objects of similar air resistance but radically different weights would fall at the same speed, Einstein decided, because gravity and inertia are the same thing.[679] "Inertia" is defined for present purposes as "the property of matter by which it retains its state of rest or its velocity along a straight line so long as it is not acted on by an external force."[680] Inertia is a resistance to force.[681] At the same time that gravity is pulling down on the object in question, the object's inertia—i.e., its natural tendency to resist being pulled toward the center of gravity—is holding the object back.[682] While this is happening, there is a direct relation between the force of gravity acting on the object and the object's inertia.[683]

Newton knew of the tug-of-war between gravity and inertia but had no way of accounting for it other than as a coincidence.[684] In recognizing that gravity and inertia are the same thing (the "equivalence" principle), Einstein took a step that had the effect of bringing accelerated motion within the principle of relativity. When motion is "accelerated," there is a change of the object's speed or direction;[685] thus, inertia is implicated. In Einstein's revolutionary 1916 formulation, there was no gravity-relativity conflict because gravity had been re-defined in a relativity-friendly (inertial) way. It could now be said that *all* motion is relative, and Einstein's theory of relativity had become a *general* theory, no longer restricted to uniform motion situations.

---

[678] See, again, the "WhereThought Experiments Lead" section of the Proving God chapter herein.

[679] Gardner, *Relativity*, 66.

[680] *Random House Webster's*, 977.

[681] Gardner, *Relativity*, 65.

[682] Ibid.

[683] Ibid., 66.

[684] Ibid.

[685] Greene, *The Elegant Universe*, 62.

# RELATIVITY

Saying that all motion is relative has an interesting effect as far as Bible interpretation (hermeneutics) is concerned: It seems to introduce the possibility of viewing the rest of the universe as being in motion around the earth, as long as the earth's equatorial bulge is explained other than as a product of centrifugal (inertial) force.[686] It's all a matter of one's choice of a frame of reference. Is it the universe or is it the earth?[687] The Bible would then no longer need to be read figuratively (nonliterally) when it appears to make Earth the center of the cosmos—as, for example, by saying that the sun rises and sets (Genesis 15:12, 17; Deuteronomy 16:6; 2 Samuel 23:4; Psalm 113:3; Micah 3:6; Nahum 3:7; Matthew 5:45; Ephesians 4:26; James 1:11). Geocentrism is affirmed by Einstein's conclusion that all motion is relative, since Earth's position is entitled to as high a dignity as that of any other celestial body. It should go without saying, further, that while the visible heavens change, God does not (Psalm 102:25-26; Malachi 3:6). In Him there is "no variableness, neither shadow of turning" (James 1:17). The relativity of the motion of the stars and planets thus evokes the comparative of God's utter stability (see Psalm 18:31).

Since "action at a distance" was no longer a correct way of viewing gravity, Einstein was faced with the necessity of framing a new definition, which he did: gravitational force, he said, exists because of a curvature of space-time in response to mass or energy.[688] What Einstein's general theory of relativity says may be easily explained: Imagine, if you will, a trampoline. A bowling ball situated at the center of the trampoline will make a depression. A marble placed at an outer edge of the trampoline will then roll toward the bowling ball, following a geodesic path that is curved but is nonetheless the "straightest" possible route gravitationally to the center of the trampoline.[689] The trampoline, bowling ball, and marble are space-time, the sun, and the earth respectively. The marble rolls toward the bowling ball because the ball has warped

---

[686] See, in the latter regard, Gardner, *Relativity*, 62-63, 72.

[687] Ibid., 72.

[688] Hawking, *A Brief History*, 30; and see the definitions of "gravitational force" and "general relativity" in Greene, *The Elegant Universe*, 416.

[689] Gardner, *Relativity*, 86-89; Hawking, *A Brief History*, 30-32.

the trampoline, not because the ball is in any way "pull(ing)" on the marble.[690] The bowling ball's sole contribution has been to create a curvature of space-time in its vicinity.

How is it determined where space-time will have the most pronounced curvature? Answer: space-time is most radically curved where matter and energy are the most concentrated.[691] We must keep in mind one of the consequences of Einstein's special relativity theory: mass can become energy and energy can become mass, and under the law of "conservation of mass-energy," none of the total mass-energy of the universe can be lost.[692] Thus, space-time can respond to *either* energy or mass in its curvature. In the words of the late John Wheeler, "Mass tells space how to curve; space tells mass how to move."[693]

When we say that space-time is curved, we're saying not just that the paths of material objects in space are influenced by gravity. Implicit in our use of the term "space-time" is the idea that gravity affects time. It's not just a matter of our making something so by nomenclature, as is the wont of the New Atheists, who for some reason like to call themselves "the brights."[694] According to Einstein's equations, strong gravitational fields have a slowing effect on time.[695] This would mean that any rhythmic process, such as the vibrations of atoms or the ticking of a specific number of ticks on a balance-wheel clock, would take slightly longer on the sun than on planet Earth.[696] Obversely, a clock on a high tower standing on the earth would gain relative to a clock at the tower's base.[697] The rule that less gravity means quicker passage of time is what is at the heart of Russell Humphreys' cosmology, which

---

[690] Gardner, *Relativity*, 86-87.

[691] Hawking, *A Brief History*, 30; Wolfson, *Einstein's Relativity*, Pt. II, lecture 14; Greene, *The Elegant Universe*, 68.

[692] Gardner, *Relativity*, 56.

[693] Quoted in Rohmann, *Ideas*, at 340.

[694] Obviously unmindful of Psalm 14, verse 1 ("The fool hath said in his heart, There is no God").

[695] Gardner, *Relativity*, 94.

[696] Ibid.

[697] Davies, *God and the New Physics*, 122.

posits billions of years passing out on the edge of the expanding universe while only a few thousand years have transpired here on Earth.[698] This not only supports the Bible's overall "young earth" cosmology; it also explains how the light from stars billions of light years away would have been able to reach Earth by the fourth day of creation.[699]

Of course, there's no need to involve general relativity theory if one believes that during history's first week God simultaneously created stars and light beams stretching from them to the earth, under what Henry Morris calls the principle of mature creation, or creation of apparent age.[700] In this scenario, God gives us a universe that's in place by the end of the week and a set of physical laws to help us understand what we thereafter see happening in that universe (operational rather than origins science).[701]

Since we're talking about stars, this would seem an appropriate time to mention Mach's Principle. Austrian physicist Ernst Mach (1838-1916) was one of the participants in a protracted debate, still going on, over whether the space-time structure in which we find ourselves (sometimes called the "metrical field") depends on there being stars in the cosmos.[702] Early writers on relativity saw the space-time structure as independent of the stars, though it was given local distortions by the stars.[703] Even if there were no other object in the universe, they argued, it would still be possible for the earth to rotate relative to the space-time structure, the curvature of which, if any, would be irrelevant.[704] The spinning Earth would still bulge around its middle, because it would still be possible for particles of its matter to be forced into paths that were not geodesics in the space-time structure, paths that were against the natural "grain"

---

[698] See the "Normal and Abnormal Science" section of my Paradigm Paralysis chapter, supra.

[699] See, again, Humphreys, *Starlight and Time*, 9-29, 37.

[700] See Morris, *Study Bible*, n. to Gen. 1: 16.

[701] See my "The Canal and Other Wonders" section in the Paradigm Paralysis chapter, supra.

[702] See Gardner, *Relativity*, 101-7.

[703] Ibid., 102.

[704] Ibid.

of space-time.[705] If the sole object in the universe was a spaceship, that craft could still switch on its rocket motors and accelerate, causing astronauts within the ship to feel inertial forces.[706]

Einstein, one of the other principals in the debate, preferred the view of Bishop George Berkeley (1685-1783) that if the earth were the only body in the universe, it would be meaningless to say that it could rotate.[707] Similar views had been held by German philosopher Gottfried Wilhelm von Leibniz (1646-1716) and Dutch physicist Christian Huygens (1629-95), but it remained for Mach to put meat on the scientific bones (a reference to Ezekiel 37:1-10; no need to look it up). Stars were indeed necessary in order to have a space-time structure relative to which the earth could spin, Mach said.[708] They were also needed as a basis for gravitational (inertial) fields capable of bulging the equator of a planet.[709]

It all came down to geodesics (gravity-directed paths), Mach explained, in a line of thinking that anticipated relativity theory.[710] Without a space-time structure such as that described above, there would be no geodesics.[711] In the absence of geodesic guidance, a light beam speeding through empty space would not know what path to take.[712] Even the existence of a spherical body such as the earth might not be possible, since planetary particles are packed together by gravity and gravity conducts particles along geodesics.[713] A planet such as Earth would therefore not know what shape to take, Mach theorized.[714] In short, if Mach's theory is correct, the lack of stars would mean the lack of everything associated with our

---

[705] Ibid.

[706] Ibid.

[707] Ibid.

[708] Ibid., 103.

[709] Ibid.

[710] Ironically, although his thinking was incorporated by Einstein in a successful theory, Mach wound up never accepting relativity. Gardner, *Relativity*, 103.

[711] Ibid., 103-4.

[712] Ibid., 104.

[713] Ibid.

[714] Ibid.

world (including us, since we would have no planet that we could inhabit and our life-giving sun would also be missing).

It takes no great leap of imagination to see in Mach's scheme the cast of characters surrounding God in the Bible. In Genesis 1:14-18, we see God putting the sun, moon, and stars in place to *rule, guide, and give light to* the earth and its inhabitants. This picture is given fuller dimension by Daniel 12:3, where it is prophesied that the resurrected saints (you and I, as believers) will "shine *as the brightness of the firmament* and they that turn many to righteousness (will also shine) *as the stars for ever and ever*"(emphasis and parentheticals added). Celestial bodies, in fact, play various roles in the Bible. In Job 38:7, for example, we see the morning stars shouting for joy when the foundations of the earth are laid.

The above leads us into the concept of the "host," in which stars and angels overlap. The word "host" (*tseba'ah* in the Old Testament Hebrew, *stratia* in the New Testament Greek) signifies among other things an army, and thus describes the angels who serve and fight for the Lord as well as give praise to Him (1 Kings 22:19; Psalm 103:20-21; Luke 2:13). The heavens similarly are under the command of God and extol Him; thus they, also, are called "host" (Genesis 2:1; Nehemiah 9:6; Psalms 33:6; Isaiah 45:12). In a word, the "host" *focuses*. God-ward orientation can happen because there is a universe of stars and angels that give direction to man's thoughts and a particular form and shape to God as we know Him. Thus, reasoning analogously with Ernst Mach, we conclude that nothing intelligible can happen without the celestial presence that God has put in place. Even the atheist has this awareness of God's provision, I contend; otherwise, he would have nothing to set himself against.

As I lie awake on my cot under an open window in a lakeside bedroom, listening to the waves scrubbing the shore with enough vigor that it drowns out the sound of Sue's breathing, it occurs to me that the most powerful thing about Mach's principle is not that it mirrors the Bible, although that is certainly the case. It's simply the idea that God has provided a way for the universe to exist. Through a physical law, gravity, He's put everything together. Even though He didn't have to do that, being under no obligation to give us the time of day or even a day for us to have the time of. I find this truly amazing.

Is there anything physical that isn't affected in some way by gravity? Einstein seems to have assumed that the answer to this question is no. After framing his "equivalence" theory, he proceeded to the question of gravity's effect on light. Since acceleration is the equivalent of gravitation, he was able to approach the above question from the perspective of the Doppler effect.[715] It was known that the pitch of an oncoming police car's siren drops when the vehicle passes you and begins to disappear into the distance.[716] This is because the car compresses the sound waves that are before it to a higher frequency (number of cycles per second) when it's coming toward you and stretches the sound waves behind it to a lower frequency when it's going away from you.[717] Essentially, it would be the difference between a boat's bow wave and its wake. Einstein reasoned that there must be a similar effect with light, which also travels in waves—specifically, that the light from an approaching body would be more toward the blue end of the light spectrum while that of a departing body more toward the red end because of the difference in their frequencies, the same "red shift" that Edwin Hubble observed.[718]

Once you've established that gravitationally-defined accelerated motion has an effect on the frequency of light waves, you're at the threshold of another gravity-related question: is light curved when it passes concentrations of matter? Einstein thought that it had to be, since photons would presumably have to follow the same spatial fabric as material bodies, but his theory needed experimental confirmation.[719]

Einstein's suggestion as to how to prove experimentally that matter curves light was that there be an observation of starlight passing close to the sun during a solar eclipse, and British

---

[715] See further, in regard to the Doppler effect, the "Big Bomb" section of my Paradigm Paralysis chapter, supra.

[716] Davies, *About Time*, 88.

[717] Ibid.

[718] See, again, the "Big Bomb" section of my Paradigm Paralysis chapter.

[719] Greene, *The Elegant Universe*, 76-77, 394 (n. 10).

astronomer Sir Arthur Eddington agreed to undertake this project.[720] The idea was that if the predictions of Einstein's general relativity theory were correct, the stars of interest would be seen to be in different positions in relation to the sun than Newtonian mechanics would have placed them.[721] The experiment was carried out, with equipment that was set up on a small island off the coast of Africa, during a solar eclipse that occurred on May 29, 1919.[722] The result, after some five months spent analyzing the photographs that were taken, was that Einstein's predictions were confirmed.[723]

*Singularities*

As suggested in previous discussion, a singularity is the ultimate space-time curvature. Stephen Hawking defines "singularity" as "(a) point in space-time at which space-time curvature becomes infinite."[724] Michio Kaku gives the broader perspective in defining "singularity:"

> A state of infinite gravity. In general relativity, singularities are predicted to exist at the center of black holes and at the instant of creation, under very general conditions. They are thought to represent a breakdown of general relativity, forcing the introduction of a quantum theory of gravity.[725]

Prior to the advent of quantum mechanics, it was widely believed that space-time came out of a singularity, this being one of the solutions to the equations of general relativity.[726] This singularity at the first moment would have consisted of the matter of an entire

---

[720] See ibid., 76-77.

[721] See ibid.

[722] Ibid., 77.

[723] Ibid

[724] Hawking, *A Brief History*, 203.

[725] Kaku, *Parallel Worlds*, 398.

[726] See Hawking, *A Brief History*, 53, 90.

universe infinitely compressed and squeezed into a single point.[727] Essentially, this would reflect a running all the way back of the tape (if there had been one) of the cosmic expansion seen by Edwin Hubble in 1929. By 1998, nonetheless, when the tenth anniversary edition of *A Brief History of Time* was published, Stephen Hawking was attempting to convince other physicists that there was in fact no singularity at the beginning of the universe, that the big bang solution could disappear once quantum effects were taken into account.[728]

The belief that in a singularity the general theory of relativity breaks down, which continues to be held by many scientists, is based ultimately on Alexander Friedmann's solutions to Einstein's theory, which uniformly arrived at the conclusion that at some time in the past, between ten and twenty billion years ago, the density of the universe and the curvature of space-time would have been infinite. Our mathematics can't handle infinite numbers, however.[729] This leaves us with no way to penetrate the hypothesized singularity. Stephen Hawking explains that even if there were events before the postulated big bang, one could not use them to determine what would happen afterward.[730] He states:

> If, as is the case, we know only what has happened since the big bang, we could not determine what happened beforehand. As far as we are concerned, events before the big bang can have no consequences, so they should not form part of a scientific model of the universe. We should therefore cut them out of the model and say that time had a beginning at the big bang.[731]

Paul Davies sketches a cosmology that sounds less like a conclusion that we simply can't know what was happening before the

---

[727] Davies, *Mind of God*, 49.

[728] Hawking, *A Brief History*, 53.

[729] Ibid., 49.

[730] Ibid.

[731] Ibid.

alleged big bang and more like the still-popular understanding that the big bang, pure and simple, was the beginning of everything. After reminding his readers that under general relativity theory the ongoing expansion of the cosmos is a matter of space itself expanding rather than the galaxies rushing outward into a pre-existing void, Davies states:

> Conversely, in the past, space was shrunken. If we consider the moment of infinite compression, space was infinitely shrunk. But if space is infinitely shrunk, it must literally disappear, like a balloon that shrivels to nothing. And the all-important linkage of space, time, and matter further implies that time must disappear too. There can be no time without space. Thus the material singularity is also a space-time singularity. Because all our laws of physics are formulated in terms of space and time, these laws cannot apply beyond the point at which space and time cease to exist. Hence *the laws of physics* must break down at the singularity (emphasis supplied).[732]

All physical laws! In both of the above formulations, it strikes me, we're allowing the limitations of science to put us in a know-nothing situation as far as the supposed big bang singularity is concerned. "After the bang" and "before the bang" (if, indeed, there was a "before") can't be connected because either (1) general relativity drags us into pernicious infinities that overwhelm our mathematics, or (2) general relativity causes time to vanish along with space when we try to look behind the bang, which in itself removes any possibility of our having a scientific theory about anything pre-bang.

Either of the above difficulties is enough to turn one to the God of Creation. Not only is it true that the Lord "by wisdom hath founded the earth; by understanding hath established the heavens (Proverbs 3:19);" it's also the case that nothing is hidden from Him. "(H)is understanding is infinite," declares the psalmist (147:5). He

---
[732] Ibid., 49-50.

"discovereth deep things out of the darkness," Job acknowledges (Job 12:22). Indeed, He *"knoweth* what is in the darkness," the prophet Daniel states (Daniel 2:22; emphasis supplied). Why trust mathematics when you can be informed by Him who made heaven and earth (Genesis 1:1), who will write all necessary truth on your heart (see 2 Corinthians 3:3)!

Michio Kaku's suggestion, that there is a gap in cosmology that may be filled by quantum mechanics, seems to me to point to an intent on the part of popular science to exclude God from the story of our origins that is more obvious than in the now-classic big bang model. With the general-relativity-based (big bang) version of the story of our origins, we can trace the history of the universe back to a point at which space, time, and matter had a definable beginning.[733] This is certainly not inconsistent with the biblical account, "In the beginning God created the heaven and the earth."[734] The conjoining of the words "the" and "beginning" makes it clear that what the first verse of the Bible is referring to is the *absolute* beginning of the temporal and material world.[735] As previously urged, the Genesis account must be understood as indicating that time, space, and matter came into existence *simultaneously*.[736] To this extent — but only to this extent — popular cosmology and the Bible agree.

There is, nevertheless, no explanation in the big bang theory of how space-time could have abruptly come into existence by operation of physical law, since the theory presupposes an initial singularity in which we are unable to apply general relativity or any other known law of physics. On this basis alone, the big bang theory should be denied acceptance. If creation by supernatural act is rejected and one is unable to wink at the problems posed by an eternal universe (not the least of them, that it ignores the second law of thermodynamics),[737] then quantum theory might be resorted to (fecklessly, in my view) as a way out of the throes of philosophical desperation.

---

[733] Davies, *About Time*, 131-32; Hawking, *A Brief History*, 9.

[734] Genesis 1:1.

[735] *King James Study Bible*, n. to Gen. 1:1.

[736] See the "Going Farther" section of the present chapter.

[737] See, in the latter regard, the "Romans 1: 20" section of my Region of Awe chapter.

Quantum theory, it will be recalled, is centered on Heisenberg's uncertainty principle, which states not only that position and velocity of a particle cannot simultaneously be measured, but also that whatever is measurable is subject to unpredictable fluctuation in its value.[738] Because of the latter fact, quantum effects can't be said to be determined strictly by preceding causes.[739] Some outcomes are more probable than others, but in a very real sense quantum events just happen.[740] This allows the quantum physicist to posit a universe that doesn't need a creator. It came into existence from nothing as a result of a quantum fluctuation, and no law of physics was violated, supposedly.[741]

Of course, the objection may be raised that the application of quantum mechanics is usually restricted to atoms, molecules, and subatomic particles, and that quantum effects are usually negligible for larger objects;[742] but this almost certainly will not be enough to deter a theorist bent on denying God's creative role. The quantum physicist will perhaps insist that whether we're constructing a relativistic or a quantum cosmology, we're still talking about a primordial universe that was compressed to minute dimensions, and that in a world where events were taking place on a very small scale, quantum processes must have been important.[743] In this way, relativity and quantum mechanics might be combined, but not convincingly as far as I'm concerned. The impression persists, in my mind at least, that this is an ill-advised attempt to apply to the universe as a whole a theory that has validity only on a micro-level.[744]

---

[738] Davies, *Mind of God*, 61; see also the "Going Farther" section supra this chapter; also the "Big Bomb" section of my Paradigm Paralysis chapter, supra.

[739] Davies, *Mind of God*, 61.

[740] Ibid.

[741] Ibid., 61-62.

[742] See ibid., 31, 61.

[743] Ibid., 62.

[744] See ibid.

## Black Holes and Beyond

Michio Kaku indicates that a singularity may be found at the center of a black hole.[745] Stephen Hawking defines "black hole" as "(a) region of space-time from which nothing, not even light, can escape, because gravity is so strong."[746] Light is both particles (photons) and waves.[747] The particle property translates into gravity being able to affect light.[748] Gravity will be especially strong where a star has used up its hydrogen and other nuclear fuel and has begun to collapse in on itself.[749] In 1939, American physicist J. Robert Oppenheimer was able to preliminarily establish that according to general relativity, the collapse of a star would have the effect of bending light cones inward, toward the star.[750] Hence, the blackness of a black hole.

Work done by Stephen Hawking and his colleague Roger Penrose between 1965 and 1970 demonstrated that in fact there must be a singularity, involving infinite matter density and infinite space-time curvature, within a black hole.[751] Hawking remarks that this is "rather like the big bang at the beginning of time," only it would be an end of time for the collapsing celestial body and any astronaut venturing past the black hole's event horizon, its boundary that is a point of no return.[752] Our hapless astronaut would suffer physical death almost instantly as a result of his being brutally crushed upon crossing the event horizon.[753]

Simultaneously, this unfortunate space traveler or his essence would have stepped outside of time in another sense, since whatever remained of him would now be in a zone where, as far as

---

[745] Kaku, *Parallel Worlds*, 398.

[746] Hawking, *A Brief History*, 199.

[747] Ibid., 83.

[748] Ibid., 83-84.

[749] Ibid., 85-87.

[750] Ibid., 87-88.

[751] Ibid., 90.

[752] Ibid., 90-91.

[753] Kaku, *Parallel Worlds*, 116.

we can tell, no physical law could define either time or space for him. As at the alleged big bang singularity, the laws of physics and science's ability to predict the future would break down.[754] Nonetheless, Stephen Hawking says, any observer outside the black hole would not be affected by this failure of predictability, since neither light nor any other signal could reach him from the singularity.[755] This separation of the universe at large and the non-world within the black hole led British physicist Roger Penrose to propose a *cosmic censorship hypothesis*, paraphrased "God abhors a naked singularity."[756] Singularities resulting from gravitational collapse will be decently hidden from outside view by an event horizon, under this supposition.[757]

The possibility of the entire universe collapsing in on itself, in a "big crunch," has already been mentioned.[758] Whether this actually happens depends on the amount of "dark matter" in the cosmos.[759] The existence of dark matter can hardly be doubted at this point, the early protestations of physicist Eric Lerner notwithstanding.[760] In the 1930s, Swiss astronomer Fritz Zwicky of the California Institute of Technology observed that galaxies in the Coma cluster of galaxies were not flying apart as they should have been doing under Newtonian rules of motion.[761] The only explanation Zwicky could think of for this, other than Newton's laws being incorrect at galactic distances, was that the Coma cluster had hundreds of times more matter than could be seen by telescope.[762]

For reasons not the least of which was the clubbiness of the American astronomers' community, Zwicky's work was rejected

---

[754] Hawking, *A Brief History*, 91.

[755] Ibid.

[756] Ibid.

[757] Ibid.

[758] See the discussion in this regard in the "Big Bomb" section of my Paradigm Paralysis chapter.

[759] See Kaku, *Parallel Worlds*, 40-44.

[760] See, again, the "Big Bomb" section of my Paradigm Paralysis chapter.

[761] Kaku, *Parallel Worlds*, 70.

[762] Ibid., 70-71.

or ignored for some thirty years.[763] Then, in 1962, astronomer Vera Rubin rediscovered the motion problem first noted by Zwicky, and over the next sixteen years, with the help of colleagues, examined eleven spiral galaxies and found them all to be spinning too fast to stay together under Newton's laws.[764] Gravity from unobserved ("black") matter was the reason for their cohesion, it was concluded. In the years that followed, various other methods were devised that effectively confirmed the presence of dark matter, which included *inter alia* measuring the distortion of starlight as it passed through the invisible matter.[765]

The metaphorical import of black holes and the big bang and big crunch singularities can hardly be missed by a Christian believer. At a black hole the traveler disappears over the horizon never to return, and the same thing happens to a person at death, since, per Hebrews 9:27, it is "appointed unto men once to die," No reincarnation, no vagabond migrating soul as in Hinduism. The difference is that believers are not consigned to a fate as horrific as the black hole; instead, their eternal destiny is wonderful beyond imagination (1 Corinthians 2:9). The singularity at the center of the black hole repeals the existing laws of physics, as does God when we enter His eternal kingdom (Revelation 21:4-5, "for the former things are passed away(b)ehold, I make all things new."). Roger Penrose's cosmic censorship hypothesis evokes Exodus 33:20, where God tells Moses, "Thou canst not see my face, for there shall no man see me and live." Our astronaut's exit from time upon crossing the event horizon is like my own anticipated going off the clock "(w)hen I have crossed the bar" (Alfred, Lord Tennyson, *Crossing the Bar*, stanza 4).

Perhaps the bit of singularity-related science that strikes me as the most relevant theologically is the idea that the big bang and the big crunch, as described, are bookends. The cosmos goes from oblivion to oblivion, unknown to unknown. In popular science's version of the story, we don't know what was there before our universe, or even that there was a "there" or a "before," and we

---

[763] Ibid., 71-72.

[764] Ibid., 73.

[765] Ibid.

likewise lack any knowledge of what will be there after the present universe, if there is an "after." In the true story of the universe set forth in the Bible, the world likewise begins and ends symmetrically, only there's no mystery concerning the first and the last; it's Jesus, who is the Alpha and the Omega.[766]

More broadly, it's the entire Godhead.[767] Scripture affirms that all three persons of the Trinity—Father, Son, and Holy Spirit—precede and survive the universe. God's eternal purpose was accomplished in Christ (Ephesians 3:11), who is "the same yesterday, and today, and for ever" (Hebrews 13:8). If the Creator's other attributes haven't changed, how can His duration have changed? The Word, God the Son, was "with God,...was God" in the beginning (John 1:1), so as to be able to create all things (John 1:3). Does this sound like a god who didn't come into existence until he was created? "I AM THAT I AM" is God's name forever (Exodus 3:14, 15). There is no statement in Scripture that any person of the Trinity began to exist at any particular time or, for that matter, ever. The Spirit, like the Father and the Son, was present at the creation (Genesis 1:2) and is likewise beginningless. It is said that the Spirit will abide with believers forever (John 14:16). Jesus likewise lives forever (John 12:34; 1 Timothy 6:14, 16), as does Father God (Isaiah 9:6; 1 Timothy 1:17).

*Reflection*

We seem to have come a ways, after more or less abandoning philosophy (which used to be the world's science) and getting mired in mysticism. Each was offered but not easily traveled as a road to truth. To a considerable extent, as things stand now, it seems to be science that resonates, to use a cliche` of the day. So where do we stand overall? What can we say we know based on clues science has given us? What must we admit we have no clue about?

---

[766] John 1:1-3 and *Henry Morris Study Bible* nn. thereto; Revelation 22:13.

[767] See Morris, *Study Bible*, "Godhead" n. to Romans 1:20.

Some things that we can affirmatively state:

(1) The "big bang," popular science's favored theory of origins, is rife with inconsistencies and relies on assumptions for which observational support is ambivalent if not nonexistent. Particularly troubling is popular science's reliance on "inflation" as a cure-all for the big bang theory's logical deficiencies. As I've previously complained, "inflation" has the feel of something jury-rigged and ad hoc. Also, although the big bang qualifies as *a* solution to the equations of general relativity, it is but one of several possible solutions, and the science gods do not appear to have been seriously considered others. The big bang theory is the main alternative offered by the scientific community to divine special creation, although there are pockets of reliance on quantum acausality (see numbered paragraph [9], following).

(2) The presence under relativity theory of relativistic invariants, such as the space-time interval and the speed of light, which give values that are not dependent on the state of motion of the observer, suggests the existence of other absolutes, such as God and His law.

(3) Any attempt to posit a natural God or supreme intelligence, an organizing principle that exists solely by reason of the laws of science, must take into account the second law of thermodynamics. Because of the law of entropy, a natural God that existed in eternity past could no longer exist, because that God would already have had an eternity to exhaust the energy needed to live and do work. On the other hand, a natural God that had a beginning would die when the universe dies at some point in the future. Both of the immediately preceding statements give the lie to pantheism.

(4) Herman Minkowski's graphs (visualizing "block time") and Stephen Hawking's light spheres (defining "past," "future," and "elsewhere"), which are artifacts of a scientific

theory of space-time, cannot dictate human eternal destiny. Salvation is a spiritual matter. It has nothing to do with space or time as we know them, except for the manner in which it (salvation) is initiated.

(5) Acknowledging that all motion is relative opens the door to a geocentric view of the universe and therefore a literal interpretation of the Bible in this and other respects.

(6) If you're unwilling to accept scenarios that feature the application of raw supernatural power, there are at least two scientific ways of explaining how we could have a cosmos as young as the Bible says it is, specifically, (a) Russell Humphreys' "white hole" cosmology, theorizing fast time passage on the periphery of an expanding universe, or (b) ongoing light speed decay, of a radical nature, from the beginning of the cosmos until now. A young universe is deadly to the theory of evolution.

(7) Mach's principle not only reflects the Bible metaphorically (the starry "host" surrounding God); it also shows that we could not have a universe without stellar gravitation, put in place by (guess who?) God.

(8) Singularities reflect a reality that's outside of time and beyond the reach of physical law. That reality has to have pre-existed the universe and necessarily will survive the demise of the cosmos. From copious Scripture and unassailable logic, similarly, it is clear that the existence of Father, Son, and Holy Spirit is from everlasting to everlasting.

(9) Even disregarding the clumsiness of "inflation" theory as a device to overcome the big bang's "horizon" problem, there remains the difficulty that no satisfactory explanation has been given of how the alleged initial explosion, the scientific Genesis, happens to have happened in the first place. What set it off? Science offers vague speculation

as to quantum effects, but nothing more. In the quantum world, things happen without identifiable causes (see the "Quantum Indeterminism" section of my Radiating Reality chapter, infra, also the "Singularities" section of my Relativity chapter, supra); thus, one explanation for the big bang would be that it "just happened." Since we seem to be able to do no better than this for an explanation, the big bang is not entitled to recognition as the standard model of cosmology.

(10) Allowing the mathematics of general relativity to control natural science in all matters relating to cosmology leads to unworkable concepts such as infinite space-time curvature. Mathematics should not be allowed to trump observation and common sense.

Some of the observations that we can make are empirical without being science-based:

(11) There is an inductively provable case for Jesus' resurrection.

(12) The life, ministry, and death of Christ were foretold in detail by the Old Testament prophets.

As suggested above, the conclusions that we have been able to draw thus far are almost exclusively products of our examination of science, in particular relativity theory. I've already complained that metaphysics and mysticism seem to have left us handfuls of air, nothing more. The questions that were posed in my Introduction and in the Fluid Reality and Essence chapters herein remain largely unaddressed. How does God inhabit eternity but still operate in time? If He's situated in time, does that mean He's not omniscient? If God has exhaustive foreknowledge, does that preclude our being able to meaningfully interact with Him? Must God be changeable, in the sense of being able to be aware of position shifts on the part of individual human beings only after they have happened, in order for man to have free will? Is exhaustive foreknowledge on God's part essential to His immutability? What did Paul and Peter mean by their supposed predestination statements? Did Augustine successfully reconcile predestination, to either heaven or hell, with free will on the part of man?

And there are certain even more basic questions that have to be answered if we're going to talk about the destination of men's souls, are there not? Is the reality of things not in what we see before us, but rather "out there?" In our imaginations? In the mind of God? In what our senses tell us? How can anything have reality in the face of change? How, exactly, is "self" defined? You've heard these questions asked before in this book.

Philosophy has been unable, without embarrassing itself, to attempt answers to the above questions, or has avoided such questions altogether. The apostle Paul warned of the danger inherent in the love of human wisdom, stressing that philosophy would cause the truth seeker to look to the world rather than Christ for answers (see Colossians 2:8).[768] Mysticism, in my estimation, likewise places us at no vantage point from which we can know with certainty any spiritual or metaphysical fact. Except, of course, that God is good. Not that we don't need to recognize and rely on God's lovingkindness every day; it's just that in the meantime we're still looking for an ultimate explanation of the universe and eternity, a "theory of everything" that satisfies both the head and the heart. Do we therefore need to continue forging ahead with science, our third major discipline, in search of answers? Do we have any choice?

---

[768] See also, once again, Morris, *Study Bible*, n. to Col. 2:8.

# 7
# RADIATING REALITY

*Gone Fishing*

In explaining why it's the individual's choice whether he or she winds up saved or condemned for eternity, I've brought in the idea of alternative realities. Heaven and hell are equally valid as projections of one's future, up to a certain point. That point is arrived at when, while you're still within your life on Earth, you either opt to accept or ultimately reject Christ's offer of salvation, which you can do up until the very last moment of your life here. The heaven/hell reality duality is part of a broader choice structure that extends into all facets of a person's life. So it's a matter not just of where you'll spend eternity and what rewards God will bestow on you in the hereafter, but also of how you'll spend the rest of your life "here below"—whether or who you'll marry, for example, what career you'll choose, what your politics will be, and so on. I now wish to explain to you how I came to the conclusion that there are alternative realities.

For all practical purposes I was an only child, as you may have guessed. My half-sister Margery was twenty-three years older than I was, and my half-brother Warren twenty years my senior. Both had finished college and left home before I was old enough to have times with them that I could remember. In 1948 or thereabouts, I was introduced to Margery's new husband, Carroll, a World War II veteran (infantryman in the Pacific theater) and a lifetime resident of the state of Maryland, in whose civil service system he secured employment after the war. Around the summer of 1950, Mother and Dad and I visited Margery and Carroll in Maryland for the first time. This week-long visit became an annual pilgrimage, which always took place in July. Dad was in charge of bus maintenance for our school district and for this reason needed to be on hand throughout the school year, even during school vacations. In August he needed to be there to train new bus drivers.

The annual trip to Maryland thus always took place in hot weather. This was before the days of car air conditioning, for our family at least (in the 1950s we had a 1949, then a 1952, then a 1956 Chevy, none of which had air). Also, there were no superhighways that were part of our route; therefore we couldn't always avail ourselves of "four-sixty" air conditioning (four windows open at sixty miles an hour). Though U.S. 15 was only two lanes wide in most places, much of it was open road; thus, we at least were able to have the hot breath of summer rushing in on us when we were between towns.

I always tried to think cool thoughts no matter what speed (and accordingly what airflow) were dictated by local traffic conditions and laws. This was relatively easy when we were motoring along the Susquehanna, which north of Harrisburg becomes wide and dotted with tiny islands. As a youngster, I would imagine myself swimming in the shallow depths of the slow-running river, whose stony shoreline is not unlike Otisco's. After I had grown older and had some experience in Boy Scouts with canoes, I would fantasize that I was paddling from islet to islet.

Thinking refreshing thoughts was more of a challenge when we were driving through downtown Baltimore, stoplight to stoplight, at 5 o'clock in the afternoon on sweltering July days. The next (and second-to-last) leg of the trip after Baltimore would be the Sandy Point ferry, which would take us across the Chesapeake Bay (Fig. 10), called "a noble sea" by Captain John Smith in James A. Michener's novel *Chesapeake*.[769] We were always more than ready

---

[769] Michener, *Chesapeake* (New York: Random House, 1978), 45.

for what the ferry offered: a chance to lower our body temperatures (or so we thought) and raise our spirits as we stood on the upper deck watching the eastern shore of the Chesapeake grow larger. The star player during the half-hour ride to the Eastern Shore was always a stiff Bay breeze, which would coerce the surface of the water ahead of us into moving ridges that the ferry would plow through with steady insistence, and which would further tousle my already disheveled hair.

The ferry ride was a respite from the merciless Maryland heat and humidity, but only a brief one. During the first couple of years of our trips to Maryland, Margery and Carroll lived in a fifth-floor walk-up apartment in Easton, a sleepy Eastern Shore town a few miles from where the ferry landed. I can remember begging Carroll and Margery to provide me with Coca-Cola after Coca-Cola. After the Penfield Apartments, my sister and her husband lived in a duplex in Easton that was more comfortable than the apartment, at least when there was air stirring outside. This place had only a first and second floor. I don't remember there being any substantial presence of fans in either place.

There would, nonetheless, be "joy in the morning" (Psalm 30:5), a period of relief from the heat of Easton,[770] since part of each summer's trip to Maryland would be time spent on the waters of the Chesapeake Bay, amid the zephyrs off Tilghman Island, where lived Carroll's mother and stepfather, Mr. and Mrs. Lednum. Mr. Lednum made his living as a fisherman and charter boat captain on the Bay and was a man of generous disposition; therefore it was the norm that one or two days of each year's side trip to Tilghman (pronounced Till'-mon) would involve our going out fishing. One summer we brought with us to Maryland my aunt Emma, a forlorn lady whose husband (uncle Earl, my father's brother) had long since deserted her in favor of the bottle and other women (he also stole money from the Mobil station and garage that Dad ran before Dad went to work for the school district). That year, we counted

---

[770] Quoted out of context in that the reference of the verse ("but joy cometh in the morning") is to the Lord's anger, which "endureth but for a moment," not to the heat of summer, but the phrase still somehow seems apropos.

some hundred and thirty fish hauled in over the side of the Bea B.[771] Aunt Emma caught many of those. It was wonderful, I thought, to see her having so much fun.

A brief sidebar will be forgiven by you the reader, I'm hoping. Uncle Earl and Aunt Emma had two children (Betty and Billy) before Earl's dereliction began. Given the circumstances, I can easily imagine my cousins asking themselves: are we so ugly and stupid that someone would not want to parent us? I'm not necessarily saying that this is what happened after Uncle Earl's abandonment, but let's just suppose it as a hypothetical. If indeed you've wantonly caused a child to have such self-doubt, it's problematic whether you fulfill the definition of either a parent or an adult, I hear myself saying. I'm tempted to say that you deserve hellfire. But now I need to cut my tirade short, I realize.

An end to speechifying was indicated because it's not up to me to decide whether someone deserves eternal punishment for having wantonly affected the lives of vulnerable young persons. That kind of judgment is in the hands of the Lord (John 25:26-27; Romans 14:10; 2 Corinthians 5:10). Anyone who undertakes the condemnation of others acts self-righteously and should himself be ashamed (see Luke 18:10-14). I therefore now admit that I spoke presumptuously in questioning Uncle Earl's humanity a moment ago, and ask forgiveness of God and you the reader. "Judge not that ye be not judged," Jesus teaches (Matthew 7:1).

As a postscript to my retracted rant, I should note that Uncle Earl made contact with Aunt Emma shortly before his death. He was in the Marcy sanatorium (near Utica, New York), a treatment facility for alcoholics and other persons with disabilities. He wanted to die surrounded by family. Too late, Uncle Earl, too late. Now I'm beginning to feel sorry for the man. Jesus both counsels and models forgiveness (Matthew 18:21-22; Mark 11:25-26; Colossians 3:13).

Back to the Bay. What I enjoyed fully as much as the fishing was exploring by boat the interwoven estuaries that complicate

---

[771] Mrs. Lednum's unshortened first name was Beatrice; hence part of the name of her husband's boat. I never did find out where the "B" came from, or maybe have forgotten.

the eastern shore of the Chesapeake. Mr. Lednum (Ollie, to adults) seemed to relish checking in with watermen he knew on various rivers, finding out what they were catching, and he often took me with him on his riparian wanderings. There were rivers named after Indian tribes, such as the Choptank (major players in Michener's *Chesapeake*), and towns that were flanked by august brick and clapboard manor houses, but what was most fascinating for me were the seemingly infinite fractal branchings-off of the various arms of the Bay (see Fig. 10).[772] A particular run of water, as it extended away from the Bay itself and into the Delmarva (Delaware, Maryland, and Virginia) peninsula, would split, then split again, then split again. The visual impression given, especially when one looked at a map, was that of a tree with branches, in this case crooked ones, that were shooting out in every direction. It was as if there was no shoreline as such because there was no end to the splitting off of waterways.

The significance of the fractal eastern Chesapeake shoreline as a metaphor for things spiritual was not evident to me as a child or youth, or even relevant as far as I was concerned during my salad days, and by the time I was in college Margery and Carroll had retired to Florida; thus, there was little reason or opportunity to think about the Eastern Shore for a long time. Then, there occurred circumstances that directed my thinking, in divers ways, to the idea of alternative realities.

My first intellectual bivouacs upon retiring from the daily practice of law were at the University of Rochester main library and the Colgate Rochester Divinity School library. The latter facility's books had not yet been shipped out to the U of R's river campus as part of a later-implemented cost-cutting plan that included the firing of Colgate's librarians. I did research at Colgate on the question of divine predestination, the parameters of which I've already

---

[772] For those unfamiliar with the work of French mathematician Benoit Mandelbrot (b. 1924), a fractal, as the term is commonly used, is a progression in which a particular shape or configuration repeats itself on a smaller and smaller scale while remaining part of a whole. Self-similarity is a key characteristic of a fractal set. See Benoit B. Mandelbrot, *Fractals: Form, Chance, and Dimension* (San Francisco: W. H. Freeman and Co., 1977), 16-17.

outlined.[773] However, what seemed like weeks of inquiry left me as dry as the paper of the books that surrounded me.

Then, one day during my first year of freedom from the law—I was sitting at my desk at home when it happened—I experienced what for me, in my dimness, was a blinding flash of insight: God's knowledge of what to us is the future logically should keep each of us from being able to affect what will befall personally down the road—poverty or prosperity, marriage or singlehood, heaven or hell, for example—but it has that effect only if personal futures come in just one version. Suppose, instead, that there are differing histories that a person could have and that God knows them all not as mere possibilities but rather as present reality. There is no preferred future as far as God is concerned. Anything can happen, and when it does, whatever it is, God foresaw it (or, to state the matter more correctly, knew of its occurrence). If I go to heaven when I depart from this veil of tears, God saw that as my very fortunate fate. If I end up in a less agreeable setting upon my exit from this life, God simultaneously saw that as my destiny also. Which outcome happens is up to me.

The above structure of being, which I've elected to call "radiating reality," is derived from what seem to me to be two irrefutable facts: (1) God is infinite and (2) He is more real than anyone or anything else. These statements being true, God's awareness of reality as regards any person or object cannot be limited to one linear past-present-future progression. He sees all possible outcomes as real, because He is infinitely real. This means that man is able to be free even though God is omniscient. God doesn't dictate your future, notwithstanding the fact that He has knowledge of the course that you will freely take in any particular area of life (for example, marriage, salvation, or employment) when you stop having multiple world lines in that area—i.e., when that part of what to you is your future finally arrives. Your reality in the meantime resembles the fractal Eastern Shore, or, perhaps even more descriptively, a bush with branches radiating outward ad infinitum,

---

[773] See the "Foreknowledge and Freedom" section of my Essence chapter, supra, also the "Reflection" section of my Relativity chapter, supra.

each parting of branch from branch showing the existence of available alternatives. The short form of the story: it's your choice and it's already true.

The scheme of things that I've just sketched took shape in my thinking before I'd ever devoted much time to studying either relativity theory or quantum mechanics. I was operating strictly on intuition at that point, and therefore, I suppose, could have been said to have been walking with C. S. Lewis in the "foothills" of mysticism.[774] As I've suggested, Lewis's mystical-mindedness consisted largely of being sensitive to correspondences between what he saw in nature and spiritual realities that he became aware of, rather than following any mystic's regimen.[775] Saying that something points to God's truth by analogy may be more powerful as persuasion than claiming mystical revelation, in any event. In my study of relativity theory, I was able to go beyond mysticism and metaphor, thankfully: I looked at science and found conclusions that *directly* pointed to the truth of the Bible.[776]

It was a few months after the above-described revelation that I actually began my study of relativity theory, sending away for a Teaching Company tape course that I had seen advertised in a political commentary magazine.[777] Although I didn't realize it at the time, it was theologically logical that I should have proceeded in this way. The thought that there is divine knowledge of the future, whether or not accompanied by divine control of the same, brings up the threshold question of God's relationship to time. Einstein's rebellion against the paradigms of physics, through his special relativity theory, involved an assault on the Newtonian view of time as an absolute. Time, for the part of the physics community that signed on to Einstein's revolution, now became the handmaiden of light, obsequiously fitting itself into the reality described by

---

[774] Lewis, *Letters to Malcolm*, 63.

[775] See the "Cottage" section of my Introduction, and the "Mysticism and Metaphor" and "Soul Searching" sections within my Region of Awe chapter, all supra.

[776] See the "Reflection" section of my Relativity chapter, supra.

[777] Wolfson, *Einstein's Relativity and the Quantum Revolution*, Pts. I and II. Prof. Wolfson's presentation was in no way political.

Einstein's special theory rather than lording it over everyone and everything in sight as before.

Time also became mortal. General relativity extended time back to a point of beginning, the alleged "big bang," and forward to a final singularity, a "big crunch." In the latter instance, it would be necessary that enough matter could be found in the universe to cause the cosmos to collapse back in on itself by reason of gravity. Time's having an origin of some description strengthens the case for a Creator, as it casts God in the role of a likely sponsor of time's origin. Time's having an anticipated end is consistent with the eschatological scheme set forth in the Bible, which provides that the present heaven and earth will one day pass away (2 Peter 3:10; Revelation 21:1), and that the setting for the world to come and the believer's existence therein will be eternity (Isaiah 57:15; Mark 10:30; John 12:25; Revelation 21:1-4). Time as we know it is inseparable from space, relativity theory tells us;[778] thus, it's not surprising to learn from the written Word of God (the Bible) that these entities in their present forms will be done away with.

Prof. Richard Wolfson, the lecturer on the Teaching Company tapes mentioned above, was also responsible for my initiation into the world of quantum physics, of which his last nine lectures were an introduction.[779] I then began reading treatments of the subject by other authors, including Fritjof Capra, Paul Davies, Brian Greene, Stephen Hawking, and Michio Kaku. Consequently, I became aware, *inter alia*, of the "many worlds" theory of Hugh Everett III, which allows the existence of conflicting quantum conditions based on an interpretation of Heisenberg's uncertainty principle. Everett's theory says that the contradictory states (a dead cat versus the same cat alive, for example) are not a problem because they're in different worlds, so that no condition is a challenge to the trueness of the opposite.[780] Thus, it has been said, "There is a universe where

---

[778] See the "Going Farther" section of my Relativity chapter.

[779] Wolfson, *Einstein's Relativity and the Quantum Revolution*, Pt. II, lectures 16-24.

[780] Kaku, *Parallel Worlds*, 167-71; Davies, *God and the New Physics*, 116-18. An example of how the status of something macroscopic can be affected by an event on the particle scale is described in the "Schrodinger's Cat" section that follows.

Elvis is still alive."[781] Some people say that the Everett interpretation was what former New York Yankees and Mets manager Casey Stengel had in the back of his head when he said, "I made up my mind both ways." Being right in two universes isn't bad.

At the same time, I learned that there is a version of the quantum theory, the Copenhagen Interpretation,[782] which says that every possible state of a subatomic particle, every possible combination of locus and speed, exists until there is an attempt to make an *observation* of the particle in order to determine either its actual location or its actual energy.[783] Observation will have the effect that it will "collapse" the particle's probability wave function. You see, quanta, the indivisible "chunks" of matter or energy of which the universe is composed, exhibit a wave/particle duality. Probability wave function collapse will cause a particle, or any larger object whose condition is dependent on the particle's condition, to go into a definite state, which will be the end of its so-called "hybrid reality."[784]

Further in the above regard, I learned (1) that American physicist Richard Feynman had devised a method, called "sum over histories" or "sum over paths," for calculating the probability that a subatomic particle will be found at a particular location, (2) that a particle is assumed, under Feynman's approach, to have possible world lines that go back in time, and (3) that Feynman's alternative world lines all exist with equal reality until the particle's wave function is collapsed.[785]

Finally, I took Cornell University mathematics professor Steven Strogatz's VCR course on chaos theory, which included lectures on fractals—a subject more familiar to the general public, I would

---

[781] Alan Guth, quoted in Kaku, *Parallel Worlds*, 169, quoting BBC-TV's *Parallel Universes*, 2002.

[782] The city of Copenhagen was the home base of Niels Bohr, the Danish physicist responsible for the model of the hydrogen atom from which quantum mechanics was developed.

[783] It is not possible to measure position and velocity simultaneously. See Wolfson, *Einstein's Relativity and the Quantum Revolution*, Pt. II, lecture 19.

[784] See Kaku, *Parallel Worlds*, 151-52.

[785] Hawking, *A Brief History*, 138, 167-69; Kaku, *Parallel Worlds*, 163-65; Greene, *The Elegant Universe*, 108-12.

suspect, than "hybrid reality" and some of the other features of quantum physics.[786] Fractalization involves things splitting apart and going separate ways, multiplying routes as in the quantum effects described above. As previously noted, fractal splitting produces *copies* of a basic structure, and on a scale that becomes *smaller with each new generation* of copies, in much the same manner as does the growth of the florets of the vegetable romanesco, a member of the mustard family that might be described as a more complicated version of its cousin broccoli. Once again, Nature informs us.

With a radiating/alternative reality template now in place to guide our thinking with respect to God's openness (vulnerability?) to change, I believe we can answer some of the questions that bedeviled Millard Erickson when he was writing *What Does God Know and When Does He Know It?*[787] Does God repent after taking an action, as if He did not know that the action would cause a bad result? Does He ever change His mind about what He intends to do? Does He in good faith test individuals or nations in particular situations, not knowing how they will react? Is He surprised, even shocked, when horrendous events such as pagan child sacrifice occur? Can statements on God's part complaining that His wishes have gone unfulfilled simply be chalked up to frustration when God should not have expected wish fulfillment? These and other questions posed by the open theists have a common strand: the Bible account, taken at face value, is inconsistent with God's omniscience. But if all possible outcomes exist in the mind of God and have equal reality until one of them becomes concrete reality, then the conflict is resolved.

---

[786] See, for the portion of the abovementioned course dealing with fractals, Steven Strogatz, *Chaos*, Pt. II (VCR) (Chantilly, Va.: The Teaching Company, 2008), lectures 13-18.

[787] See, by way of review, the "More Openness" section of my Fluid Reality chapter, supra.

## Schrodinger's Cat

"Superposition" is a term used to describe the relationship of alternative quantum states prior to collapse of the probability wave function.[788] "Hybrid reality," a term that I used in the preceding section of this chapter, is another name for this relationship.[789] As stated in Michio Kaku's *Parallel Worlds* glossary, "wave function" refers to a mathematical description of the probability of finding a given subatomic particle at a particular location.[790] To speak of a particle's probability aspect as a wave is to describe it accurately, as English physicist Thomas Young demonstrated in 1800 with his famous two-slit experiment. Young's test of the idea that light is a wave involved his shining beams through parallel openings in a screen and observing the sequence of light and dark bands that appeared on a screen behind the slit screen.

Young's light band pattern eventually came to be recognized by the physics community as reflecting wave interference, which would have been a result of the peaks and troughs in the beams from the two openings being out of step, then in step, then out of step, and so on.[791] What is somewhat startling about the two-slit experiment is that a similar pattern is shown when only one photon (light particle) at a time is allowed to pass through the system: once again bright and dark bands are in evidence. This phenomenon (a single light particle interfering with itself), as you may have guessed, is a reflection of the superposition of quantum states, of the hybrid nature of reality.[792] In effect, two worlds are present together.[793]

As is the case with light particles, matter particles are also waves, as was shown by the Frenchman Louis de Broglie in 1923 and shortly thereafter confirmed by Americans Clinton Davisson

---

[788] See Wolfson, *Einstein's Relativity and the Quantum Revolution*, Pt. II, lecture 21.
[789] See Davies, *God and the New Physics*, 109-10.
[790] Kaku, *Parallel Worlds*, 401
[791] Davies, *God and the New Physics*, 108-10.
[792] Ibid., 109-10; see also Greene, *The Elegant Universe*, 97-103.
[793] Davies, *God and the New Physics*, 109-10.

and Leslie Germer using a version of the two-slit experiment.[794] In 1926, the final piece of the puzzle fell into place when German physicist Max Born advanced the theory that the wave function of an electron (a particle) must be interpreted in terms of probability: places where the magnitude of the wave is relatively large are places where the electron is *more likely* to be found, and places where the wave amplitude is smaller are places where the electron is *less likely* to be located.[795]

A thought experiment to show how quantum superposition might translate into events in the macroscopic world, in this case with respect to the physical well-being of one of God's furry creatures, was devised by German physicist Erwin Schrodinger (1887-1961).[796] In the hypothetical constructed by physicist Schrodinger, a cat would be placed in a closed box that contained a quantum system, in this case a radioactive atom that had a fifty percent chance of decaying every hour. The decay of the atom, if and when it occurred, would be a random event; quantum physics could predict only the probability of decay. Also in the box would be a Geiger counter, which would be able to detect decay of the particle. The Geiger counter would be connected to an apparatus that would dispense cyanide into the cat's box if and when the counter sensed decay.

In addition to being easily unsealable to allow placement or removal of the animal, the box would have a window that could be unshuttered so as to let an observer peer inside. Checking on the status of the cat after it has been in the box for an hour, done by looking through the window, would constitute observation or measurement for purposes of this quantum system. According to the Copenhagen Interpretation, a quantum system is in a superposition (hybrid reality) state until observation or measurement brings about collapse of the relevant probability wave function. Thus, until someone looks in the box, Schrodinger's cat is in a

---

[794] Greene, *The Elegant Universe*, 103-4.

[795] Ibid., 105.

[796] See Wolfson, *Einstein's Relativity and the Quantum Revolution*, Pt. II, lecture 21; also Kaku, *Parallel Worlds*, 158; and Davies, *God and the New Physics*, 114-15.

superposition of dead and alive![797] The poor feline's life status is equivocal because no one has looked in on him or her.

At this point in the discussion, someone may ask, shouldn't the Geiger counter be considered a measuring system that would have collapsed the wave function the instant it registered any decay? This would mean that the human eye in the cat box's window, checking on the animal inside, would play no role in achieving concrete reality. Prof. Richard Wolfson replies that the answer to the above question is yes only in classical physics, defined as systems not incorporating quantum theory's uncertainty principle.[798] In classical physics, the measuring system is thought of as distinct from what it's measuring.[799] In quantum physics, on the other hand, we must consider the quantum state of the measuring apparatus as well. According to the Copenhagen Interpretation, the cat's box and all the things in it are in a quantum superposition state until we look into the box.[800] The Geiger counter is therefore both showing decay and not showing decay prior to human observation of the quantum system.

The question of who or what is allowed to do the measurement that's needed to collapse the wave function is not an issue in the Hugh Everett "many worlds" scenario described earlier, since the universe simply splits at every juncture at which a choice of an alternative creates quantum states that disagree.[801] The parallel worlds thus created, physicist/writer Michio Kaku assures us,

> are not ghost worlds with an ephemeral existence; within each universe, we have the appearance of solid objects and concrete events as real and as objective as any.[802]

---

[797] Wolfson, *Einstein's Relativity and the Quantum Revolution*, Pt. II, lecture 21; see also Kaku, *Parallel Worlds*, 158; and Davies, *God and the New Physics*, 114-15.

[798] Wolfson, *Einstein's Relativity and the Quantum Revolution*, Pt. II, lecture 21; and see Kaku, *Parallel Worlds*, 384.

[799] Wolfson, *Einstein's Relativity and the Quantum Revolution*, Pt. II, lecture 21.

[800] Ibid.

[801] See Kaku, *Parallel Worlds*, 167-71.

[802] Ibid., 168.

Schrodinger's cat is thus alive in one Everett universe and dead in another, and no collapse of the wave function is necessary for the purpose of establishing a final concrete reality.[803] Schrodinger's cat is both dead and alive as soon as he or she is placed in the box, but not before that. In Everett's interpretation, as summarized by Kaku, wave functions "never collapse, they just continue to evolve, forever splitting into other wave functions."[804] What we're presented with under the "many worlds" theory, then, is the somewhat unwieldy image of a tree constantly sprouting new branches, each branch of which represents an entire universe.[805]

The differences between Everett's theory and other multiple-universe hypotheses include not only the circumstance that the number of Everett universes increases while other schemes assume a pre-existing infinity of universes.[806] Since Everett holds that the number of "you"s increases every time you make a choice that could have gone a different way, your acceptance of Christ as Lord and Savior would have the effect of creating a "you" who is going to hell.[807] Your intention in trusting Jesus for salvation would thereby be frustrated, since in choosing heaven you would be denied the ability to avoid hell. Such an arrangement would reflect at least passive cruelty on God's part in allowing the would-be believer to act self-destructively. It would essentially be Calvinism with the blame for hell shifted from God to man (cf. the "'Predestinarian' Footnote" section of the present chapter, infra). Suffice it to say, authoring or permitting such a thing would not be within the character of a merciful God (Psalm 103:17; Luke 1:50)

With "radiating reality," my personal view, there is only one universe and all possible things that might happen within that universe are contained in the mind of God. The psalmist says of the Lord that "his understanding is infinite" (Psalm 147:5), and Isaiah declares that there is "no searching of his understanding" (Isaiah 40:28). Thus, we can count on God not only to be aware

---

[803] Ibid.

[804] Ibid.

[805] Ibid.

[806] See Davies, *God and the New Physics*, 172-73.

[807] See ibid., 141-42.

of everything that occurs or could occur in His creation, but also to assure that there will not be outcomes such as the generating of alternative human selves who will suffer condemnation as a result of alter ego decisions.

There is no scientific evidence of other universes, certainly none that I know of. As Paul Davies noted in *God and the New Physics*, the many-universe theorists have conceded that the "other worlds" of their theory "can never, even in principle, be inspected."[808] Multiple universes have been seen as "a purely theoretical construct, a belief (that) must rest on faith rather than observation."[809] At best, as Prof. Davies later explained in *About Time*, these alternative worlds would be "*contenders* for reality" (emphasis Davies').[810]

Nor does the Bible in any way hint that other universes might exist. From Genesis to Revelation, "heaven" and "heavens" appear to be used interchangeably (see and compare, for example, Genesis 1:1; 1:15; 2:1; Revelation 12:12; 20:9). A particularly strong indication that there is but one universe is found in Revelation 21:1, where the apostle John states that in his vision he "saw a new heaven and a new earth: for the first heaven and the first earth were passed away;." This statement, mentioning only an old and a new universe, is conclusive as far as I'm concerned.

The "many worlds" formulation is an extension of the theory of "decoherence," under which there are splits between quantum wave functions that are contradictory to each other. "Decoherence," like "many worlds," does away with the need for observational collapse of the wave function. The basic idea, as conceived by German physicist Dieter Zeh and translated into Schrodinger's thought experiment, is that you can't separate a cat from his or her environment.[811] The cat is continually in contact with things like molecules of air, its box, and even cosmic rays that pass through the experiment area.[812] This contact, it is said, will inevitably dis-

---

[808] Ibid., 173.

[809] Ibid., 174.

[810] Davies, *About Time*, 277.

[811] Kaku, *Parallel Worlds*, 166-67.

[812] Ibid.

turb the wave function and cause it to split into the functions of a live cat and a dead cat.[813] The splitting shows that the live cat and dead cat wave functions, having fallen out of synchronization and thus lost their coherence, can no longer interact; thus, they must be permanently separated from each other.[814]

The cat in a "decoherence" scenario, keep in mind, will already be dead or alive before you open the box, rendering observation superfluous.[815] But while decoherence makes it unnecessary to consider the effect of consciousness in quantum mechanics except to the extent that human decisions precipitate quantum wave function splits before environmental factors can do the job, this still doesn't resolve the question that troubled Einstein: how does nature (what Einstein would have called God) "choose" the final state of the cat?[816]

As I've suggested above, not all theorists have been ready to exclude animate consciousness from the process by which existence is established under quantum theory. Among the scientists who have viewed mind as having a critical role as regards matter was American physicist Eugene Wigner (1902-95), author of the saying, "consciousness determines existence."[817] Wigner wrote, concerning his deliberations on this subject, that it

> was not possible to formulate the laws of quantum mechanics in a fully consistent way, without reference to the consciousness [of the observer]...the very study of the external world led to the conclusion that *the content of the consciousness is the ultimate reality* (emphasis supplied).[818]

---

[813] Ibid., 167.

[814] Ibid.

[815] Ibid.

[816] Ibid., 168.

[817] See Kaku, *Parallel Worlds*, 165.

[818] Eugene Wigner, quoted by Kaku, *Parallel Worlds*, 165, quoting Robert Crease and Charles Mann, *The Second Creation: Makers of the Revolution in Twentieth-Century Physics* (New York: Macmillan, 1986), 67.

Or, as English poet John Keats once stated, "Nothing becomes real until it is experienced."[819] Wigner's view was that it is the entry of information concerning a quantum system into the mind of an observer that collapses the quantum wave function and converts a schizophrenic, hybrid, ghost state into concrete reality.[820] But what about the reality of the observer? Doesn't someone have to have collapsed the observer's wave function in order to make the observer real? This person is sometimes called "Wigner's friend."[821] And don't we need someone to observe Wigner's friend? Aren't we falling into a "turtles all the way down" trap, heaping contingency on contingency? Or is there a noncontingent "superturtle" at the bottom of it all?[822]

A variation on Eugene Wigner's "consciousness determines existence" thesis is the "it from bit" idea that physics patriarch John Wheeler was toying with at the time of his death in 2008.[823] Wheeler began with the assumption that information, defined as data, is the basis of all things.[824] When we look at the moon, a galaxy, or an atom, he said, what we're seeing is something the essence of which is contained in the information stored within that object.[825] According to Wheeler's theory, necessary information ("bit") sprang into existence at the beginning of the universe ("it") when the information underlying the cosmos was observed.[826] The observer was the universe itself, Wheeler explained, in the sense that the physical world at some stage generated observers who were able to look back toward the origin of the cosmos and by their acts of observation make everything concrete.[827]

---

[819] John Keats, *Letter to George and Georgiana Keats [March 19, 1819]*, quoted in *Bartlett's Familiar Quotations*, 440.18.

[820] Davies, *God and the New Physics*, 115.

[821] Kaku, *Parallel Worlds*, 165.

[822] See in this regard Davies, *The Mind of God*, 223-24.

[823] See Kaku, *Parallel Worlds*, 171-72.

[824] Ibid., 171.

[825] Ibid., 171-72.

[826] Ibid., 172.

[827] Ibid.; and see Davies, *God and the New Physics*, 110-11; also Davies, *The Mind of God*, 224-26.

Retroactive creation of reality had already been recognized as a possibility in quantum physics, it should be noted, in that after collapse of a probability wave function the surviving part of a previously hybrid reality is seen as having been *the* reality all along.[828] The picture sketched by Wheeler was thus one of a reality created by a consciousness that is dependent on that reality.[829] Michio Kaku remarks that under the line of reasoning implied in the above, "the universe does have a point: *to produce sentient creatures like us who can observe it so that it exists*"(emphasis Kaku's).[830] This is what's been termed a closed loop, in which everything is supported by something else within the loop rather than by anything or anyone external to the loop.[831] There's no explanation of how the materials and principles necessary to create the loop came into being, other than that they're quantum effects.

Wheeler's theory of an observation-based universe, as summarized above, goes farther than merely implying that it is our presence that makes the universe possible;[832] it speaks negatively regarding the necessity of God for any purpose. If Wheeler's thesis is indeed seen as ipso facto exclusion of God from the picture, then it seems to me that what we're left with amounts to solipsism, the belief that your self is all that can be known by you to exist, that the only reality you can rely on is what's in your head.[833] Wheeler himself seems to have on occasion leaned toward this interpretation of his position, since he once stated:

> I confess that sometimes I do take 100 percent seriously the idea that the world is a figment of the imagination and, other times, that the world does exist out there independent of us.[834]

---

[828] See Davies, *God and the New Physics*, 39, 110-11.

[829] See ibid., 110-11.

[830] Kaku, *Parallel Worlds*, 351.

[831] See Davies, *The Mind of God*, 44, 224.

[832] See Kaku, *Parallel Worlds*, 172.

[833] See Rohmann, *Ideas*, 370-71.

[834] John Wheeler, quoted in Kaku, *Parallel Worlds*, at 171, quoting Jeremy Bernstein, *Quantum Profiles* (Princeton, N.J.: Princeton University Press, 1991), 132.

I find it intriguing that Prof. Wheeler, said by Richard Feynman to be the last of the physics "monster minds,"[835] should exhibit such indecisiveness regarding the location of reality. Eugene Wigner, also, would seem to be a solipsist, by reason of his statements that "consciousness determines existence" and that "the content of the consciousness is the ultimate reality."[836]

I'm trusting that Mssrs. Wigner and Wheeler would eventually have come to what I believe is the correct conclusion on the measurement question: observation can establish concrete reality only if it can be said at some point that "(t)he buck stops here."[837] There must be a true end-of-the-line observer/measurement taker, someone whose existence is not subject to the vagary of infinitely regressing layers of observation. It's difficult for me to see how anyone but God would be able to fill this role. The observer cannot be the universe itself, since this would allow matter to be what only mind can be, rank materialism in my opinion. "Decoherence" and "many worlds" get around the need for animate participation, as previously noted, but—like Wigner and his friends and Wheeler with his bits—are unable to explain why one reality rather than another is actualized in the universe that we inhabit. While unpredictability may be "the quantum way," the prospect of dealing with quantum indeterminism on a macroscopic level is hardly a reassuring one, if results matter.

Much of the above discussion, it hardly needs saying, tempts one to return to the original probability wave function collapse explanation and, yes, think in terms of the primacy of the human mind. The mind has "a special status" as regards quantum physics, Paul Davies has observed.[838] Indeed, Davies says, the central paradox of the quantum theory is "the unique role played by the mind in determining reality."[839]

---

[835] Kaku, *Parallel Worlds*, 150.

[836] See supra.

[837] A reference to the sign on President Harry Truman's desk, "The buck stops here." *Bartlett's Familiar Quotations*, 705.12.

[838] Davies, *God and the New Physics*, 210-11.

[839] Ibid., 141.

At first glance, it would seem a salutary development that science is now able to venture away from the mechanistic determinism of earlier centuries, so that the universe in our era "seems to be nearer to a great thought than to a great machine," in the words of British astrophysicist James Jeans (1877-1946).[840] But, it must be asked again, what ultimately is the foundation of the observing consciousness? Is our ability to observe or measure attributable to the natural God imagined by Paul Davies, who presides over a cosmos that could not be otherwise?[841] In that case, inevitability rather than conscious intention would be the dominant principle, quantum uncertainty aside, and no one other than Nature—necessity under the laws of science—would be in control of anything. Or is it solipsistic man who is in charge, ruling (blindly, it may be said) over a world that exists only within his head? What place is there for the God of the Bible or a super-intelligent organizing principle, we ask, where the only things that have reality are the thoughts of men? Absent the assistance of the Word, we wrestle with such questions the whole night through, as Jacob is said to have contended with an angel, without achieving an outcome (cf. Genesis 32:24-30).

*Speaking to God*

Having grasped intuitively, and also by analogy to happenings in the subatomic world, that there are alternative realities the existence of which is one factor that releases us from determinism, we're left with what I believe is persuasive metaphorical support for at least two of the corollaries of human free will. The first of these is that prayer makes a difference.

"Prayer," Webster's Unabridged says, involves petitioning God and having communion (discourse) with Him.[842] Webster's also allows prayerful acts to be directed toward "an object of

---

[840] Quoted at Kaku, *Parallel Worlds*, 350, quoting Gary Kowalski, *Science and the Search for God* (New York: Lantern Books, 2003), 19.

[841] Davies, *God and the New Physics*, 223; and see the discussion in the "Science and Religion Generally" section of my Paradigm Paralysis chapter, supra.

[842] *Random House Webster's*, 1519.

worship,"[843] which I understand as referring to material icons, but that's not the meaning attached to the word "prayer" by Bible-believing Christians, nor should it be. Prayer and worship directed to idols was an issue at Ephesus in the first century AD, when the silversmith Demetrius caused a great stir among his fellow craftsmen by publicizing the threat to their livelihood that was posed by the apostle Paul's preaching that the tin household deities that they daily pounded out were not gods at all (Acts 19:23-41). Some five hundred years earlier, the scribe Ezra, likewise stating the obvious, had said regarding the Israelites and their graven images, "For all the gods of the people are idols, but the Lord made the heavens." 1 Chronicles 16:26. Prayer must be directed toward the Creator of the universe, not at any of His creatures or things made by them.

Prayer is not a one-way street, as some seem to think it is. There are innumerable instances in the Old Testament of God speaking to believers, and much monastic prayer practice over the centuries since the advent of Christianity has consisted of the discipline of *hesychasm*, approaching God in stillness or from a contemplative stance rather than proactively asking Him for things.[844] Indeed, so-called contemplative prayer and the discipline of silence are at the heart of Emergent Church mysticism.[845]

Quakers, who traditionally gather in meetinghouses and sit silently until moved by the Spirit to speak,[846] are another example of believers hearkening to God rather than hounding Him. That prayer may be defined to include receiving communication from the Almighty without directing anything verbally to Him was something I had known even as a teenager, since during one of our family's trips to Maryland, my sister Margery took us to the old Quaker meeting house at Easton, a plain wooden building with inward-facing benches all around its walls, where a docent described for us the Quaker way of contacting God.

The voice of the Lord must be carefully listened for, unquestionably. "Be still, and know that I am God," the psalmist quotes

---

[843] Ibid.

[844] See, in the latter regard, Johnson, *Mystical Tradition*, lecture 17.

[845] See Gilley, *Out of Formation*, 41-62, 81-100.

[846] Rohmann, *Ideas*, 321.

the Lord as saying (Psalm 46:10). In the book of 1 Kings, the prophet Elijah, being pursued by the wicked queen Jezebel, is told to go and stand before the Lord on Mount Horeb, whereupon the Lord passes by and there are in succession wind, earthquake, and fire, but the Lord is not in any of these (1 Kings 19:11-12). Instead, He is in the "still small voice" that follows the fire (19:12), which tells Elijah what he must do and what will happen next (19:13-18).

Whichever way the communication runs, how is it able to happen? God inhabits eternity (Isaiah 57:15) and thus, for the present at least, is in a milieu entirely different from mortal man's habitation. Man is a resident of time, where evening follows morning and morning evening, where effect follows cause, and where events sally forth in an inexorable procession. "Eternity," on the other hand, has been defined as "(o)utside of time."[847] This simple definition, framed by a professor of philosophy,[848] could be taken to mean not only that there are no days in eternity, but also that eternity lacks sequentiality, since one thing leading to another is a feature of time. Time is likewise treated as something totally separate in at least one Webster's dictionary definition of eternity ("the *timeless* state into which the soul passes at a person's death [emphasis supplied].").[849]

Nontheological Webster definitions of eternity ("infinite time;" "an endless or seemingly endless period of time")[850] make eternity and time *coextensive*—i.e., equal in duration.[851] This has the effect of making the biblical "for ever" mean, as *The Scofield Study Bible* puts it, "eternal in the sense of unending,"[852] and seems to leave in place most of the deficits associated with temporality. A God who is in time would presumably be subject to change, most importantly.[853]

---

[847] Nichols, *Free Will and Determinism*, glossary, 120.

[848] Shaun Nichols teaches at the University of Arizona, where he has a joint appointment in Philosophy and Cognitive Science. As indicated above, he is also a lecturer for The Teaching Company.

[849] *Random House Webster's*, 665.

[850] Ibid.

[851] See Random House Webster's definitions of "coextend" and "coextensive," at 399.

[852] *Scofield Study Bible* n. to Rev. 20:14.

[853] Davies, *God and the NewPhysics*, 133.

But if the triune God is the cause of all things, as the Bible seems clearly to say He is (Genesis 1:1-2; John 1:1-3), then it makes no sense to think of Him as changeable.[854] What is your starting point for everything created, if the First Cause is typically in a state of flux and thus not secure as far as being is concerned?

Moreover, if time can come to an end, as Einstein's theory of relativity suggests by its linkage of time and matter, what is there that's left at that point for a time-encumbered eternity to attach itself to? God's habitation, eternity (Isaiah 57:15), if defined as endless time, must also disappear. Does this not put the very existence of the Ruler of the Universe, who would then be homeless, in jeopardy?[855] How can God be omnipotent if He can be overcome in this way by operation of the laws of physics?

From the above, it should be clear that getting eternity tangled up with time brings with it some serious hazards. But entanglement of some description seems necessary if we are to have a God to whom we can pray and from whom we can expect communication in return, in a sequential (time-implicating) relationship.

Perhaps C. S. Lewis can help us through our dilemma. In his apologetic *Mere Christianity*, Lewis addresses the popular notion that it is not possible that God could attend to millions of human beings who are all praying to Him at the same moment.[856] If men's applications came one by one and God had endless time to give to them all, there would be no problem, Lewis notes; however, the life regarding which we seek God's aid comes to us moment by fleeting moment, which, Lewis says, is simply "what Time is like."[857] This potential difficulty is easily resolved, as far as Lewis is concerned: God "(a)lmost certainly" is not in time, he declares; therefore, God's life does not consist of moments following one another.[858] Instead, God lives in an eternal present, a realm such as that described by Augustine, which allows Him to see at once all moments—past, present, and future—but react to individual

---

[854] Ibid.

[855] See ibid.

[856] See generally Lewis, *Mere Christianity*, 166-71.

[857] Ibid., 166-67.

[858] Ibid., 167.

concerns unhindered by time constraints. Thus, Lewis explains, God "has all eternity in which to respond to the split second of prayer put up by a pilot as his plane crashes in flames."[859] God has the time because time does not have Him.

Joining in Lewis's appeal to Augustine's "eternal present" gets us only part of the way to establishing the efficacy of prayer, to showing that "(t)he effectual fervent prayer of a righteous man availeth much" (James 5:16).[860] We must persuade the skeptic not only that God has the time (or, rather, the eternity) to answer all prayers, but also that no particular answer or lack thereof to any specified prayer is a foregone conclusion. Happily, our burden of persuasion on the latter point is eased by the manner in which God has chosen to construct the subatomic world, which suggests metaphorically that different outcomes may flow from any given prayer to the Almighty. As noted above, the Copenhagen Interpretation of quantum physics declares that all possibilities for the motion of a subatomic particle remain open until there is an attempt by an observer to establish definitely the particle's position or velocity, at which time the particle's state will become something other than hybrid, something other than a collection of probabilities.[861]

Richard Feynman's "sum over histories (paths)" method of calculating likely particle position, previously discussed in this chapter, is even more emphatic in its advocacy of multiple-reality thinking, since it posits world lines for particles that go back in time and declares as an article of faith that all world lines exist with *equal* reality until the particle's wave function is collapsed. The wave-collapse part of the process, which occurs when the probability wave function is suddenly no longer needed, may be treated as a metaphor for God's acting on the believer's request. God's making a decision with respect to a prayer in the macroscopic

---

[859] Ibid. As Kathleen Norris notes in her Foreword, *Mere Christianity* was a collection of radio talks that C. S. Lewis gave during World War II, when the battle for air supremacy over Britain was raging. See *Mere Christianity*, xvii, xviii. It is not surprising, then, that Lewis used this as an example.

[860] This principle comes with the additional qualifier that the supplicant must not "ask amiss" (James 4:3)—i.e., must not pray for something that Jesus would not have asked the Father for. See Morris, *Study Bible*, n. to James 4:3.

[861] Kaku, *Parallel Worlds*, 151-52.

world would accompany His assessment (observation) of the situation prayed about. Prior to that, the doors to all outcomes would remain open for the supplicant.

"Prior" is meaningful only in speaking from the petitioner's point of view, however. The term is meaningless in speaking of God's knowledge of the action taken by Him on the proffered prayer, since God knows His response as a matter of present knowledge rather than foreknowledge. Because God doesn't have foreknowledge, He can't be said to be acting deterministically, to be predestining results for His creatures. He sees Himself doing what He had the option but not a plan in advance to do. Moreover, He doesn't provide His assistance without there being a request for it. "(Y)e have not, because ye ask not," early church leader James tells fellow Christians in his general epistle (James 4:2). Obviously, asking God for something would be meaningless if God knows *in advance* whether the asker will receive the thing requested. But if God knows His response as knowledge obtained by Him in His eternal present and as a choice that was possible along with others at the time He was prayed to, then the above verse, "ye have not-because ye ask not," is something other than a smoke screen.

The intermeshing of time and eternity, it should be emphasized, is a core principle in God's being able to give ear to all persons praying to Him. The same principle is involved in God's ability to see alternative realities, which gets Him past the limitation that having foreknowledge rather than present knowledge would impose on His ability to respond to prayer. Also emblematic of God's position astride the boundary between time and eternity are the incarnation of God the Son and the Son's atoning sacrifice for sin before the world ever came into being.[862]

Regarding the incarnation, the apostle John's gospel, referring to Christ as "the Word," announces at 1:1, "In the beginning was the Word, and the Word was with God, and the Word was God." That Word, that perfect Godness, John says,"was made flesh, and

---

[862] In referring to the physical universe, the New Testament writers used Greek words that have been translated "world." See, e.g., Ephesians 1:4; 2 Timothy 1:9; Titus 1:2; 1 Peter 1:20; with *The New Strong's Exhaustive Concordance of the Bible* (Nashville: Thomas Nelson Publishers, 1984), at 1236, 1237; and Greek Dictionary of New Testament therein, at 9, 43.

dwelt among us" (John 1:14). The beloved apostle emphasizes the unity of the Father and the Son by quoting Jesus in this regard: "I and my Father are one." John 10:30. The life that the Father and the Son share, John indicates, is an eternal life (1 John 1:2), and yet it is a life that was manifested while the Son was on earth (1 John 1:1-3). It is in this way that time and eternity are joined in the person and life of Jesus Christ.

Elsewhere, John refers to Jesus as "the Lamb slain from the foundation of the world" (Revelation 13:8). The apostle Peter writes that Christ's sacrifice was "foreordained before the foundation of the world" (1 Peter 1:20). God's salvation plan necessarily had to be in place from the very beginning because, as Paul tells Titus, eternal life for believers was "promised before the world began" (Titus 1:2). Paul in the same vein writes to Timothy that God's grace was "given us in Christ Jesus before the world began" (2 Timothy 1:9). What we see, then, is an event (Christ's atoning sacrifice) that takes place at a particular point in history (33 AD) even though in practicality it has already happened—God's gracious gift was actually given before the world began, we're told—and therefore benefits all believers for all time, both after and before the year 33. Here, as elsewhere in Scripture, any distinction between time and eternity is blurred. Communication traversing the boundary between the two thus seems fully and gloriously possible!

The reader will note that I have not referred to any of the many-universes theories as either a metaphor or direct proof of anything having to do with prayer. To be a road sign pointing to an interpretation of the Gospel, a theory that is offered as scientific must be what it purports to be: science. As I pointed out earlier, "many worlds" and its kindred many-universes theories have no empirical support and are in fact untestable.[863] They're pure speculation, fueled by an obvious need to avoid a theological explanation for a universe which, viewed from a scientific perspective, would appear to have come into being against some very long odds. My own personal, arguably mystical insight regarding the radiating nature of reality, as I indicate, is supported by Scripture and by analogies found in the teachings of the Copenhagen school of

---

[863] See the "Schrodinger's Cat" section of this chapter; also, Rohmann, *Ideas*, 355.

quantum physics. I offer my faith as belief consistent with science; Hugh Everett and his "many worlds" coterie offer nonscience as science.

Worth thinking about is not only how prayer is able to work, but what it consists of. It has been said that we pray to God through our actions, an idea discussed at some length by Irish divinity professor John Oulton in Volume 2 of his and Prof. Henry Chadwick's *Alexandrian Christianity*.[864] Commenting on the "mystical side" of early church theologian Origen (185-254), Oulton interprets the apostle Paul's command at 1 Thessalonians 5: 17 to "pray without ceasing" as including—i.e., counting as prayer—the performance of good deeds and the fulfilling of commandments.[865] When words and actions coincide in this way, as they do for Origen, we can correctly say that the whole life of the saint is "one unbroken prayer," Oulton concludes.[866] It then no longer makes sense to think in terms of balancing the devotional and the practical in one's life, since the two are so interrelated as to form an integrated whole.[867] Viewing one's orison in this way would seem to render a mystic or would-be mystic immune to the familiar charge that he or she is "too heavenly-minded to be any earthly good."

Equally as important as the underlying philosophical structure and the how and when of prayer is the question of intensity. As an immanent God whose glory fills the earth (Isaiah 6:3), our Lord exhibits a revealed will that roars through and throughout our habitation like a rushing passenger train. We need to match our will, our spiritual momentum, to His if we are to be able to board this rapidly moving conveyance. This principle is consistent with the biblical view of the Word of God as an active factor, "quick, and powerful, and sharper than any twoedged sword, piercing even to the dividing asunder of soul and spirit, and of the joints and marrow, and a discerner of the thoughts and intents of the heart" (Hebrews 4: 12). A casual attitude toward prayer will not get the job done,

---

[864] John Ernest Leonard Oulton and Henry Chadwick, transls. and eds., *Alexandrian Christianity*, Vol. 2 (Philadelphia: Westminster Press, 1954).

[865] Ibid., 208-9.

[866] Ibid., 209.

[867] Ibid.

therefore, any more than anything less than God the Son's sacrifice would have been sufficient to purchase our redemption (see 2 Corinthians 5:21; also, 1 Peter 1:18-19).

*Prophecy*

So far in this chapter, we've been focused on scientific theories pointing to spiritual reality analogously. In the position taken by Cambridge University professor Mary Hesse, as noted above, scientific theories are themselves metaphors.[868] But even if we accept this idea, we should still also be able to view the hypotheses of science as indirect proof of the truth of Scripture. The usefulness of science as metaphor is not diminished by the status of science as metaphor, as far as I can see. Accordingly, we'll now approach prophecy using the same general methodology as we applied with respect to prayer and Schrodinger's cat. Thus, our inquiry in the next few pages will be aimed at discovering ways, if there be any, in which the tenets of quantum physics *by analogy* support the co-existence of fulfilled prophecy and personal freedom.

The previously-stated physical principle having the most relevance as far as prophecy is concerned, it seems to me, is that part of quantum theory that allows an observer to retroactively create reality.[869] Upon collapse of the probability wave function — probability waves abruptly ceasing to exist because of observation or attempted measurement of the velocity or position of a subatomic particle — you go from a hybrid reality (A, B, C, etc., are all true) to there being a single, concrete reality (let's say A). That single reality, now that it's been validated as *the* reality, is in effect recognized as having been in place all along, as we said earlier. It was *never anything but* the whole story, quantum theory says. It has to be given this kind of recognition in order to serve as the foundation for the next step in any progression of which it's a part.

The retroactivity feature described above is time-based — you're in a hybrid reality but you're still moving linearly (moment followed by moment) — thus, we're not worrying at this point in

---

[868] See the "Mysticism and Metaphors" section of my Region of Awe chapter, *supra*.

[869] See Davies, *God and the New Physics*, 39, 110-11.

our exercise about how we factor in the eternality of God, even though this is critical as far as questions such as heaven or hell are concerned. Thinking metaphorically and not taking God's timelessness into account, let us assume that the macro-scale prediction made by a Bible prophet describes a single concrete reality such as in the quantum world of particles would be a result of a wave function having been collapsed as part of a sequence of events in time. "Phantom" reality would not be a feature of our prophecy. It would have vanished at the moment of particle observation or its macroscopic analogue, which would have reduced the possibilities to the one already retroactively established.

Let's see if we can say the above another way: Macro-world prophecy that would parallel a quantum observer's knowledge of a particle's state after collapse of the wave function is not possible until the reality focused on has become nonhybrid. Thus, the above scenario, without additional or different assumptions, would not allow any outcome that was not arrived at in advance of the relevant prophecy. Concretized reality would dictate prophecy because it in effect comes before prophecy. We would call this determinism, since it involves the past (a reality retroactively concretized by observation with respect to a particular variable) dictating the future as described in a prophecy. The outcome—John Jones getting saved or not getting saved, for example—would never vary from the prophecy, and our subject citizen of the universe, Mr. Jones, would have had no say regarding this outcome.

But what if we purged our minds of the notion that accurate prophecy can *only* happen in a sequence such as I've just described, consisting of a prediction attaching to a retroactively-created reality? What if, instead, we saw prophecy and its fulfillment as happening in the context of God's eternity? What if, also, we plugged in the phantom realities/alternative world lines assumptions of the Copenhagen Interpretation of quantum physics and Richard Feynman's "sum over paths" formula? What if, finally, we compared the above presuppositions to what Scripture allows? Applying the above method of approach, we might arrive at a line of reasoning that looks something like the following:

(1) God sees alternative realities. The Copenhagen Interpretation and Feynman's "sum over paths" point to this on a micro- scale. The micro- can affect the macro-. Accordingly, as an example, Schrodinger's cat can be both dead and alive until someone looks in on the poor creature.

(2) The alternative realities that God sees are not limited in any way. Scripture states that "with God all things all things are possible" (Matthew 19:26; Mark 10:27). Thus, nothing is beyond His power (Genesis 18:14; Revelation 19:6).

(3) With regard to prophecies, alternative realities therefore include both outcomes and fulfilled prophecies concerning outcomes.

(4) God inhabits eternity (Isaiah 57:15).

(5) God's position in eternity means that considerations associated with the passage of time (what comes before what, in particular) are not controlling as far as prophecy issuance and fulfillment are concerned.

(6) God is the ultimate observer. His eyes "run to and fro throughout the whole earth" (2 Chronicles 16:9; see also Zechariah 4:10). They are "in every place" (Proverbs 15:3). His power as an observer is not limited to microscopic particles. He is, after all, "the Lord God omnipotent" (Revelation 19:6). Thus, He can observe anything in the universe into concrete reality, in effect substituting His eyes for the tongue that He used to speak things into existence whenever He said "(l)et there be" during the six days of creation (see Genesis 1:3, 6, 14).

(7) Since God is not limited by time (normal chronology), scale (micro- or macro-), or number (of alternative realities, in this instance), He can make anything (i.e., any event) true, regardless of when it would have happened if man were the

concretizing observer. Thus, we can imagine a scenario in which God, from His position outside of time, sees a freely-chosen outcome (voluntary acceptance by John Jones of His gracious offer of salvation, for example) and in effect journeys back in time, like one of Richard Feynman's particles, to put the outcome in the mouth of the prophet.

Our God-as-observer cases only resemble the man-as-observer cases up to a point, I caution. In the God- cases, as in the man-cases, outcome matches prophecy. The difference is that in the God-as-observer cases the prophecies potentially affecting any individual are multiple. If each of them is true, then our cosmic citizen has freedom of choice re the prophesied outcome (for example, saved or not saved, health or lung cancer from smoking, slenderness or obesity). Whatever the outcome that freely occurs, it happens because God conforms the prophecy to the event, not the event to the prophecy.

*Quantum Indeterminism*

The alternative pre-collapse realities referred to above exist in a world, the quantum world, where indeterminacy is the rule. *Determinism*, as the word is generally used, is something distinct from *predestination*, which is the idea that God has foreordained everything, including world history and human destiny.[870] The determinist view holds that the course of events is solely a result of the operation of physical laws in a universe that is seen as self-perpetuating and driven by an endless sequence of cause and effect.[871] As noted earlier, the determinist position was articulated in the nineteenth century by French astronomer and mathematician Pierre Laplace, who speculated that if one had knowledge of the status of each of the universe's elementary particles, it would be possible to predict the entire future of the cosmos and all human behavior.[872]

---

[870] Rohmann, *Ideas*, 100, 313.

[871] Ibid.

[872] See the "Determinism" section of my Essence chapter, supra; also, Rohmann, *Ideas*, 100; and Hawking, *A Brief History*, 55.

Although it had held sway since at least the time of Isaac Newton, determinism was discredited in the twentieth century due to the advent of scientific theories one of which was quantum mechanics.[873] Werner Heisenberg's uncertainty principle, which is the core of the quantum theory, warned that it was not possible to know precisely both the position and momentum of a subatomic particle; thus, gone was the perfect predictability that had been a central theme of classical mechanics.[874] Laplacians could no longer hope, even in their wildest dreams, to know what every particle in the universe was doing position- and velocity-wise, and therefore they no longer had any prospect, even in principle, of establishing exactly what the future held in store. The most that could be hoped for, in the brave new world brought into being by the quantum revolution, was to know the respective probabilities of various particle behaviors.[875] Since subatomic particles were found to appear and disappear unpredictably, even the assumption that every event had a cause was called into question.[876]

The suggestion that events can be causeless raises as a topic of debate the role of God, who has long been thought of as the "first cause" of the universe and its inhabitants, based on the idea that while every event has a cause, there cannot be an infinite chain of causation.[877] The reasoning in this regard is similar to that which is applied with respect to quantum measurement or observation: the buck has to stop somewhere.[878] But if there can be something that lacks a cause, doesn't this jeopardize the idea of God?

The answer to the above question is no for at least two reasons. The first of these reasons is that we cannot rule out *hidden*

---

[873] Rohmann, *Ideas*, 100, 327-28.

[874] Ibid., 327-28, 412; and see the "Big Bomb" section of my Paradigm Paralysis chapter, supra; also Hawking, *A Brief History*, 57-58.

[875] See Hawking, *A Brief History*, 57-58; also, Wolfson, *Einstein's Relativity and the Quantum Revolution*, Pt. II, lecture 19; and Davies, *God and the New Physics*, 108.

[876] Davies, *God and the New Physics*, 34-35, 102.

[877] See ibid, 33-43; also, the "WhereThought Experiments Lead" section of my Proving God chapter.

[878] The assumption apparently underlying the sign on President Harry Truman's desk, "The buck stops here." *Bartlett's Familiar Quotations*, 705.12.

*variables*, physical properties unknown to man at present but nonetheless able to act as causes of effects. Consistently with Dietrich Bonhoeffer's prediction that human knowledge in general would steadily increase and thus reduce the importance of the "God of the gaps" as a default explanation,[879] we should anticipate developments in physics that will have the effect of reducing the area of the unknown. Bonhoeffer made his prediction not only with respect to human problems such as death, suffering, and guilt, but also with respect to the relationship between scientific knowledge and God.[880] Thus, we can't say that there are effects in physics that will be forever causeless. Thus, we can't rule out a "first cause" God.

A second consideration, the *anthropic principle*, was mentioned in another context earlier.[881] Under the "strong" version of AP, the universe's properties must be such as to at some point permit intelligence and life. If there were not multiple universes or cosmic regions that increased the likelihood of sentient consciousness occurring *somewhere*, then intelligent design was required in order to fine-tune nature's constants so as to allow intelligent life to exist.[882] Under the "weak" version of AP, the universe became tuned for intelligence and life by a series of happy accidents, but no one is suggesting that there was any intelligent design. As the cheerleaders used to sing on the bus, "We're here because we're here," and there's nothing more to it than that. In either case, the universe and our being able to live in it suggest a confluence of physical factors that the odds would have been overwhelmingly against.[883] Call it what you will, intelligent design or an anthropic principle in nature, what one sees in the overall structure of the

---

[879] Bonhoeffer, *Letters and Papers*, 311; and see the "Induction and Deduction" section of my Proving God chapter, supra.

[880] Ibid.

[881] See the "Whence Cometh Life" and "Dogs and Computers" sections of my Essence chapter, and authorities cited therein.

[882] Hawking, *A Brief History*, 130; Kaku, *Parallel Worlds*, 242; Davies, *God and the New Physics*, 188-89; Rohmann, *Ideas*, 21, 163-64.

[883] See, in this regard, Davies, *God and the New Physics*, 186-89; also, generally, Michael J. Behe, *Darwin's Black Box: The Biochemical Challenge to Evolution* (New York: Simon & Schuster, 1996).

universe bears little resemblance to a quantum arena in which things "just happen."

Having established that quantum acausality does not require a causeless universe, we can seek the true message presented by the uncertainty principle. In this regard, Michio Kaku may have said it all. Comparing the classical theory of relativity to the wide-open world of quantum physics, Prof. Kaku says of the quantum theory that it

> gives us a picture of the universe much stranger than the one given to us by Einstein. In relativity, the stage of life on which we perform may be made of rubber, with the actors traveling in curved paths as they move across the set. As in Newton's world, the actors in Einstein's world parrot their lines from a script that was written beforehand. But in a quantum play, the actors suddenly throw away the script and act on their own. The puppets cut their strings. *Free will has been established* (emphasis supplied).[884]

Thus, Kaku sees not so much the absence of God as human personal freedom in quantum theory's disconnect between cause and effect at the subatomic level.

A "free will" understanding of quantum acausality is consistent with the Copenhagen/Feynman interpretation of quantum physics, which postulates that alternative realities exist for any particle until its probability wave function is collapsed. Clearly, the Copenhagen/Feynman interpretation by analogy supports a faith system in which, as far as a person's eternal destiny is concerned, all options are open during his or her earthly life. The individual in question may accept God's gracious offer of salvation by believing the Gospel and coming to the Lord in repentance (John 3:16; Acts 3:19; Romans 3:21-22; Galatians 3:26), or may decline to do this, to his or her eternal detriment (Matthew 23:37; 2 Thessalonians 1:7-8; Hebrews 2:3; Revelation 21:8). A believer, by faithful service, may

---

[884] Kaku, *Parallel Worlds*, 149.

earn rewards that will be given by God in the next life, or may fail to amass much in the way of rewards even though salvation has been secured (Matthew 5:12; 6:1, 4; 1 Corinthians 3:8; 2 Timothy 4:14). Contrary to Calvinist teaching, nothing is foreordained.[885]

The alternative is a cosmology which holds that there is no "first cause" of the universe because there is a lack of causes at the subatomic level. This in effect recognizes quantum indeterminism as an excuse for a shoulder shrug regarding the origin and purpose of the cosmos. Essentially, you're denying the existence of the God of the Universe because of something that's happening in particle physics. This gives the quantum acausality metaphor more respect than it deserves, in my opinion.

*A "Predestinarian" Footnote*

Many pages ago, in a preliminary exploration of the question of God and change, I introduced the Reformed doctrine of predestination, an aspect of Calvinism.[886] Much later, I attempted to give a view of doctrinal predestination from the perspective of relativity theory.[887] Most recently, as I suggest above, the seeming indeterminism of quantum mechanics has given me further occasion to search for and examine clues on the question of whether we inhabitants of the macro- world have free will.[888] In each instance, I've sought to create in the reader an understanding that a predestinarian view is required neither by Scripture nor by careful observation of nature, the latter involving the theories of modern physics referred to above. I nonetheless believe that more needs to be said on this subject. Calvinism is supposedly on the rise among evangelical Christians,[889] a development that should be of concern

---

[885] See the discussion in the section that follows.

[886] See the "Foreknowledge and Freedom" section within my Fluid Reality chapter, supra.

[887] See the "Block Time and Eternity" section within my Relativity chapter, supra.

[888] See the immediately preceding section herein.

[889] See David W. Cloud, "Calvinism on the March Among Evangelicals," 8/22/13 Calvinism, cited in *http://www.wayoflife.org/index_files/calvinism_on_the_march. html*.

to all believers who see their God as a compassionate, merciful, and loving deity.[890] Therefore, we should not give short shrift to this matter.

The Calvinist position is stated in the form of five points. Those five points came into existence as a response to five articles of faith drawn up by the followers of Dutch seminary professor Jacobus Arminius a year after his death in 1609, which challenged the teachings of Protestant Reformation leader John Calvin (1509-64).[891] The Arminian "Remonstrance" (i.e., protest) was examined in the course of a synod held in 1618 and 1619 at the Netherlands city of Dort (Dordrecht), which produced a lengthy writing defending Calvin's position. Five chapters of this document became known as the five points of Calvinism.[892] Thus far, mention has been made herein of only two of the five points, "total depravity" and "unconditional election," the latter by implication. The first letters of the five points as they're normally stated make up the initialism "TULIP." The complete five "TULIP" points, with Arminian responses, are as follows:[893]

*Total depravity.* Man is spiritually helpless, unable on his own to effect his salvation by believing the Gospel, according to Calvinist teaching. Adam's fall, abetted by Eve, caused a fatal condition in man, sin, that can only be overcome through complete regeneration by the Holy Spirit, in effect producing an entirely new person, which is necessary before anyone can believe and be saved. Therefore, faith is not something man contributes toward his salvation. It is God's gift to the sinner, not the sinner's gift to God.

---

[890] Regarding the abovementioned attributes of God, see 1Chronicles 16:34; Psalm 23:4; 36:5; 78:38; Isaiah 40:1; Jeremiah 31:3; Lamentations 3:22; Micah 7:19; Mark 5:19; Luke 1:50; Romans 5:8; 2 Corinthians 1:3; 2 Peter 3:9; and 1 John 3:1; 4: 7, 10. See also, generally, Dave Hunt, *What Love Is This? Calvinism's Misrepresentation of God*, 4th Ed. (Bend, Ore.: The Berean Call, 2013).

[891] Steele and Thomas, *The Five Points*, 13-15.

[892] Ibid., 14-15.

[893] Calvinist and Arminian arguments, identified herein as such, are as summarized in Steele and Thomas, *The Five Points*. My apologies for the duplication of "depravity" material already set forth in the "Foreknowledge and Freedom" section of my Fluid Reality chapter. It seemed necessary given the large number of intervening pages.

The Arminian replies that there is no scripture that says that man is *unable* to respond to God's call. The prophet Jeremiah's pronouncement that the human heart is "deceitful above all things and desperately wicked"[894] does not equate to spiritual helplessness. Human nature was seriously affected by the fall, certainly, but man still retains enough of "the image of God," in which he was originally made (Genesis 1:27), to be able to choose good over evil. The Holy Spirit, the third person of the Trinity, merely *helps* man to believe. The Spirit enabling but not replacing man spiritually is the true import of Paul's statement to the Ephesians that they "by grace are saved through faith; and that not of (themselves): it is the gift of God."[895] Thus, Paul can write to the Christians in Galatia, "ye are all the children of God by faith in Christ Jesus."[896]

*Unconditional Election.* Calvinists say that before the foundation of the world, God chose certain individuals to be saved. His "election" of these persons was not based on anything they might do, nor was it based on foreknowledge that they of their own free will would repent and believe the Gospel. Faith is required for salvation, Calvinists admit, but it is a gift that comes from God; if it were something exercised or generated by man, it would be a work. Those whom God sovereignly elects He brings to a willing acceptance of Christ as savior—i.e., to faith.

Arminians hold that the "gift of God" referred to in Ephesians 2:8 is not faith, but salvation, which must be claimed through freely exercised faith on the part of the sinner. Faith and works are distinguished in both the Old Testament and the New Testament;[897] thus, faith is a condition of salvation but not a work. Man's exercise of faith is done freely because God does not have foreknowledge of man's decision, which would preclude any outcome but the one foreknown. When Peter addresses his fellow believers as "(e)lect according to the foreknowledge of God" (1 Peter 1:2), it is a reflection not that God has foreordained the salvation of "the elect,"

---

[894] Jeremiah 17:9.

[895] Ephesians 2:8.

[896] Galatians 3:26.

[897] See Genesis 15:6 and *Henry Morris Study Bible* n. thereto; also Romans 4:2-5 and nn. to Rom. vv. 4:2 and 4:3; also James 2:17-18.

but rather that God from His position in a timeless eternity sees a positive response by believers to His call. Those answering yes to the call by coming to repentance (remorse for sin) and trusting in Christ's work on the cross, since they do so in what is God's past (a part of His eternal present, along with man's future) are already deemed "elect" or "chosen" (1 Peter 1:2; Ephesians 1:4). But respond they must; otherwise, exhortations such as "Repent ye therefore, and be converted" (Acts 3:19) would be meaningless.

*Limited Atonement.* Christ's atoning sacrifice was only for the benefit of those persons whom God had chosen for salvation before the beginning of the world, Calvinists say. To assure that the elect have the faith needed for their salvation, God sends it to them as a gift through the Holy Spirit. Christ's work on the cross is all that's needed for salvation, therefore. Faith, exercised by the believer on his or her own, is not required.

Arminians counter that Christ's redeeming work made it possible for *anyone* to be saved but did not actually *secure* the salvation of anyone. Although Jesus died to redeem everyone, only those who choose to repent and believe the Gospel are saved. Stating the matter another way, lack of human repentance and lack of belief are the only factors limiting the scope or effectiveness of Christ's atonement. That the atonement of Christ was intended to apply to all believers is repeatedly shown by Scripture—for example, John 3:16 (God gave His only begotten Son "[t]hat *whosoever* believeth in him should not perish"), Romans 3:22-23 (salvation comes "unto *all* them that believe; for there is no difference, [f]or *all* have sinned, and come short of the glory of God"), Hebrews 2:9 (Christ "taste[d] death for *every* man"), and 2 Peter 3:9 (God is "not willing that *any* should perish, but that *all* should come to repentance").[898]

*Irresistible Grace.* The Calvinist position in this regard is that there are two calls to salvation. There's an *outward*, general call that's made to everyone who hears the Gospel preached. The call that gets sinners saved, however, is the *inward* call that's made exclusively to the Calvinist elect. God's internal call cannot be rejected; it never fails to bring the sinner to salvation. But since the elect sinner has no choice but to accept God's offer, a question

---

[898] Emphasis supplied in all quotes.

arises as to whether the love directed by him toward God is what C. S. Lewis would call a "love worth having."[899] Calvinist teaching assures us, in this regard, that the Holy Spirit will put the elect sinner in a frame of mind that will allow him to cooperate with God's plan.

The Arminian position is that the Holy Spirit calls inwardly all who hear a valid presentation of the Gospel, but that what happens after that is up to the sinner. Some will resist the Spirit's call. Until their resistance is overcome, such persons cannot be regenerated ("born again"). The Holy Spirit will do all that He can to overcome the resistance, but people are still free to reject salvation. God's gracious call can be and often is successfully resisted, therefore. Citing the Jewish leaders' killing of the prophets who have urged them to turn from their wicked ways, Jesus laments over Jerusalem, at Matthew 23:37, "how often would I have gathered thy children together, even as a hen gathereth her children under her wings, and ye *would not* (emphasis supplied)!" This doesn't sound like irresistible grace. At John 5:40, we find Jesus saying to the Jews, "ye *will not* come to me, that ye might have life." Again, we see the Lord's detractors called but not accepting the call.

*Perseverance of the Saints.* Not surprisingly, Calvinism teaches that the elect—i.e., all who were chosen by God, redeemed by Christ's sacrifice, and given faith by the Holy Spirit—are eternally saved. They're kept in faith by the God who chose them, supposedly, and thus persevere in their Christianity throughout their earthly lives. This is not the same as the "eternal security" spoken of in non-Calvinist denominations. Calvinist "perseverance" is a corollary of predestination—the elect person was always saved and by definition cannot be or become unsaved—whereas non-Calvinist "eternal security" is the product of a step of faith taken at a specific time in the believer's terrestrial life.

How does the Calvinist know that he or she is among the elect? From his or her performance as a Christian, comes the answer[900]. But what may seem to one person to be conduct probative of

---

[899] Lewis, *Mere Christianity*, 47-48, 183; see also the discussion in this regard in the "OpennessTheology" section within my Fluid Reality chapter, supra.

[900] See Boettner, *The Reformed Doctrine*, 308-12.

salvation might not appear so to God. Our Lord counseled against behavior that was all for show—for example, visible alms-giving or public prayer (Matthew 6:1-6). Those who behave with such ostentation already have their reward, He said (vv. 2, 5). As the apostle Paul stated, we should be conducting ourselves "(n)ot with eyeservice, as menpleasers, but as the servants of Christ, doing the will of God from the heart" (Ephesians 6:6). But, as the prophet Jeremiah notes (17:9), the heart of man is "deceitful above all things, and desperately wicked: who can know it?" What it boils down to, then, is that there's no solid assurance of salvation after the Calvinist begins supposing that he's saved; he just has to wait and see how it all plays out. As a practical matter, this would seem to encourage the idea of salvation by works, which is what Calvinists accuse Arminians of inasmuch as they say that faith is a work.

Arminians, defined as opponents of five-point Calvinism, are not in agreement among themselves regarding eternal security. Some say that one's salvation can be lost by subsequently doubting the Gospel message. Others, notably Baptists, maintain that once a believer is regenerated, he can never be lost. Scripture would seem to support the Baptist position. At John 6:37, Jesus states, "him that comes to me I will in no wise cast out." At John 10:28, speaking of the sheep of which He is the good shepherd, Jesus says, "I give unto them eternal life; and they shall never perish, neither shall any man pluck them out of my hand."

As suggested in the preceding discussion, the footing on which Calvinist predestination rests, without which the doctrinal cornerstone Unconditional Election would have no scriptural support whatsoever, is God's "foreknowledge," so called in the New Testament. At Acts 2:23, we see Christ delivered to be crucified "by the determinate counsel and foreknowledge of God," First Peter 1:2 describes believers as "(e)lect according to the foreknowledge of God." As indicated above, a way of reconciling God's foreknowledge and human freedom is to take into account the fact that God and man operate in different milieus, God in eternity and man in time. Thus, it has been theorized that God not only sees man freely conducting his business within time; He also from His vantage point in eternity sees the future of each individual, be it heaven or hell.

If God knows everything, this would have to mean that He knows not only individual man's fate for eternity, but also what decision(s) any person will make on his or her way to eternity—i.e., that he knows whether any particular individual will accept or reject the Gospel. Man twists his own arm, essentially, and God watches as His creature makes the fateful decision. Since God will invariably allow man's decision re salvation to stand (He cannot do otherwise, since to alter man's future would be to render His foreknowledge faulty), this is not unlike saying that man's decision as to how he will spend eternity is foreordained. Foreordination accompanies foreknowledge.

"Alternative realities," I submit, is a better way than reference to God's position in eternity to get around predestination, which seems to be an unavoidable concomitant of God's knowing (what to us is) the future. The idea that with respect to any particular there is more than one world line for any traveler in the macro- world, suggested by analogy to quantum physics, not only allows us to envision prayer that works and prophecy that doesn't foreclose free will; it also defeats Calvinist "election." Predestination is logically impossible if for an individual there are differing futures each qualifying as an object of divine "foreknowledge." But— this is the kicker—the alternatives in question must exist in one cosmos only and in the mind of one god only. There can be no infinity of universes each with its own deity, such as Edgar Allan Poe visualized.[901] Otherwise, our god would not be God. His mind and knowledge would not be infinite (see Psalm 147:5 ["his understanding is infinite."]; also Isaiah 40:28 [ "there is no searching of his understanding."]). He would not be the ruler of everything (see Deuteronomy 10:14 [ "the heaven and the heaven of heavens is the Lord's"] and Psalm 97:9 [ He is "exalted far above all gods."]). In short, he would not be the God of the Bible.

---

[901] See Edgar Allan Poe, "Eureka: A Prose Poem," in *The Complete Works of Edgar Allan Poe*, 17 vols. (New York: Fred de Fau & Co., 1902), vol. 16, pp. 275-76 (orig. publ., New York: G. P. Putnam, 1848).

Educator/author Norman Geisler remarks that the Five Points "more or less stand or fall together, particularly the first four."[902] Thus, if we've convincingly disposed of Unconditional Election (by in effect removing the "fore" from "foreknowledge"), it would seem unnecessary to separately attack the remaining four points. Nonetheless, some parting shots are in order as far as I'm concerned.

Total Depravity, as the late Dave Hunt noted, involves "the irrationality of blaming the non-elect for failing to do what they can't do."[903] In his sermon on Mars Hill in Athens, the apostle Paul announces that God "commandeth all men every where to repent" (Acts 17:30). Concerning this announcement, Bible scholar Hunt elaborates:

> To say that God commands men to do what they cannot do without His grace, then withholds the grace they need and punishes them eternally for failing to obey, is to make a mockery of God's Word, of His mercy and love, and is to libel His character.[904]

Hyper-Calvinists hold that God *actively* elects not just to heaven, but also to hell.[905] This would in the most unmistakable terms make Him the author of evil, which the Bible declares absolutely He cannot be (see, e.g., James 1:13). Other Calvinists, including C. H. Spurgeon, have rejected the hyper-Calvinist view that God has no love whatsoever for the nonelect and have instead ascribed to Him a "general redemptive love" for all men, as contrasted to His "electing love" for the chosen;[906] however, this doesn't appear to change the ultimate destination of the nonelect.[907] Passive nonelec-

---

[902] Norman Geisler, *Chosen But Free: A Balanced View of Divine Election*, 2nd Ed. (Minneapolis: Bethany House Publishers, 2001), 57.

[903] Hunt, *What Love*, 115.

[904] Ibid., 117.

[905] Geisler, *Chosen But Free*, 215-16.

[906] Ibid., 218-19.

[907] See ibid., 215-19.

tion, as it's called, would presumably lead to the same place in the end as active nonelection: hell.

Denigration not only of God the Father, but also of God the Son's sacrifice, is a fundamental problem as far as Limited Atonement is concerned. Christ was sent to be the savior of all men (1 Timothy 2:4; 4:10; Hebrews 2:9; 2 Peter 3:9; 1 John 4:14). Any suggestion to the contrary cheapens what He did for us all on the cross. Was Christ's precious blood insufficient to cover the sins of all mankind? Both hermeneutics and honor are involved here, it should be noted. Since there's no Scripture supporting an atonement limited in its scope, Calvinists are forced to resort to the questionable tactic of ascribing restrictive meanings to words such as "world."[908] The world thus becomes the "elect."[909] They justify their rewriting of the Gospel as a matter of practical necessity if they are to maintain their position in favor of Limited Atonement, concerning which Dave Hunt comments:

> In considering the scriptures bearing on this subject, it becomes clear that the only way Limited Atonement can be defended is to assign, arbitrarily, a *restrictive Calvinist meaning* to key words (emphasis supplied).[910]

Strangely, the love of God for mankind in its full dimension is left out of the equation. God *is* love, it must be remembered (1 John 4:16), and thus the scope not only of His love but also of His very being is called into question by the doctrine of Limited Atonement.

The "two calls" formulation with respect to Irresistible Grace is also totally unscriptural, an obvious contrivance made necessary by the Unconditional Election point of five-point Calvinism. Why would a God of truth issue a sham call! It's impossible for Him to lie (Hebrews 6:18), and yet such a deception, leading persons to trust Him for a blessing not actually available to them, would clearly be a form of lying. Why would a loving, compassionate,

---

[908] Hunt, *What Love*, 298-300, 337-38.

[909] Ibid., 300.

[910] Ibid.

and merciful God behave so disingenuously? The name "Faithful and True" (Revelation 19:11) would not apply to Him were He to act in this manner.

Calvinist Perseverance of the Saints and Baptist "eternal security" reach the same final result. Therefore, the only argument is as to the road taken to reach that result. If we assume that the sinner made a sincere commitment to Christ as his Lord and Savior (see Romans 10:9), the only possible remaining issue, whether for the Calvinist or the non-Calvinist, is God's faithfulness, which is emphatically declared by the Bible to be a nonissue. "Faithful is he that calleth you," Paul tells his followers (1 Thessalonians 5:24). "(T)he Lord is faithful, who shall stablish you, and keep you from evil," he assures them (2 Thessalonians 3:3). And yet man's steadfastness is always a potential matter of concern. "Most men will proclaim every one his own goodness," the Book of Proverbs states at 20:6, "but a faithful man who can find?" Even the high and mighty are subject to lapses. See, e.g., Numbers 20:7-13 (Moses smiting the rock at Meribah rather than speaking to it as instructed by God, and as a result being denied entry into the promised land). Thus, only God's fidelity is a foregone conclusion.

# 8
# IDENTITY AND ORIENTATION

*Smith Hollow*

"Identity" is defined *inter alia* by Webster's Unabridged as "the sense of self, providing sameness and continuity in personality over time (although) sometimes disturbed in mental illnesses, as schizophrenia."[911] "Orientation" is defined in the same lexicon as "the ability to locate oneself in one's environment with reference to time, place, and people."[912] "Orientation" is also said to be "one's position in relation to true north, to points on the compass, or to a specific place or object."[913] The relevance of these meanings to our discussion will shortly become apparent.

A body of water other than Otisco that played a role in my childhood was Fish's Pond, also known as Smith Hollow because it was located in a dell of the same name just west of Navarino. There weren't any lakes or ponds that I remember other than Smith Hollow that were visible from U.S. Route 20 between Cazenovia and Skaneateles, which are some thirty miles apart. The road went over some high hills and dipped into some deep valleys, but there just wasn't any natural or man-made repository of water other than

---

[911] *Random House Webster's Unabridged*, 950, def. 5 (parenthetical supplied).

[912] Ibid., 1366, def. 4.

[913] Ibid., def. 5.

Fish's Pond to be seen on that stretch. The north end of Otisco Lake was several miles to the south of Route 20 and therefore tucked out of sight behind intervening hills.

The Pond would come upon you as a surprise when you were driving west out of Navarino on Route 20, since the road sliced downhill through a steep hillside immediately upon leaving the village (see Fig. 11), and the embankment on the south side of the road was high enough to hide the pond from your view until the last minute. "The cut," as it was called, was a likely place for juvenile-delinquent Navarino teens to find entertainment, since we could throw rotten vegetables down on passing cars and then flee into woods that adjoined the field from which the squash, gourds, and other organic projectiles that we threw were gathered.

Not all of the activities associated with Smith Hollow were destructive. During my pre-teen years, Uncle Leslie, Mother's brother, would take me fishing on the Pond. Uncle Leslie had only one child, a daughter, and so I seemed to be regarded by him as the son he never had. I don't remember our catching a single fish, but it was pleasant sitting in a gently rocking rowboat and looking up at steep wooded hillsides. In winter, kids and adults alike would ice skate on the Pond. I remember one time when there was an ominous sound, sort of like a great sigh, which one of the older boys said was the ice settling, but we nonetheless continued our game of "fox and geese," basically "tag" restricted to the circle and spokes of a giant wheel shoveled out of the snow.

In the summer, there would be the annual Navarino Volunteer Fire Department clambake, held in some rickety pavilions at the south end of the Pond, an event that would give us as teenagers the chance to see who the most prodigious beer drinkers among the locals were, folks who as an unsaved youth I might have wished to emulate. Navarino had a noteworthy history as far as alcohol was concerned: in the 1950s, an attempt to establish a tavern there was stopped dead in its tracks, largely through the efforts of the women of the Navarino Methodist church. Therefore, the nearest drinking establishment was Smitty's, some three miles to the west on Route 20. Thus, the firemen's clambake was likely seen by the drinking men of our immediate community as an opportunity to cut loose closer to home.

In the summer of 2010, I returned to Onondaga County for my fiftieth high school reunion, minus Sue because she was getting ready for hip replacement surgery. I stayed at a motor inn just outside Syracuse, in the vicinity of the State Fairgrounds; thus, I came in by the New York Thruway and Thruway feeder I-690. After attending the reunion, which was held at a restaurant on a different side of Syracuse from where my high school was located, I spent part of the next day with my friends John and Carol and some others who had worked on the reunion. I then determined that I would return to Rochester via the "long way"—i.e., Route 20. I had not been in or near Navarino since Mother passed in 1995, which was thirty years after Dad's homecoming (entry into heaven).

Coming down the hill into the village from the east, I could notice changes, which were not unexpected. The house that I grew up in now had vinyl siding which replaced the asbestos shingles that had been so easy to chip with a baseball. The gas pumps were gone from Mullin's Atlantic, and in their place were boxes piled all the way up to the overhang. The biggest surprise—in fact, the only surprise—came as I emerged from "the cut" heading west: Fish's Pond was dried up. Where there had been gentle waters that supported enjoyable activities, there was now a low bed of nondescript vegetation. Saplings and broad-leafed weeds stood shoulder to shoulder, in cacophonic competition for what little moisture remained. Apparently, the dam on the other end of the Pond, which allowed this body of water to exist, had given way.

Seeing something emptied of meaning has caused me to consider, once again, the postmodern condition, which can be summed up as follows: we don't know who we are or where we're going; indeed, we don't know anything at all.[914] We're wandering around in the fog morally, in a medium the lack of transparency of which is not a concern for us inasmuch as we've decided that there's nothing beyond the fog to see anyway. Having to again stare into the expressionless face of postmodernism has made me realize

---

[914] See my remarks at the end of the "Scripture and Spirit" section of my Region of Awe chapter, supra.

that I haven't escaped from the snares of philosophy as cleanly as I thought I had. I'm now right back in the fray with the old Truth Trader, trying to deal with the confusion to which the linguistic turn of philosophy has ultimately led. Poetic justice, perhaps. In the meantime, I've found myself slipping back into a nature mysticism such as that practiced by C. S. Lewis, arguably more a matter of simply recognizing correspondences than meeting with the Lord. Ironically, the correspondences that have most recently sustained me on this journey have come from a discipline with which religion has for many years been seen as being at odds: science.

*The Essence of Postmodernism*

Since the philosophy we're about to discuss is "post-", it must have come after something else. That something else, one may easily surmise, is *modernism*. Modernism was a way of thinking that in large measure developed during the Enlightenment of the eighteenth century, which I've already commented on, when "reason" gained ascendancy over divine revelation in the world-view of the so-called educated classes.[915] Rejecting the idea that Western civilization reached its highest point in ancient Greece and Rome, modernity envisioned further human progress, which would supposedly be achieved through rationality and technological advances.[916]

It was somewhat surprising, then, that modernism would mingle the irrational with the rational, particularly in the arts. What happened in the areas of painting, literature, and music has been seen, more than anything, as a reaction to the nineteenth century's romanticism (nature-oriented free expression of feelings) and its realism (presentation of the world as it appears to the ordinary senses). The modernist response in the arts included such developments as cubism (Pablo Picasso, Georges Braque), surrealism (Salvador Dali, Rene Magritte), stream-of-consciousness writing (James Joyce, Virginia Woolf), and dissonance in music

---

[915] See the "Where We Stand Now" section of my Proving God chapter, supra; also Rohmann, *Ideas*, 115-16.

[916] See Rohmann, *Ideas*, 265.

(Igor Stravinsky, Arnold Schoenberg).[917] In the meantime, church leaders of the early twentieth century were struggling to reconcile religious doctrine with what was being called "science"—in particular, the theory of evolution—and also with the social change that Charles Darwin's theory brought.[918] In Catholicism, this resulted in charges of heresy being brought by Pope Pius X, in 1907, against certain liberal theologians. In Protestantism, in the 1920s, the challenge of "science" to the Bible's authority sparked a revival of fundamentalism.[919]

If modernism consisted in drinking from the cup of license, then postmodernism might be seen as a hangover thus produced. Pretenses of relevance are nowhere to be found in the postmodern world and also lacking, it seems, is any hope of restoring a sense of order in our commercialized, trivialized culture.[920] It's difficult for someone such as myself who is not a social scientist to know where to begin in explaining postmodernism, but Prof. James K. A. Smith of Villanova University, in his book *Who's Afraid of Postmodernism?*[921], has provided what seems to me a useful starting point. Dr. Smith does not deny that postmodernism is a problem for contemporary society. In his opening chapter, he states:

> While we might not name it as such, our experience of cultural shifts and changes can be traced to the advent of postmodernity and the trickle-down effect of postmodernism on our popular culture. The transition calls into question all our previously held sureties and rattles a faith that has been too

---

[917] See ibid.

[918] Ibid.

[919] Ibid., 152-53, 265.

[920] See ibid., 310.

[921] James K. A. Smith, *Who's Afraid of Postmodernism? Taking Derrida, Lyotard, and Foucault to Church* (Grand Rapids, Mich.: Baker Academic, 2006).

easily equated with Cartesian certainties, sometimes issuing in a kind of vertigo.[922]

Smith nonetheless indicates—and this may come as surprise to some observers—that the reception postmodernism has received in the churches has not been uniformly negative. He reports:

> To some, postmodernity is the bane of Christian faith, the new enemy taking over the role of secular humanism as object of fear and primary target of demonization (footnote citation omitted). Others see postmodernism as a fresh wind of the Spirit sent to revitalize the dry bones of the church (footnote citation omitted). This is particularly true of the "emerging church" movement, which castigates the modernity of pragmatic evangelicalism and seeks to retool the church's witness for a postmodern world.[923]

Professor Smith's "castigates to retool" statement is an obvious reference to the abandonment of modernism after some two hundred years of its failure to deliver on its promise of human progress through the application of reason and technology.[924] Intellectual and technological props for his sense of well-being having been snatched away from him, late-twentieth-century man either sank into hopelessness or turned to a version of Christianity that long preceded either modernism or postmodernism: mysticism, primarily in the form practiced by medieval Christians and modern Roman Catholics.[925] This is where things stand at present.

---

[922] Ibid., 17. The reference is to Rene Descartes, who sought to establish a solid foundation for philosophy with his maxim, *cogito, ergo sum* ("I think, therefore I am").

[923] Ibid., 18.

[924] See immediately preceding discussion; also Rohmann, *Ideas*, 265; and Gilley, *This Little Church Stayed Home*, 22-24.

[925] Gilley, *This Little Church Stayed Home*, 112-13; and see Gilley, *This little church had none*, 36-62.

The mysticism referenced above, while far from being anything new, has become the religion of the nascent "emerging (emergent) church," which is described by evangelical pastor and writer Gary Gilley as a loose-knit configuration of believers that is "coming out of the more traditional understanding of the church and *emerging* into a postmodern expression of the church (emphasis supplied)."[926] As such, it is, Gilley says, "defined by its culture rather than by Scripture."[927] The dangers of such a departure from *sola scriptura* (scripture alone) are explained elsewhere in *Water*, as are various other details with respect to the mysticism spoken of above.

Doctor Smith brings to the discussion the assumption that postmodernism "is something that has come slouching out of Paris," that it "owes its impetus to French philosophical influences."[928] He then names names, saying that he will engage "something of an unholy trinity of postmodern thinkers"—specifically, Jacques Derrida (1930-2004), Jean-Francois Lyotard (1924-98), and Michel Foucault (1926-84).[929] He will consider, he announces, three slogans of postmodernism associated with these philosophers:

(1) "There is nothing outside the text" (Derrida).

(2) "Postmodernity is incredulity toward metanarratives" (Lyotard).

(3) "Power is knowledge" (Foucault).[930]

Smith admits that these slogans have been seen as antithetical to Christianity. He states:

> Generally, these three slogans are invoked as being mutually exclusive to confessional Christian faith. How could someone who takes the sweeping

---

[926] Gilley, *This Little Church Stayed Home*, 142-43.

[927] Ibid., 143.

[928] Ibid., 19.

[929] Ibid., 21.

[930] Ibid., 21-22.

> narrative of the Scriptures as the Word of God reject metanarratives? How could someone who believes in the existence of a transcendent God and his creation deny that there is a reality outside texts? How could someone who worships the God who is love participate in a Nietzschean celebration of the will to power as the basis of reality?[931]

How, indeed! Dr. Smith nevertheless goes on to say that his goal is "to demythologize postmodernism by showing that what we commonly think so-called postmodernists are saying is usually not the case."[932] The reader will forgive me, I hope, if I respectfully disagree with various points in Prof. Smith's defense of Mssrs. Derrida, Lyotard, and Foucault, and will permit me the observation that a book that says that postmodernism is not as bad as it seems will get you attention but not necessarily nods of agreement.

Jacques Derrida's strategy in approaching texts (anything written, printed, or spoken) is one of "deconstruction"—i.e., of exposing the multiplicity and changeability of the meanings of the words therein. True, says Smith, there is apparently in Derrida's view no objective structure into which we can fit words in determining their meanings.[933] Nonetheless, he explains, it is not the case that Derrida is counseling the reader to adopt the stance of a linguistic idealist (someone who thinks there are only words, no corresponding things).[934] The existence of God is not necessarily denied by the statement that "there is nothing outside the text," Smith urges, since our Bible is a text and does refer to something rather than nothing.[935] What Derrida is really saying, according to Smith, is that when you read a text, you give the words an interpretation which in turn incorporates other interpretations, which then

---

[931] Ibid., 22.

[932] Ibid.

[933] See Stokes, *Essential Thinkers*, 189.

[934] Smith, *Who's Afraid*, 35.

[935] Ibid.

draw in still other interpretations, and so on, so that you never get past the interpretation level and arrive at "pure" reading.[936]

Saying that everything is interpretation doesn't seem to resolve the problem, however. Just before making his claim that "there is nothing outside the text," Derrida says that a reading or interpretation

> cannot legitimately transgress the text toward something other than it, toward a referent (a reality that is metaphysical, historical, psychobiographical, etc.) or toward a signified outside the text whose content could take, could have taken place outside of language, that is to say, in the sense that we give here to that word, outside of writing in general.[937]

My translation: it's wrong to interpret text as referring to anything outside itself. But this is the same as saying that there's "nothing outside the text," isn't it? Elsewhere, according to Dr. Smith, Derrida attempts to clarify matters by saying that "nothing outside the text" really means "nothing outside context."[938] What you see or hear from others in connection with an event or idea determines what you understand about it. In this way, the community in which you're situated plays a role.[939]

If everything is interpretation or there's nothing outside of context, where does that leave Jesus and the Gospel? If, for example, we lay aside all the accounts of miracles recorded in the Old Testament and New Testament, which would seem to me definitely to qualify as "context," thus reducing our belief system to simply "God exists and He loves you," can we still say that we have a Gospel? Does it matter, anyway, that what we think we know about God is "interpretation (instead of turtles) all the way down"? Does it matter that different individuals in the area of Jerusalem would have seen and

---

[936] See ibid., 37-38.

[937] Jacques Derrida, *Of Grammatology*, Corr. Ed., transl. by Gayatri Chakravorty Spivak (Baltimore: Johns Hopkins Univ. Press, 1974, 1976, 1997), 158.

[938] See Smith, *Who's Afraid*, at 52.

[939] See discussion at Smith, *Who's Afraid*, 51-53.

heard different though not inconsistent things on the day of the crucifixion, as long as at least one of them would have been able to "put it all together" and rationally conclude that Jesus was the Son of God? The centurion who was watching Jesus and the other soldiers who were with him were informed that the veil in front of the Holy of Holies in the temple was torn in two and that graves in the vicinity of Jerusalem had been opened by an earthquake, allowing their occupants to arise and come out, and it was this combination of circumstances, along with a prolonged darkening of the sky, that led the soldiers to recognize Christ's divinity (Matthew 27:51-54; Mark 15:38-39; Luke 23:45, 47-48). Is it fatal to a finding that there is indeed something "outside the text," a risen Savior, that persons in the Jerusalem area other than the centurion and his detachment would have been denied access to critical facts relevant to Jesus' identity? Does true knowledge have to be objective in the sense of being not based on interpretation and able to be "universally known by all people, at all times, in all places?"[940]

The above are questions that need to be answered before anyone can say definitely whether Derrida's teaching is friendly, neutral, or hostile toward Christianity. For me they already have been, each a divine extra-textual presence and person who is describable in rational terms.

Needless to say, a milieu in which truth is almost totally relative is not a healthy society, since there is no knowledge base that can be relied on either to govern interpersonal relations or to support the gathering of essentials. And yet this is the situation in which we find ourselves if we hold, as the most straightforward reading of Derrida seems to suggest, that there is nothing of substance outside our words that they can attach to in order to give them meaning.

Proceeding to Jean-Francois Lyotard's claim, Smith prefaces his analysis with the statement that postmodernism "can be understood as the erosion of confidence in the rational as the sole guarantor and deliverer of truth, coupled with a deep suspicion of science—particularly modern science's pretentious claims to an

---

[940] Ibid., 48.

ultimate theory of everything."[941] Smith notes in this regard that Lyotard was one of the first philosophers to attempt a definition of postmodernism.[942] Writing a "report on knowledge" commissioned by the government of the Canadian province of Quebec, Lyotard opened his analysis with the statement: "Simplifying to the extreme, I define *postmodern* as incredulity toward metanarratives."[943] Smith notes that the French term that's translated as "metanarratives" is *grand recits*, big stories.[944] Postmodernism, then, is the suspicion and disbelief of "big stories."[945] In this regard, Smith observes (correctly):

> Now, if ever there was a big story, it is the grand narrative offered in Scripture, spinning a tale from before creation until the consummation of time (and beyond). Thus, if postmodernism is incredulity toward metanarratives, and Christian faith as informed by the Scriptures is just such a metanarrative, then postmodernism and Christian faith must be antithetical: postmoderns could never believe the Christian metanarrative, and Christians should not participate in postmodernism's incredulity.[946]

The conclusion that postmodernism and Christianity are mutually exclusive by reason of Lyotard's "incredulity toward metanarratives" does reflect "a reading suggested by even the most nuanced Christian commentators on postmodernity," Smith further

---

[941] Einstein unsuccessfully sought this holy grail throughout his lifetime. See the discussion in this regard in the "Science and Religion Generally" section of my Paradigm Paralysis chapter, supra.

[942] Smith, *Who's Afraid*, 63.

[943] Jean-Francois Lyotard, *The Postmodern Condition: A Report on Knowledge*, transl. by Geoff Bennington and Brian Massumi (French original, 1979; Minneapolis: University of Minnesota Press, 1984), xxiv.

[944] Smith, *Who's Afraid*, 63.

[945] Ibid.

[946] Ibid.

concedes.⁹⁴⁷ But he insists that this judgment is "a bit hasty, not informed by a careful understanding of what Lyotard means by 'metanarrative' and of what it would mean to no longer believe in metanarratives."⁹⁴⁸

For Lyotard, according to Dr. Smith, metanarratives not only tell a big story but also claim to be "able to legitimate or prove the story's claim by an appeal to universal reason" (Smith's phraseology).⁹⁴⁹ It is the supposed rationality of modern scientistic (science-like) stories about the world that makes them metanarratives, Smith says.⁹⁵⁰ In Lyotard's view, Homer's *Odyssey*—although it tells a grand story and makes universal claims about human nature—is not a metanarrative because it does not claim to legitimate itself by an appeal to a supposed universal, scientific reason; instead, it is simply in the nature of preaching, nothing more, which demands only a response of faith.⁹⁵¹ On the other hand, the scientific stories told by modern rationalism, scientific naturalism, or sociobiology are metanarratives insofar as they claim to be demonstrable by reason alone.⁹⁵²

It is the issue of legitimation that is at the heart of the distinction between modernity and postmodernity, in Smith's (and Lyotard's) analysis.⁹⁵³ Modernity appeals to science for legitimation, while postmodernity looks to narratives, not-necessarily-true accounts of events or experiences, inasmuch as it is unwilling to accord "truth" status to scientific observations.⁹⁵⁴ But when judged by the criteria of modern science, narratives or stories are seen as little more than fables.⁹⁵⁵

---

[947] Ibid., 63-64.

[948] Ibid., 64.

[949] Ibid., 65.

[950] Ibid.

[951] Ibid.

[952] Ibid.

[953] See ibid., 65.

[954] Ibid.

[955] Ibid.

When pushed, however, science must still legitimize itself.[956] This raises the problem, Smith says, that postmodernism has suggested that the emperor of modernity has no clothes—i.e., has relied on an illusion to substantiate its proffered view of reality. The gist of the postmodern critique of modernity, as Smith explains it, is that science—which has been critical of the "fables" of narrative—is itself grounded in narrative.[957] Narrative knowledge, when based on the *custom* of a culture, normally does not need legitimation. The problem, however, is that in the postmodern era we have individuals attaching different meanings to any given word; thus, we can no longer establish the legitimacy of a narrative by reference to custom.[958] The words of an account tell a story that varies depending on who the hearer is. Therefore modern scientific knowledge, when called on to legitimate itself, cannot avoid appealing to narrative, cannot help resorting to storytelling. This "return of the narrative in the non-narrative" is inevitable, according to Lyotard.[959] Thus, we must not believe metanarratives because science ultimately relies on narratives and in turn is the stock in trade of modernity, which postmodernists have come to distrust. Guilt by association, it would seem.

As an amateur at best with respect to scientific matters, I cannot say much regarding the underpinning of science, but as a Christian I find incomprehensible the suggestion that my faith is founded on fables. I see the beauty of the earth and the universe, the very livability of this our terrestrial home, the existence of life and all the miracles of both the Old Testament and the New Testament, including the resurrection of Jesus Christ and the promise of everlasting life made manifest by His return from the dead, all inductively proven and all pointing to a loving God who has a plan of being Father to children that are able to commune with Him both now and forever, and seeing these things wonder how there could ever be "incredulity" toward the narrative that underlies the Christian faith. The legitimation of our metanarrative is before our

---

[956] Ibid.

[957] Ibid., 66; and see Lyotard, *Postmodern Condition*, 27-31.

[958] Smith, *Who's Afraid*, 66-67.

[959] Ibid., 67; and see Lyotard, *Postmodern Condition*, 27-28.

eyes and too obvious to be ignored, too direct and powerful to be dismissed as attaching to a set of made-up stories.

Michel Foucault's contribution to postmodernism, as noted above, is the axiom that "power is knowledge." Prof. Smith explains that in the world-view of this third person of the "unholy trinity," there is a network of power relations that controls our major institutions—hospitals, schools, businesses, and prisons—so that what counts as knowledge is merely the information that is assembled within connecting systems of social, political, and economic power.[960] Foucault states, near the beginning of his signature work *Discipline and Punish*:

> We should admit that power produces knowledge (and not simply by encouraging it because it serves power or by applying it because it is useful); that power and knowledge directly imply one another; that there is no power relation without the correlative constitution of a field of knowledge, nor any knowledge that does not presuppose and constitute at the same time power relations.[961]

Unfortunately, what a power-as-knowledge system produces is not necessarily good for society. Smith characterizes Foucault's effort in *Discipline and Punish* as an attempt to show that modernism's claims to scientific objectivity or moral truth are "fruits of a *poisoned* tree of power relations" (emphasis supplied).[962] He refers to Foucault's example of a criminal's confession as demonstrating the symbiotic relationship of power and knowledge. Rationalizing the long survival of public torture and executions in France, Foucault states: "If torture was so strongly embedded in legal practice, it was because it revealed truth and showed the operation of power."[963] Writing in the late twentieth century, Foucault goes on to say that

---

[960] Smith, *Who's Afraid*, 85.

[961] Michel Foucault, *Discipline and Punish: The Birth of the Prison*, transl. by Alan Sheridan (New York: Vintage Books, 1977, 1995), 27.

[962] Smith, *Who's Afraid*, 87.

[963] Foucault, *Discipline and Punish*, 55.

the truth-power relation "remains at the heart of all mechanisms of punishment and is still to be found in contemporary penal practice."[964]

In the course of his analysis, additionally, Foucault refers to what he calls "the formation of a disciplinary society."[965] This society, a mercantile economy, is made up of individuals each of whom is "a reality fabricated by specific technology that (Foucault had) called 'discipline.'"[966] In this regard, Foucault warns the reader:

> We must cease once and for all to describe the effects of power in negative terms: it 'excludes', it 'represses', it 'censors', it 'abstracts', it 'masks', it 'conceals'. In fact, power produces; it produces reality; it produces domains of objects and rituals of truth.[967]

But what is the actual nature of the reality that power produces? In this regard, Prof. Smith observes:

> (H)ere we must make an important distinction: we can distinguish good discipline from bad discipline by its *telos*, its goal or end. So the difference between the disciplines that form us into disciples of Christ and the disciplines of contemporary culture that produce consumers is precisely the goal they (convents and monasteries versus factories and prisons) are aiming at. Discipline and formation are good insofar as they are directed toward the end, or telos, that is proper to human beings: to glorify God and enjoy him forever (citation omitted).[968]

---

[964] Ibid.

[965] Foucault, *Discipline and Punish*, 193; and see Stephens, *Philosopher's Notebook*, 174-75.

[966] Foucault, *Discipline and Punish*, 194.

[967] Ibid.

[968] Smith, *Who's Afraid*, 102.

I could not agree more. Regarding the disciplinary mechanisms implicit in modern consumerism, Smith complains: "Nothing frustrates me more than the 'label idolatry' evident in my children."[969] Again, I find myself sharing Prof. Smith's sentiments. The main fault that I see in his handling of Foucault is that he does not seem to be troubled by the Frenchman's apparent readiness to cast aside normal procedural restraints in favor of the expedient of obtaining proof of questionable reliability for use in the court system. This in spite of the fact that Dr. Smith appears to have a world-view very different from that of the atheist Foucault.

Gary Gilley suggests that postmodernists have not always been as accepting of the effect of power relations as Foucault apparently was. Indeed, as Dr. Gilley reports, the standard postmodernist thinking seems to be that to claim to have found out the truth about anything is to deceitfully employ a tool by which one attempts to manipulate or control others.[970] This is the case, Gilley indicates, even where the truth claim is made in the context of a supposedly objective discipline such as science or medicine.[971] Such a conclusion follows unavoidably from the postmodernist aphorism that there is no such thing as absolute truth. Thus, it is said, the only reason anyone would claim to know anything, given that real knowledge is impossible, would be that, as Gary Gilley puts it, they want to "empower themselves and enslave others."[972]

The themes stated by the "unholy trinity" philosophers were articulated by Friedrich Nietzsche (1844-1900) some sixty or seventy years before the mid-twentieth-century advent of postmodernism, in *The Will to Power*, an assemblage of his writings first published by the enigmatic German's sister Elisabeth the year after his death.[973] Thus, for example, we find Nietzsche expressing "(p)rofound aversion to reposing once and for all in any one total view of the world," which anticipates Lyotard's "incredulity toward

---

[969] Ibid., 105.

[970] See Gilley, *This Little Church Stayed Home*, 28, 37.

[971] See ibid., 37.

[972] Ibid.

[973] See, generally, Friedrich Nietzsche, *The Will to Power*, transl. by Walter Kaufmann and R.J. Hollingdale, ed. by Walter Kaufmann (New York: Random House, 1967).

metanarratives."[974] Also foreseen by Nietzsche are the interpretational convolutions occasioned by Derrida's announcement that there is "nothing outside the text." Nietzsche states that "(in) so far as the word 'knowledge' has any meaning, the world is knowable; but it is *interpretable* otherwise, it has no meaning behind it, but countless meanings."[975] Thus, Nietzsche continues: "There are no facts, everything is in flux, incomprehensible, elusive; what is relatively most enduring is—our opinions."[976] Consequently, he says that the wisest man "would be the one richest in contradictions, who has, as it were, antennae for all types of men—as well as his great moments of *grand harmony*—a rare accident even in us!"[977] Foucault's alliance of force and knowledge is anticipated by Nietzsche's conclusion that "(i)t is our needs that interpret the world," which for Nietzsche means that "(e)very drive is a kind of a lust to rule; each one has its perspective that it would compel all the other drives to accept as a norm."[978] The association of postmodern thinkers with Nietzsche—especially, of Foucault—is potentially a sinister one, it must be said, since Nietzsche's philosophy has had the reputation, rightly or wrongly, of supporting Nazism.[979]

In the century just past, the skeptics' banner was raised high by the existentialists, as indicated earlier. In the twentieth-century version of existentialism, a person's character is not a product of the influence of God; therefore it's easy to find roots of atheistic postmodernism in existentialism. Thus, it appears, there was more that was involved in the spread of the cold chill of postmodernism over the warm body of modern man than just Nietzschean nihilism (rejection of any total world-view, "knowledge" consisting of multiple meanings, and our personal needs interpreting the world)

---

[974] Nietzsche, *Will to Power*, 470 (1885-1886). The first digits for each passage cited herein are a number assigned to such writing for identification, while the numbers in parentheses are the year(s) in which the passage is believed to have been written.

[975] Nietzsche, *Will to Power*, 481 (1883-1888).

[976] Ibid., 604 (1885-1886).

[977] Ibid., 259 (1884).

[978] Ibid., 481 (1883-1888). Further regarding the Nietzschean will to power, see Stephens, *Philosopher's Notebook*, 123.

[979] Stokes, *Essential Thinkers*, 146-47.

or the failure of modernism (reason plus science) to deliver on its promises of societal well-being. The way had been paved by another philosophy—existentialism—that recognized no absolutes and accordingly saw no reason for the universe to exist.[980]

Is my theory of "alternative (radiating) realities" merely no-absolutes postmodernism in disguise? Answer: certainly not. The alternative realities in my formulation are different but existent courses of action for God or His subjects to take. Saying that these alternatives exist side by side, in parallel as it were, is not the same as saying that none of them exists, which I see as a bottom-line conclusion required by postmodernism's "nothing is real" credo. The existence of God cannot be established under postmodernist thinking, since that would constitute imposing an absolute; however, my theory, by giving all possibilities reality within God's infinite mind, encompasses absolutes and at the same time allows God to exist changelessly, as the Bible says He does (Malachi 3:6; James 1:17). Since God's mind content never changes, in that there is nothing in the way of new information that can be added to it, and since beings are defined by their information, God himself never changes. "Alternative realities" and postmodernism therefore have nothing in common worth talking about, since in the one theory there is reality and in the other there is not.

*The Effluvia of Postmodernism*

The problem presented by postmodernism is not just that its slogans tend toward nihilism. The difficulty is quantitative as well as qualitative. We have already spoken of the effect of postmodernism on Christianity, where centuries-old orthodoxies are being supplanted in a headlong rush to mysticism. Further reading on the subject of postmodernism fallout suggests that the malaise associated with "the postmodern condition" is widespread and deep and has affected many areas of life. Jean-Francois Lyotard in his report states:

---

[980] See Rohmann, *Ideas*, 127-28.

> What is new in all of this is that the old poles of attraction represented by nation-states, parties, professions, institutions, and historical traditions are losing their attraction. And it does not look as though they will be replaced, at least not on their former scale. Identifying with the great names, the heroes of contemporary history, is becoming more and more difficult (footnote citation omitted).[981]

A professional group that has been particularly affected in a negative way, Lyotard says, is the scientific community, within whose ranks the "narrative proof" challenge is said to have produced demoralization of both researchers and teachers.[982] More than bad methods is involved, as Lyotard analyzes the problem; it's also a matter of bad motives, particularly as related to the acquisition of power. Lyotard explains:

> The production of proof, which is in principle only part of an argumentation process designed to win agreement from the addressees of scientific messages, thus falls under the control of (a) language game in which the goal is no longer truth, but performativity—that is, the best possible input/output equation. The State and/or company must abandon the idealist and humanist narratives of legitimation in order to justify the new goal: in the discourse of today's financial backers of research, the only credible goal is power. Scientists, technicians, and instruments are purchased not to find truth, but to augment power.[983]

This sounds more than a little like Foucault, does it not?

But even more is involved in the postmodern condition than nihilism, godlessness, the loss of heroes, and the corrupt acquisition

---

[981] Lyotard, *Postmodern Condition*, 14.

[982] Ibid., 7-8.

[983] Ibid., 46.

of power. Professor Emeritus Arthur Asa Berger of San Francisco State University has put together a compendium of essays on various other aspects of postmodernism, each of which is accompanied by his own comments, which he has entitled *The Portable Postmodernist*.[984] For one thing, Berger indicates, our lives have typically become a mishmash that lacks any core belief structure. Pastiche—the throwing together incongruously of forms and motifs from assorted sources—is how we give the past its due, he says.[985] The postmodernist penchant for pastiche evokes Jean-Francois Lyotard's observation that a postmodernist listens to reggae music, watches a western, eats McDonald's food for lunch and local cuisine for dinner, wears Paris perfume in Tokyo, dons 'retro' clothes in Hong Kong," and thinks that knowledge is (nothing more than) "a matter for TV games."[986] Eclecticism would not be so bad, in Prof. Berger's view, if we did not "become lost in a rapid succession of images as we try on and cast off one identity after another and lose any sense of self we have—assuming that a self involves some kind of coherent sense of one's identity."[987] This should not matter, anyway, Berger suggests, since, unlike existentialism, postmodernism does not consider authenticity to be of any importance.[988]

More than sartorial tackiness or indigestion is potentially involved when someone applies a "pastiche" approach to theology. There may be a jumping around between different parts of the Bible or variant theologies that is distracting or confusing for the reader. There may be outright misquoting of the Word, whether intentional or inadvertent. A sticky-note treatment of the Bible, grabbing at verses as they're stumbled across, may in reality be the pursuit of an agenda that is not God's. Surely the inspired Word

---

[984] See, generally, Arthur Asa Berger, *The Portable Postmodernist* (Lanham, Md.: Rowman and Littlefield Publishers, 2003).

[985] Berger, *Portable Postmodernist*, 5, 47.

[986] Lyotard, *Postmodern Condition*, 76.

[987] Berger, *Portable Postmodernist*, 11.

[988] Ibid., 73, referring to Mark Gottdiener, *Postmodern Semiotics: Material Culture and the Forms of Postmodern Life* (Cambridge, Mass.: Blackwell, 1995), 233-34.

of God deserves—indeed, requires—handling that is more honest and careful than this.[989]

It's not surprising, then, that a postmodernist will frequently be heard to say that nothing is real. In an apparent reference to "reality" television programming, postmodern society has been described as "a cinematic, dramaturgical production."[990] Reality is nothing more than a play that one puts on for the entertainment (but not necessarily the edification) of the rest of mankind.[991] Not only does art imitate life; life now imitates art.[992] Berger says that we are the same as Hollywood film actors, always taking on new identities and casting them off, except that we play out our roles in malls and shopping centers.[993] Quaere how the public's shift to on-line shopping will affect our choice of a stage.

"Image" is therefore very important. The generation brought up in the 1960s and 1970s, it is said, was not educated to use words with precision.[994] What you see as the image presented to your eyes by this generation is not necessarily what you get, either. After quoting French philosopher Jean Baudrillard, Prof. Berger tells us:

> This selection by Baudrillard deals with his analysis of the power of images in generating simulations and a *hyperreality* (emphasis supplied) in postmodern societies—one of the dominant themes in his writing on the subject. Just as fish do not realize they are in water, we do not realize that we are living in a giant hyperreality, a world of images that reflect other images endlessly and that,

---

[989] See John 5: 39; James 1: 21; and Revelation 22: 19; also, see Gilley, *This Little Church Stayed Home*, 152.

[990] Norman K. Denzin, *Images of Postmodern Society: Social Theory and Contemporary Cinema* (Thousand Oaks, Calif.: Sage, 1991), ix-x, quoted in Berger, *Portable Postmodernist*, at 14.

[991] See Berger, *Portable Postmodernist*, 14.

[992] Ibid.

[993] Ibid., 15.

[994] Gilley, *This Little Church Stayed Home*, 32.

ultimately, in a process he describes, bear no relation to reality.[995]

Given the unreality of everything in our immediate surroundings, it might not be unreasonable to suggest that even God is a simulation, reducible under postmodernism to, and nothing more than, the bits of evidence that are said to show His existence.[996]

On the subject of simulation, we think of filmmaker/entertainment mogul Walt Disney, whose synthetic world included witches, sorcerers, and all other manner of imaginary creatures and things. Need a lift out of the world of the everyday? Magic can take you wherever you want to go, in the Disney formulation of truth—including, perhaps, from death to life. Professor Berger notes that Disney has been cryogenized, suspended in liquid nitrogen, in the hope that in the future he can be brought back to life.[997] (Perhaps Ted Williams, another cryogenic, will also come back, and bat .400 again.) What this suggests, Berger says, is that "nothing is real, nothing has meaning anymore—even death can be countered, if you can afford to be cryogenized."[998]

In any event, cryogenizing is a silly idea, it seems to me, or at least is a concept that is not well thought out. If indeed this technique will allow a physically dead you to be restored to the ranks of the living, this will still presumably be with the same normal effects of aging and mortality as afflicted you before. All you'll be doing is subtracting from your total health picture the disease that killed you, for which we assume a cure will now have been found (otherwise, the medical people wouldn't be waking you up). Will a cryogenic return from the dead restore your youth? That doesn't seem to be part of the script, any more than ultimate immortality. What we're looking at instead, it seems, is returning you to service

---

[995] Ibid., 57, referring to Jean Baudrillard, "The Evil Demon of Images and the Precession of Simulacra," in *Postmodernism: A Reader*, ed. by Thomas Doherty (New York: Columbia Univ. Press, 1993), 194.

[996] See Berger, *Portable Postmodernist*, 57.

[997] Ibid., 67.

[998] Ibid.

as a vehicle that's no less high-mileage than it was before you were flash-frozen.

And where does the road lead after your so-called resurrection? To new freezings when future physical maladies prove sufficient to kill you? What is the ultimate destination of your being? You should be looking toward God and His eternal kingdom, in my humble believer's opinion, but you probably aren't. Do you think, seriously, that you can put off the final outcome of your life forever? With the world continuing to deteriorate and the people and things that you hold dear disappearing during your long slumber(s), there would seem to be no point in being frozen and woken up. Unless, of course, you value life no matter how valueless.

Also not real or permanent is romantic love. Berger quotes supposed amour expert Eva Illouz for the proposition that the romantic "love of my life" has been replaced by the *affair*, an experience that is briefer and repeatable with one or more other partners.[999] Berger provides the following comparison of the postmodernist and modernist approaches to love:

> The postmodern perspective on romantic love is considerably different from the modernist one. Marriage and its narrative of fidelity have been replaced, in a consumerist culture, with the affair, an attempt, our author suggests, to relive the passion and emotional excitement we felt when we first had sex with someone…(s)ex has become one more lifestyle choice—an expression of autonomy not too far removed in nature from buying a pair of jeans.[1000]

As regards professional sports, also, there has been a perception that nothing is real. This is particularly so with respect to major league baseball, in which the use of anabolic steroids by players

---

[999] Eva Illouz, "The Lost Innocence of Love: Romance as a Postmodern Condition," *Theory, Culture, and Society* 15, nos. 3-4 (1998), 175-76, quoted in Berger, *Portable Postmodernist*, at 96.

[1000] Berger, *Portable Postmodernist*, 97.

to enhance performance had become extensive by the beginning of the twenty-first century, as detailed in Ken Burns's *Baseball: The Tenth Inning*.[1001] The Burns documentary follows *inter alia* the Mark McGuire/Sammy Sosa home run race, which culminated with both players finishing the 1998 season with totals in excess of Roger Maris's single-season mark of 61 (70 and 66, respectively).[1002] Suspicions grew when a newly muscular Barry Bonds later appeared on the scene and finished the 2001 season with 73 home runs, notwithstanding a temporary suspension of major league games following the 9/11 attacks on the United States.[1003] In 2002, retiring player Jose Canseco told the press of widespread steroid use in the majors.[1004] In 2004, a Congressional subcommittee began looking into the matter.[1005]

In 2005, after Jose Canseco published a tell-all biography that named other alleged steroid users, including Mark McGuire, Major League Baseball finally implemented a drug testing plan that reversed years of resistance by MLB Commissioner Bud Selig and players union representative Donald Fehr.[1006] The question then became one of whether unprincipled chemists could come up with performance-enhancing drugs whose presence in the blood would be undetectable. The year 2007 saw a joyless eclipse by Barry Bonds of Hank Aaron's career home run record of 755.[1007] Three years earlier, Bonds had admitted "inadvertent" steroid use.[1008] By 2008, pitcher Roger Clemens and slugger Alex Rodriguez had also been accused of doping, along with Sammy Sosa.[1009]

To the extent that I watch baseball nowadays, the question that remains for me is that of whether I'm watching human beings or

---

[1001] Ken Burns and Lynn Novick, *Baseball: The Tenth Inning* (DVD) (Florentine Films, 2010).

[1002] Ibid.

[1003] Ibid.

[1004] Ibid,

[1005] Ibid.

[1006] Ibid.

[1007] Ibid.

[1008] Ibid.

[1009] Ibid.

chemicals perform. It all goes back to the question of selfhood, as I see it. The Barry Bonds who as a 42-year-old player with the San Francisco Giants surpassed Hank Aaron's home run record: is he the same Barry Bonds as signed with Pittsburgh some fifteen years earlier? If he's now gone beyond the natural limits of his ability to improve, is he somehow transcending the metaphysical boundaries of self? Is there a teaching, somewhere in all of this, about being born again? Or are we simply looking at an unheeded lesson about working with what God has given you? Where does respect for God's handiwork end and hubris begin?

In the past, I would not have been giving our national pastime the cold shoulder as I am now. This author, like millions of American boys, grew up not just watching and reading about but also playing baseball, albeit in a meadow where cow pies could easily be mistaken for bases. Some of my fondest memories are of occasions when I was able to "give it a ride"—i.e., get ahold of a pitch and send the ball deep into the outfield, maybe even over the single-strand barbed wire fence that enclosed left field of our rude ballpark. Naturally, I had dreams of making it to the major leagues, "the show." I would see Mickey Mantle's picture on the cover of a magazine and think, "If he could do it…" Today, Christie Wyckoff's pasture; tomorrow, Yankee Stadium.

Dreams of personal success in sports or music or other endeavors are first cousins to one of the other grand narratives of Western culture: progress in general. Referring to yet another Baudrillard essay, Prof. Berger states:

> One of the great metanarratives that sustained us—the idea of progress—has been tossed, (Baudrillard) suggests, onto the ash heap of history. We've gone as far as we can go, Baudrillard asserts, and history as we know it has come to an end. Postmodernisn represents, then, the "real" game of *Survivor* in a world where everything has, so it seems, been emptied of meaning.[1010]

---

[1010] Berger, *Portable Postmodernist*, 23, referring to Jean Baudrillard, "On Nihilism," *On the Beach* 6 (spring 1984), 38-39.

# IDENTITY AND ORIENTATION

Emptied of meaning just like Fish's Pond.

With the disappearance of metanarratives, we seem to have lost our ability to talk intelligibly about the things on which progress and the other grand narratives are premised. On page 42 of *The Portable Postmodernist*, Dr. Berger sets forth an essay from a K. Wilhelm Abian. I will not quote any part of this writing, since it's classic pomospeak and you wouldn't understand it. In fact, nobody would. That's just the point, as it turns out. On page 43, Berger admits that the essay the reader has just read is completely meaningless and was in fact generated using the Postmodern Generator, a program that was available on the Internet when Berger downloaded it.[1011] This program takes a number of terms used by postmodernist writers and allows the user to generate essays that include postmodernist jargon but, as Berger describes them, "are nothing but gobbledygook, meaningless essays that look like postmodernist writings but don't make any sense."[1012] Some people will say that all postmodernist writing is nonsense, Berger acknowledges, while others will assert that this style of communication is merely a way for the so-called intellectual elite to "snub their noses at middle- and working-class people who love Disneyland and shopping malls."[1013]

The above calls to mind another hoax, described by Lehigh University professor Steven Goldman in one of his *Science Wars* lectures, in which the joke was similarly on the intellectuals.[1014] The background of this deception was that by the 1990s, science-technology-society studies were pointing to the possibility that supposed scientific knowledge was socially or anthropologically based, and therefore created rather than discovered.[1015] In the mid-1990s, New York University physicist Alan Sokal submitted, to the social constructionist journal *Social Text*, a paper purporting to characterize the quantum gravity theory as socially constructed and

---

[1011] Berger, *Portable Postmodernist*, 43.

[1012] Ibid.

[1013] Ibid.

[1014] Goldman, *Science Wars*, lecture 22.

[1015] Ibid., lecture 20.

ideologically based.[1016] In fact, this paper, which was ultimately published, was physics nonsense interwoven with terminology that Sokal had acquired through a brief immersion in postmodernist literature.[1017] As such, it was a spoof of the way postmodernists had been talking and writing about scientific matters. Postmodernists were outraged, but others saw Sokal's hoax as pointing to the conclusion that postmodernism was intellectually bankrupt.[1018]

This might be an opportune time to review and reflect on what we've just learned (or perhaps already knew) about postmodernism:

(1) Pastiche, defined as incongruity in how elements are combined, is big. Reality is a mishmash. How would you expect it to be otherwise when (according to postmodernists) there's no God to fit things together!

(2) Life is lived for the camera or its equivalent (the eyes of as many other persons as possible). Under these circumstances, human existence is nothing more than a dramatic production. What with all the playacting, where is the real you to be found?

(3) Visual imagery dominates, since people no longer know how to use words. How, then, do you go about describing God? Is He spirit? Superman? Statuary?

(4) Reality is hyperreality (everything simulated, maybe even God). But how do you conduct your life if you, rather than God, are responsible for the purported reality of things?

(5) Plastic is pleasing (a la Disney). Sadly, this seems to apply to people as well as amusement parks.

---

[1016] Ibid., lecture 22. Quantum gravity, touched on elsewhere herein, is gravity that obeys the quantum principle. See, in this regard, Kaku, *Parallel Worlds*, 396; also the "Big Bomb" section of my Paradigm Paralysis chapter, supra.

[1017] Goldman, *Science Wars*, lecture 22.

[1018] Ibid.

# IDENTITY AND ORIENTATION

(6) Death is no longer an absolute, since rich folks can now get cryogenized. But can they elude the Grim Reaper forever?

(7) The "love of my life" is only the affair of the moment. Who, then, do you count on for emotional support in the long haul?

(8) Cheating in sports (use of performance-enhancing drugs, principally) is acceptable, as long as you don't get caught. How about taking shortcuts in other areas? Do you really want to live in a world in which there's no operative merit principle?

(9) Human progress is an illusion. The advancement of society is usually brought about by heroes and heroines, is it not? Where have all these folks gone?

(10) There is and probably will continue to be phony erudition in philosophy, science, and other disciplines. Does this signal an abandonment by Western thinkers of serious truth seeking? Stephen Hawking and Ludwig Wittgenstein have suggested as much.[1019]

Are you beginning to see a pattern, gentle reader? A unifying theme that we've already touched on? Something that we can identify, for the benefit of Christians or prospective Christians with whom we're working, as the most basic error in postmodern thinking that must be overcome? Yes. We've already said it, repeatedly. It is that NOTHING IS REAL.

Derrida, Lyotard, and Foucault have put it in slightly different terms, but the basic message is still the same: what you see before you is not necessarily something you can rely on, certainly not in support of any grand philosophical narrative. All you can put your trust in is the exercise of human power. Corrupt human power.

---

[1019] See Hawking, *A Brief History*, 190-91; also the discussion in this regard in the "Transcendentalism" section of my Essence chapter.

That is, unless you're leaning on the everlasting arms of God (Deuteronomy 33:27).

*Out of the Mist*

The lack of personal identity and the moral disorientation that are being suffered every day by citizens of Postmodernia need remedying if ours is to be a viable society. Not surprisingly, if you've caught my drift over the past however many pages, the remedy I'm offering is the triune God and everything in the way of aids to understanding that He has provided. So, we'll begin the final section of this chapter by returning to the slogans of the postmodernist philosophers whom professor James K. A. Smith has referred to as "something of an *unholy* trinity,"[1020] then we'll see how Christianity exposes their watchwords as almost total misrepresentations of reality, and finally we'll demonstrate through Scripture (the only unassailable source of knowledge, given the unreliability of science and human reason) that the attributes of the God of the Bible are such that He can deliver trusting souls from the state of mental confusion that is postmodernism.

Jacques Derrida has told us, "There is nothing outside the text."[1021] What you hear is what you get. Hearing is subjective, however; what you hear could mean something else to someone else, and it could mean something different to the same person tomorrow. Also complicating the task of understanding the text, according to Professor Smith, is the problem that every interpretation of the text relies on other, foundational interpretations; thus, it's said, you'll never get to the point where you can simply read a text without having to worry about varying interpretations.[1022] As previously noted, the problem doesn't seem to be resolved by the nostrum, "there is nothing outside *context*."[1023]

The principal text that we need to be concerned with is the Bible. This magnificent document, consisting of sixty-six separate

---

[1020] Smith, *Who's Afraid*, 21.

[1021] Ibid., 21; see also "Essence of Postmodernism" discussion supra this chapter.

[1022] See Smith, *Who's Afraid*, 37-38.

[1023] See, again, the "Essence of Postmodernism" section supra this chapter.

books or epistles, written by forty or forty-one different human authors[1024] over a period of almost sixteen hundred years, shows a consistency in its content that for me conclusively proves that the writings collected therein had only one author, God in the person of the Holy Spirit. This being the case, what need do I have to resort to philosophy, mysticism, or science to help me understand spiritual things.? Wouldn't unitary divine authorship imply that all the truth that's needed for man's spiritual wellbeing is contained within the Bible's covers? Would the Spirit of God have given me less than I need in order to be able to walk with Him (Galatians 5:16, 25)?

Have I therefore been wasting my time trying to understand, for example, how the Eleatic philosophers could have said there's no such thing as change and how their "no change" would affect my ontological underpinning, my understanding of the nature of my being? Can metaphysical truth really be breathed more audibly by the mystic crashing surf than by the written Word of God? Do I really need to know that particles in the quantum world exist in multiple states in order to understand how God's omniscience and human autonomy can co-exist? It has been said that the best aid to understanding the Bible is the Bible. The idea that Scripture should be read against Scripture would of necessity agree with Derrida's "nothing outside the text," would it not? So what exactly is the problem with postmodernism as far as biblical interpretation is concerned?

The above queries can best be answered, I believe, with a simple, personally heartfelt observation: If we candidly look at the natural world as an aid to understanding the beyond, we must conclude that God has not left us clueless. I do not mean to denigrate Scripture by saying this. My point is simply that it is inconceivable that God would have created a home for us, the earth and the cosmos, and not have left something of Himself in His creation. That is simply not how a master builder, a true artisan, operates. The construction always reflects the constructor. The closeness of the relationship between the Creator and His creation is evident even in the first

---

[1024] The authorship of Hebrews has never been conclusively established, although there is reason to believe that the writer was the apostle Paul. See Morris, *Study Bible*, Intro. to the Epistle to the Hebrews.

chapter of Genesis, where it is stated that God "saw everything that he had made, and behold, it was very good" (1:31).[1025]

In this case, more than pride of workmanship is involved: God has hinted through His *world* what the explanation of His *Word* is (Romans 1:20). But what He has shown us in some instances is not exactly what we would have expected. We see truth staring out at us from underneath the counterintuitive. We entertain the thought that inanimate water may be able to speak to us. We are not perplexed by the idea that the state of a subatomic particle can imply simultaneously-existing opposites on a macroscopic scale — for example, that a cat in a poison-dispensing quantum system such as Erwin Schrodinger's is both dead and alive until observed, or that, similarly, a person is simultaneously saved and condemned to hell until a final decision is made by that person to accept or reject Christ's gospel. The paradoxical melds with the plain, with direct revelation, and informs our understanding of the latter. What it amounts to is a kind of reverse revelation: God has placed in His universe some phenomena that are so striking that we're compelled to conclude that they must be clues regarding spiritual things, but the significance of the preternatural can be fully grasped only with the aid of Scripture. Thus, there *is* something outside the text, but it leads back into the text.

Jean-Francois Lyotard, as we know, has defined postmodernism as "incredulity toward metanarratives."[1026] Metanarratives, *grand recits*, are "big stories."[1027] A narrative fitting this description is not to be believed, Prof. Smith tells us in his interpretation of Lyotard, unless it can legitimate itself by an appeal to universal reason.[1028] The story must not only be grand; it must be believable.[1029]

---

[1025] For further discussion regarding the revelation of a skilled Creator in nature, see Randy J. Guliuzza, "Engineering Principles Point to God's Workmanship," *Acts & Facts* 46, no. 6 (June 2017), 16-19.

[1026] Smith, *Who's Afraid*, 22; Lyotard, *Postmodern Condition*, xxiv; and see, again, the "Essence of Postmodernism" discussion supra this chapter.

[1027] Smith, *Who's Afraid*, 63.

[1028] Ibid., 65.

[1029] Ibid.

## IDENTITY AND ORIENTATION

Without question, the Bible presents a "big story." The narrative begins with a man and a woman being created as innocents and placed by God in a perfect environment, the Garden of Eden; however, their failure to obey the Creator's command (to not eat the fruit of the tree of the knowledge of good and evil) makes it necessary for the Lord to impose consequences that include expulsion from the Garden (Genesis 1:26—3:22). Next, God places mankind under a rule of conscience, which unfortunately does not keep one of the first couple's sons (Cain) from killing his brother (Abel) because the brother's sacrifice is acceptable to God while his is not. God's ruling that the only acceptable offering is a blood animal sacrifice foreshadows the coming of Jesus, who will be an atoning sacrifice in behalf of all believers (John 3:14-15; Romans 5:9; Colossians 1:14, 20; 1 Peter 1:19).

The rule of conscience also does not prevent widespread evil on the earth; therefore, God sends a worldwide flood, which only Noah and his family survive, along with pairs of every species of land animal then existing. (Genesis 3:23—8:22.) Following the flood, the surviving humans are directed to repopulate the earth and are permitted to eat animal flesh for the first time, subject to certain restrictions. Also authorized for the first time is capital punishment, characterized as an essential feature of human government. All the families of the earth speak the same language until men presumptuously attempt to build a tower with a top that "may reach (up) unto heaven," at which point the human race is scattered by God and its speech is "confused" by Him into a multiplicity of tongues. (Genesis 9:1—11:9.)

After that, the patriarch Abraham appears on the scene and is promised that he will be the progenitor of a great and blessed nation, provided he and his descendants stay in the land where God has put them. A famine in their land causes the Israelites to migrate to Egypt, nonetheless, and they are eventually treated poorly by Egyptian overlords, which brings about the return of the entire nation to the land of Canaan. (Genesis 12:1—Exodus 19: 2.) Ostensibly to preserve and protect the Israelites, God now gives them detailed law regarding spiritual and public health matters, but they fail to meet the requirements of the (Old Testament)

Law in spite of the exhortations of prophets, judges, and patriarchs. This demonstrates that God's people cannot save themselves; they need a Savior.

The Savior will eventually come, in the person of Jesus, and by his sacrifice (crucifixion) satisfy the Law. The so-called Age of Law therefore ends at the cross. (Exodus 19:3 — Matthew 27:35.) On the day of Pentecost, following Christ's resurrection and ascension to heaven, the Holy Spirit comes and empowers the believers (Acts 2:1-4), thus initiating the Church Age, which will continue until the faithful are taken up from the earth in the rapture of the church (1 Thessalonians 4:13-17). Finally will come the Kingdom Age, when Christ will return in power, rule over the earth for a thousand years, defeat the forces of evil at the battle of Armageddon, seal the doom of Satan, and judge both believers and nonbelievers (Joel 2:1-12; Zechariah 14:1-5; Revelation 19:11 — 20:15). Believers will then live eternally with the Lord in a new heaven and new earth, where there will be no more sin and God will "wipe away all tears from their eyes" (Revelation 21:1 — 22:3).

The legitimation of the above story can hardly be said to be an issue, given the witness of God's written Word. The Bible contains eyewitness accounts not only of the events by which God preserved the Israelite nation, the sacrificial milieu into which the Lamb of God would be born, but also of the miracles by which Jesus of Nazareth established himself as Jesus the Son of God. These events had been spoken of in detail by the Old Testament prophets. The Jews, as it happened, were awaiting a Messiah who would lead them militarily in shaking off Roman rule, rather than a king whose kingdom was not of this world (John 18:36). The Roman government likewise did not favor attributing deity status to Jesus, since this would draw worship away from the Emperor and because the Roman rulers were bent on placating the Jews whenever they could. Under these circumstances, the truth about Jesus would probably not have been given a warm reception by any but those persons who were in Jesus' circle of influence. Thus, it does not seem likely that New Testament believers would have attempted to show that Jesus was God incarnate by *making up* stories of miracles, including His resurrection from the dead. Such claims, which almost certainly would have brought harsh intervention by the Roman rulers, would not have been lightly made. Thus, the legitimation of the narrative that would later become the Holy Bible consists in the fact that this *grand recits* was told at all.

Again, we hear Michel Foucault, the third member of our "unholy trinity," proclaiming, "power is knowledge."[1030] Might makes right, to borrow an expression. A claim to have knowledge of something arises because power relations are lined up in such a way as to allow that claim to be made. We should nonetheless not bemoan the application of power, Foucault says, since power gets things done.[1031]

The above-described way of doing business is not counseled in the New Testament for obvious reasons. An "every man for himself" morality is not the way to a livable world, which will follow only if there is a universal application of the "golden rule" (Matthew 7:12; Luke 6:31). More is involved than the expedient of treating others well so that they will do the same toward you, however. Christlikeness, which is proof that a person is fit for eternal life with God, is a mandate that entails sharing our Lord's suffering (Matthew 16:24-25; 1 Peter 4:13).

Thus, in the Sermon on the Mount, Jesus describes as "blessed" (happy), not the powerful but rather the "poor in spirit" (Matthew 5:3), the meek (5:5), and those who are reviled and persecuted (5:10-11). This hardly sounds like the possession of truth being claimed by someone after eating "fruits of a poisoned tree of power relations."[1032] When the mother of James and John the sons of Zebedee approaches the Master with a request that her sons be permitted to sit on His right and left side in His kingdom, Jesus characterizes such privilege seeking as the way of the princes of the Gentiles, and states, "whosoever will be chief among you, let him be your servant" (Matthew 20:25-27). After all, Christ himself took on the form of a servant in coming to Earth and dying for our sins (Philippians 2:7-8). He did not come as a sovereign.

---

[1030] Ibid., 22; and see Foucault, *Discipline and Punish*, 27.

[1031] Foucault, *Discipline and Punish*, 194.

[1032] Smith, *Who's Afraid*, 87.

The power that's promised to the disciples just prior to the Lord's post-resurrection ascension, it should be emphasized, is "from on high" (Luke 24:49). It's not political or in any way garnered through human strategy, and it's not a matter of "our needs (interpreting) the world."[1033] Instead, it accompanies our giving to God "glory in the church by Christ Jesus throughout all ages, world without end" (Ephesians 3:21). The focal point is God, not man.

The word "effluvia," used earlier in this chapter, applies to exhalations that are "disagreeable or noxious."[1034] This would certainly describe some if not all of the byproducts of postmodernism identified above, whose unhealthiness hardly needs spelling out. Lack of metanarratives means there's no role for heroes.[1035] Lacking heroes, we have no one to emulate with society-benefitting behaviors. If there's no underlying idealism, the legitimation of knowledge will be solely in terms of power and profit.[1036] There will be no reason to know anything except as it serves one's need to command and control. No need to think great thoughts or, for that matter, to think any thoughts at all. If there's "nothing inside" as far as human character is concerned, then it will not be surprising to find people preoccupied with "image" and living lives that look like staged productions.[1037] If our talk cannot be said to refer to anything outside itself, then what or who is left that's real? Entitlement to belief would no longer extend even to our Lord, who according to the Bible holds the keys to life and death (Revelation 1:18), if in the era of cryogenics death as a fact of life loses its reality.[1038] If records are all that matters in professional sports, then cheating will be prevalent and proffered performance will not be real or available to inspire us toward our own achievements.[1039] If romantic partners are tried on like a pair of jeans, then marriage and fidelity

---

[1033] Nietzsche, *Will to Power*, 481 (1885-1888).

[1034] See "effluvium" (sing.), *Random House Webster's*, 622.

[1035] See Lyotard, *Postmodern Condition*, 14.

[1036] Ibid., 46.

[1037] See Berger, *Portable Postmodernist*, 14-15, 57.

[1038] Ibid., 67.

[1039] See, generally, Burns, *Baseball, Tenth Inning*.

will no longer have any reality as features of our lives.[1040] On and on goes the list of ways in which postmodernism has emptied our lives of meaning.

Even the most mundane occurrences, the things that happen to us every day that are not life-changing, reflect the postmodern condition. I think of the jet skiers, who infest every body of water of any size in upstate New York and whose cacophonous noise I'll have to listen to all summer. These guys are first cousins to the yahoos who are always hollering "GET IN THE HOLE!" at golf tournaments the instant the ball leaves the tee. Your personal worth, in both instances, is measured by the volume of sound you produce, it seems. It doesn't matter that what you're doing is annoying or distracting; the important thing is you. To heck with the other guy, who has no rights; let the poor schmuck plug his ears if he doesn't like what he's hearing.

You see, not only have we lost our ability to be genuine (what you see is what you get); we've also ceased being capable, if we ever were, of seeing things from the perspective of others, of getting outside our innate mental texts so as to be able to hear from a transcendent authority (God? conscience?) with respect to our behavior. Ironically, the inconsiderate folks referred to above, if pressed, would most likely mouth a "live and let live" philosophy as far as the other guy's religion is concerned, since postmodernism's "no truth" maxim validates any particular religious persuasion as against every other system of belief.

We've already recounted most of the negatives above, you'll note. What we haven't talked about is the ability and willingness of God to overcome the deficits (let's be bold and call them evils) of postmodernism. I bring the Almighty into the equation at this point, asking Him to lift the curse of postmodernism, first and foremost because of Paul's counsel to the church at Philippi (Philippians 4:6): "Be careful (i.e., anxious) for nothing, but *in every thing* by prayer and supplication with thanksgiving let your requests be made known to God" (emphasis added). There are things about God that make Him more than adequate to fulfill the role of prayer answerer, specifically:

---

[1040] See Berger, *Portable Postmodernist*, 96-97.

## WATER

(1) His power is unlimited, as stated in Matthew 19:26 and Mark 10:27 ("with God all things are possible."). Nothing is too difficult for Him (Genesis 18:14). His power would have to be without limit to allow Him to be the creator of the universe (Genesis 1:1; John 1:3). Everything cannot come from less than everything. His power is recognized by the heavenly host throughout the universe.[1041] The apostle John, regarding his vision on the island of Patmos, records that he "heard as it were the voice of a great multitude, and as the voice of many waters, and as the voice of mighty thunderings, saying, Alleluia: for the Lord God omnipotent reigneth" (Revelation 19:6). Infiniteness of power implies that power's transcending all time limits; thus, God's power is eternal (Romans 1:20) and He is "our God for ever and ever" (Psalm 48:14), who "ruleth by his power for ever" (Psalm 66:7).

(2) God's power is a power that is applied with wisdom. His thoughts are "higher than" our human thoughts (Isaiah 55:9). Indeed, He is referred to as "the only wise God" (1 Timothy 1:17). As stated by Hannah the mother of Samuel, the Lord "is a God of knowledge, and by him actions are weighed" (1 Samuel 2:3). His wisdom is "manifold" (many-faceted), Paul observes at Ephesians 3:10. "(H)is understanding is infinite," declares the psalmist (147:5). His foolishness "is wiser than men," Paul says (1 Corinthians 1:25).

(3) Thankfully, God "is not a man that he should lie" (Numbers 23:19). Indeed, it is impossible for Him to lie (Titus 1:2; Hebrews 6:18). We can trust Him wholeheartedly not only because His complete integrity has been declared and demonstrated to us in his Word, but also because of His great love for us, made known through the substitutionary sacrifice of God the Son on the cross to pay our sin debt (1 Corinthians 15:3; 2 Corinthians 5:21; 1 John

---

[1041] "Heavenly host" is here used to signify God's angels. See Morris, *Study Bible*, nn. to Daniel 8:10, Luke 2:13, and Acts 7:42.

4:10). Through Christ's suffering and cruel demise we as believers are set free from the law of sin and death (John 3:16; Romans 5:12; 6:23). Would a God who would thus die for us lie to us?

(4) He knows everything that's happening in His universe. All creatures are "manifest in his sight," all things "naked and opened to the eyes of |God|" (Hebrews 4:13). There is not a sparrow that falls to the ground without His knowing of it (Matthew 10:29). His eyes "run to and fro throughout the whole earth, to shew himself strong in the behalf of them whose heart is perfect toward him" (2 Chronicles 16:9). Thus, David can acknowledge, "there is not a word in my tongue (i.e., not yet spoken), but lo, O Lord, thou knowest it altogether" (Psalms 139:4).

(5) He brings to pass the things He has willed. See Isaiah 46:11, containing words of God uttered through the prophet: "I have spoken it, I will also bring it to pass; I have proposed it, I will also do it." Language to the same effect is found at Numbers 23:19, where the prophet Balaam, querying rhetorically regarding God's veracity, states, "hath he said, and shall he not do it? or hath he spoken, and shall he not make it good?"

(6) He is an immanent God, in that He indwells the universe and time while He also transcends them. We have access to Him simply by faith (Romans 5:2). This access is possible, as noted earlier, because when God the Son "gave up the ghost" on the cross, the curtain in front of the Holy of Holies, God's inner sanctum, was torn in two (Matthew 27:50-51; Mark 15:37-38). Symbolically, the tearing of the veil meant that Christ, by the shedding of His blood, had opened the way for all believers to enter directly into the presence of God.[1042]

---

[1042] See Morris, *Study Bible*, n. to Matthew 27:51; also, Hebrews 10:19, 20.

Therefore, we as individuals can approach with confidence the enterprise of restoring personal identity and orientation to the extent that the same are lacking in our lives and in the lives of others. How, you ask, is this seemingly gargantuan task undertaken? With the understanding, first of all, that no effort toward achieving spiritual wholeness will succeed without God's integral involvement. Basically, what this translates into is traditional prayer and Bible study, two activities eschewed by the New Age/Emergent Church camp, whose members seem to be geared toward listening to their own heartbeats rather than hearkening to the "still small voice" of God (1 Kings 19:12).

The technique of the Emergents, touched on elsewhere, is contemplative prayer, which involves *inter alia* repeating key phrases or words. However, Jesus counseled against using "vain repetitions" in prayer (Matthew 6:7), and nowhere in Scripture is the simple repeating of spiritual words or mantras encouraged. Thus, it's hard for me to believe that Elijah, when he heard the "still small voice," was pursuing contemplative-spirituality solitude and silence, as has been suggested.[1043] At that moment Elijah appears instead to have been running for his life from the wicked queen Jezebel (1 Kings 19:1-12).

What is needed rather than the New Age falderal described above, if someone is to be of use to Christianity, is simply a mouth that will say words acceptable to God and a heart that is open to the Lord's leading (Psalm 19:14). This does not equate to a spirituality that allows certain persons to claim insider (privileged) information concerning things spiritual. I'm speaking of the so-called spiritual directors who are showing up more and more in the evangelical church to teach mystical prayer.[1044] These people are outsiders as far as the process of personally contacting God is concerned, in my estimation.

Spirituality should be informed by Scripture rather than spiritual directors. As I've suggested elsewhere, New Age/Emergent Church folk appear to regard the Bible—the historic Bible—as a

---

[1043] See Gilley, *Out of Formation*, 85-86.

[1044] See Yungen, *A Time of Departing*, 50-51, 67, 83, 90, 172, 205.

challenge to their personal spiritual hegemony.[1045] With constant reference to the Word, the believer or would-be believer must continually be reminded of who God is, what He's done, and why He's done it (the Father of all, who has given us the potential Savior of all, because He's the lover of all). With this awareness in the forefront of his consciousness, the committed Christian must be prepared to approach persons who are of a postmodern bent and "(speak) the truth in love" to them (Ephesians 4:15).[1046]

*In the Book*

Before we proceed farther—which is to say, before we complete the final leg of this our earthly journey together—I would like to give you some counsel for persons who are not yet regular readers of God's written Word. It should help you as you try to sort out where you stand personally. My first piece of advice is that a person should not assume that his or her reading habits have anything to do with salvation. You heard me right. What we're talking about at the moment is one's daily walk with the Lord, not the leap of faith that initially gets you into God's heavenly kingdom. An important part of a person's spiritual walk as a confessing Christian, which should be happening every day, is Bible reading.

Methodism founder John Wesley did much of his study of Scripture on horseback as he rode from one English town to another.[1047] I'm not saying that our new Christian needs to buy a horse, but you get the idea: a believer should use any available time to enlarge his or her knowledge and understanding of the written Word. While daily Bible reading won't necessarily make you a Christian, it will make you a better Christian once you've accepted God's gracious offer of salvation. "All scripture is given by inspiration of God," the apostle Paul tells his follower Timothy, "and is profitable for doctrine, for reproof, for instruction in righteousness:

---

[1045] See, again, Smith, *"Wonderful" Deception*, 100-103, 105-07, 118-19. 125-26, 157-60, 187-88.

[1046] See Gary Gilley's discussion in this regard in *This Little Church Stayed Home*, at 45-51.

[1047] See Shelley, *Church History*, 337.

that the man of God may be perfect, thoroughly furnished unto all good works." 2 Timothy 3:16, 17. In a preceding verse, 2 Timothy 3:15, Paul states to his protégé that the holy scriptures "are able to make thee wise unto salvation," but redemption is still "through faith which is in Christ Jesus."

Indeed, the heavy lifting, allowing one's initial entry into God's eternal kingdom, may have been done before an unsaved person has read any Scripture at all, or at least anything beyond John 3:16. Salvation requires only that you believe the basic Gospel message and confess Jesus as Lord (see John 3:14-16, 36; 5:24; Romans 4:3-5; 10:9; and 1 John 4:15), that you make a simple avowal of Christ as the heaven-sent Redeemer. For some, nevertheless, salvation may not be a quick-and-easy acceptance of God's grace. Conversion to saving belief may be the result of a long, laborious process, as it was in the case of C. S. Lewis, who step by mincing step went eventually from atheism to a career as Christianity's foremost man of letters.[1048]

In the end it was not Lewis's mental labors that brought him the prize, but rather God's work, gratefully accepted by Jack (Lewis's nickname). In his first letter to the believers at Corinth, Paul states, "(N)o man can say that Jesus is the Lord, but by the Holy Ghost." 1 Corinthians 12:3. So if you've chosen to welcome Jesus into your heart, perhaps even after having considered the intellectual arguments for accepting Christ as savior), kindly refrain from trying to take the credit for your salvation. You've become a member of the family of God not because of your own cleverness or cleanliness but rather by reason of God's freely-given grace, as mentioned in the discussion above (Romans 3:24; Ephesians 2:8-9).

The depth of true faith, it turns out, is found at the surface of things. One sense arouses another, and the latter in turn arouses the intellect. Who could listen to Edvard Grieg's music without enjoying visions of spring's streams cascading down into Norwegian fjords, thereby describing the miracle of winter melting into spring and summer, a metaphor for the newness of life that we enjoy through Christ Jesus (Romans 6:4)? Who indeed.

---

[1048] See Brown, *A Life Observed*, 111-58; also, C. S. Lewis, *Surprised by Joy: The Shape of My Early Life* (New York: Harcourt, Inc., 1955), 197-238.

# 9
# CONCLUSIONS

It's now time to sum up, as a lawyer customarily does in his or her closing argument at the conclusion of a trial. Summing up goes most smoothly, I've found, when there's an adjournment between the final offer of proof and the arguments of counsel, which maximizes the opportunity afforded the attorneys to organize the facts and arguments on which they're relying. It's a different story with respect to personal evangelism. We do have the Master's direct assurance that the Holy Spirit will tell you what to say when you're brought before various kinds of tribunals (Luke 12:11-12), but apostle-generated Scripture advises that as a believer you should *"be ready always to give an answer"* to anyone who asks you about your faith (1 Peter 3:15, emphasis supplied) and that you should *"(s)tudy to show [yourself] approved unto God, a workman that [needs] not to be ashamed, rightly dividing* (i.e., interpreting and teaching) *the word of truth"* (2 Timothy 2:15; emphasis supplied, parentheticals substituted or added). It is in the spirit of the above verses, gentle reader, which taken together seem to counsel a sort of study-supported abandon in presenting the Gospel, that I offer you the summary that follows.

Our final focus in the foregoing discussion, you'll recall, was postmodernism. Reduced to its simplest explanation, postmodernism is about the appearance of things, as communicated through language, never quite matching reality. Knowledge is never complete because we never hear what newsman Paul Harvey would have called "the rest of the story." And even if we penetrate the linguistic haze, we find no grand narrative waiting to lift us into a clear blue sky from which we can see a sharply delineated horizon. Moreover, it may be the case that although we're able to see and hear with a measure of objectivity, the knowledge that we possess has been contaminated by being too closely associated with power. As Prof. James K. A. Smith has indicated, the above are all conditions that are inimical to Christianity.

Since clarity and completeness can hardly be said to be watchwords of postmodernism, it's no surprise to find that historic Christianity is not by and large the religion that's being offered to the postmodern world as a cure for its nihilism. A studied approach to faith would require too much in the way of attention span, I'm surmising. Nor, any more, does the nod go to the slickly packaged "entertainment" church devised by Rick Warren and Bill Hybels, in which you "do church" by sitting back and enjoying a non-challenging concert, stage production, or other holy amusement.[1049] Cultivating a crop of spectators, no matter how good the show, is simply not sufficient when we're seeking to give believers the ability to make a clear and forceful presentation of the Gospel.

The answer now, we're told, is the Emergent Church, whose strategy is to vault back over a number of centuries and adopt early church practices, mystical in nature, that were (and still are) uncounseled by Scripture.[1050] The Emergent Church movement assumes the legitimacy of questionable observances engaged in by the "desert fathers," monks who fled to Egypt to escape the worldliness of the medieval Roman church. If the standard is that of whether any particular ancient/modern practice is provided for or encouraged by Scripture, then the questionable observances (my phrase) of the Emergent Church are myriad. A number of these practices have been detailed in sources already referred to in this book.

---

[1049] See, generally, Gilley, *This Little Church Went to Market*, 9-117.

[1050] See Yungen, *A Time of Departing*, 155, 159, 174; also, Gilley, *Out of Formation*, 7-13, 17-40.

## CONCLUSIONS

Two of the more useful resources regarding New Age ancient/modern mysticism are the late Ray Yungen's *A Time of Departing*, previously mentioned herein, and pastor Gary Gilley's *Out of Formation*. These summaries describe the practices of

(1) *lectio divina*, so-called sacred reading, described in detail earlier, which involves taking a single word or short phrase from Scripture and repeating it over and over,[1051]

(2) reciting sacred words in the context of *contemplative* or *centering prayer*, after supposedly emptying the mind in order to make room for God and entering into *"the silence,"*[1052]

(3) saying *"breath" prayers*, by taking a single word or short phrase and repeating it in synchronicity with one's breathing,[1053]

(4) saying to oneself or reciting aloud *mantras*, consisting of a word or words repeated again and again to produce a trance-like state,[1054]

(5) resort to the services of a *spiritual director*, one who trains others in the *spiritual disciplines*, including "the silence,"[1055]

(6) contacting an imaginary *spirit guide*, said to already reside within the mystic, with regard to specific matters of concern,[1056] and

---

[1051] See Yungen, *A Time of Departing*, 204; also, Gilley, *Out of Formation*, 63-80.

[1052] See Yungen, *A Time of Departing*, 33, 37, 40-41, 43-46, 64-65, 81-84, 130, 142-45, 152, 172-73, 181, 202-3; also, Gilley, *Out of Formation*, 41-62.

[1053] See Yungen, *A Time of Departing*, 33, 75, 79, 145-46, 148-50; also, Gilley, *Out of Formation*, 59.

[1054] See Yungen, *A Time of Departing*, 33, 42, 86, 98, 108, 149-50, 202, 204.

[1055] See ibid., 41, 50-51, 67, 90, 205.

[1056] See ibid., 20, 95, 98, 103, 118.

(7) *creative visualization* — imagining a desired result and expecting it to happen — basically, creating one's own reality.[1057]

The ancient/modern mystic's life, as best it may be pictured from descriptions of the practices mentioned above, would seem to contrast sharply with the existence enjoyed by a nature mystic like C. S. Lewis or by a seashore habitue such as one of the individuals quoted in the "Breathing Water" section of my Region of Awe chapter. The disconnect between the ancient/modern practitioner and Nature is not a complete one, however. Emergent Church mysticism consists not only in a set of practices assiduously carried out; it also includes a world-view of either pantheism (God *is* all things, including Nature) or panentheism (God is *in* all things). Such a *weltanschauung* is necessary if the New Age believer is to maintain his or her position that the Divinity resides in us all. If the Divinity resides in us all, nobody gets judged.

As might be expected and as noted earlier, New Age thinking puts very little stock in the Bible. It is apparently the thought of the Emergents that Holy Scripture would stand in the way of "unmediated contact with ultimate power," which Prof. Luke Timothy Johnson has identified as the object of all mystic endeavor.[1058] Lacking mysticism's "unmediated contact," man has to (and should) be content with Christ's advocacy for us with God the Father (1 John 2:1) and with the revelation of God through his Word, which, it happens, is "quick, and powerful, and sharper than any twoedged sword" (Hebrews 4:12). Understanding and approaching the Father under the terms of the revelation given through the Son is nonetheless thought by the typical New Ager to be inadequate as a way of proceeding. It is better, our postmodern mystic seems to be saying, that he or she not have to deal with any constraints that a book of doctrine such as the Bible might impose on his or her thoughts or actions.

Other than the fact that there may be no readily apparent outward-directed reason for a mystic's being, what can be said about

---

[1057] See ibid., 19, 203.

[1058] See Johnson, *Mystical Tradition*, lecture 1.

the life that he or she presumptively possesses and hopes will ultimately be synonymous with God's existence (after purgation, illumination, and the dark night of the soul)? Let's postulate that it's not going to be sufficient merely to say that the subject's life is defined by God's life. Once you're in a state of union with God, after all, who or what is going to be left for you to worship or serve? Is there, at that point, any "why" that's connected with your existence? The purpose of your existence aside, what exactly is the *nature* of your being? What is it that will be left of you physically or mentally after you die? Let's assume that the puzzles of the purpose and nature of our existence can't be solved simply by your dispatching all the demons and snuggling up to the Divinity. Let's assume, to put it another way, that you can't seem to obtain the desired depth of knowledge through New Age ancient/modern mysticism, which appears to be capable of delivering good feelings but not much else, certainly not reliable metaphysical truth. What then? Where do you go from there?

At such a point as the above, may I be allowed the temerity to suggest, it would seem necessary to ponder the central question (what is life?) without regard to what might come out of "the silence," lectio divina, or whatever other ancient/modern mysticism technique is practiced. Is there really something fundamentally wrong with asking our familiar Judeo-Christian God for wisdom by means of prayer that's understandable as language spoken in sentences? The intelligibility of "normal" prayer, I'm convinced, helps focus the asker on the need at hand, which in many if not most situations should be beneficial. More beneficial, it strikes me, than vain repetition of words that fail to identify the problem (in this case, lack of intellectual understanding of things natural and supernatural) for which the Lord's counsel or intervention is needed.

As should be apparent from prior discussion, it's not always easy to fix the boundaries of the natural and the supernatural, particularly where scientific theories are characterized as metaphors.[1059] For example, consider consciousness and its object(s). A scientific saying that's widely accepted but seldom given spiritual

---

[1059] See the "Mysticism and Metaphor" section of my Region of Awe chapter.

application is the maxim that sentience supports existence. We saw this idea again and again in our study of quantum physics. Under the Copenhagen Interpretation, reality that is plural exists, albeit as phantom reality, until there is a conscious observation of the subject particle.[1060] Thus, mind has a special status as far as reality is concerned, and, as we noted earlier, even retroactively creates reality. What is seldom acknowledged, however, is that the need for sentience in the scheme of things, which allows the phantom to become real, points to the existence of God. Without His awareness of things, there would be no way, ultimately, of calling anything into being. The natural gives insight into the supernatural, in this as in many other instances.

The sentience that allows the universe to exist is associated with life. You'll recall that we considered whether the Geiger counter in the box occupied by Schrodinger's cat was an observer that could have effected a collapse of the critical probability wave function in favor of the cat's being either alive or dead but not both. There can be no existence-creating sentience without actual underlying life, we decided; therefore, a Geiger counter could not be a quantum observer, since it is not alive. This evokes the apostle John's statement concerning the pre-incarnate Christ, "In him was life; and the life was the light of men" (John 1:4). My reading of John's statement is that because Christ had life (i.e., was sentient, *inter alia*), men were able to exist rather than be consigned to darkness. Christ, as the creator of all things (John 1:3; Ephesians 3:9), was in effect the ultimate observer, to describe the situation in quantum mechanics terms.

As the product and necessary reflection of the sentience of God, life enjoys an existence independent of the material medium in which it resides—in the present example, human flesh. You can't both exist and be the thing that causes you to exist, since that would constitute a joining of cause and effect, which works only with respect to the actuality of God, the self-existent One (see, again, Exodus 3:14). Stating it another way, God's existence, which is

---

[1060] Richard Feynman, you'll recall, held that the multiple quantum states that exist prior to collapse of a particle's probability wave function each exist with equal reality.

## CONCLUSIONS

life itself, is not subsumed in (i.e., incorporated within) the flesh that He holds in being.

In its eternal state, life has a specific identity just as does temporal life; otherwise, human existence would have nothing to attach to and would at best be little more than an amorphous principle, neither particular nor personal and therefore inconsistent with any definition of life stated or implied in Scripture (see, e.g., Genesis 2:7; Job 12:9-10; Acts 17:28). The life-giving breath of God, of which the human self is the encapsulation (Genesis 2:7; Acts 17:25), is specifically directed toward individual man because it must be so directed if the human personality is to bear any resemblance to the personal God by whom all individual things consist (Genesis 1:27; 1 Corinthians 8:6; 15:48-49; Colossians 1:12-17).

There's more that we can glean from Scripture in this regard, particularly from John's gospel and his first epistle. The life that Christ has is the same life as God the Father has (see John 1:1, 4; also 1 John 1:1-2). For this reason, it was and is sufficient ontologically to be the basis of all other life. Under the physical law of biogenesis, there is no life except that which comes from life; therefore, all life ultimately comes from God the Father, since as the possessor of the same eternal life as is in Jesus He would be the only source that could always be counted on to support other life. We exist because God cannot not exist.

When we're considering the respective existences of God and man, Einstein's theory of relativity comes into play. Einstein concluded, for reasons summarized previously herein, that time is a relative concept. It will pass more slowly or more quickly depending on how fast an observer is traveling in relative motion or how great the gravity is in a particular region. Thus, we find man in a medium that's subject to change, that can be corrupted by influences outside itself. Imperfection implies perfection, Plato tells us; therefore, a changing state implies a changeless one. For now, man lives in time, but God inhabits eternity (Isaiah 57:15). The latter is consistent with God's changelessness (Malachi 3:6; James 1:17), since change defines time, which is the opposite of eternity. Someday saved man and the Lord will have the same habitation (Psalm 23:6; John 14:2-3), although the respective views of God and man from that vantage point may reflect a distinction between eternality and everlastingness (see discussion infra).

More specifically regarding God's and man's dwelling places, we note that the big bang theory and the Bible are in agreement that time had a beginning, although they are obviously at odds as to how that beginning came about. "(T)he beginning" referred to in Genesis 1:1, when "God created the heaven and the earth," necessarily was the beginning of everything, since Scripture, in at least five translations from the Hebrew that include the King James Version and the New American Standard Bible, attaches a restrictive time term, "*the* beginning," to similarly restrictive space terms, "*the* heaven" and "*the* earth." Use of the article "the" makes it clear that only one "beginning," only one "heaven," and only one "earth" are referred to by Genesis 1:1. The verse's declaration that "heaven" *and* "earth" came into being "(i)n the beginning" unites space and time origin-wise; neither was present before "the beginning" (if, indeed, there was a "before") and both were present from "the beginning" on.

Relativity theory, the basis of the big bang theory, similarly holds that time and space are inextricably joined; thus, under popular science's view as well as according to the Bible, the beginning of "the heaven and the earth" had to have been the beginning of time. God had to have been in existence when He thus created space-time. The first verse of Genesis assumes His preexistence. It does not state that God and space-time sprang into existence simultaneously; instead, it indicates that God was there to give "the heaven" and "the earth" existence that He already had. As Thomas Aquinas noted, something cannot come from nothing. To say that the universe and all that is within it were created *ex nihilo* is not to say that everything did not come from God. The Creator has to have always been in existence because otherwise there would have been nothing from which something could have come. The above statements about God are consistent with the conclusion that His habitation is a timeless eternity (Isaiah 57:15).

Life necessarily implies an environment suitable for life. The circumstance that man has a home where he can enjoy his God-given life defies some very long odds. Like the wonderfully efficient human body, man's location in the "Goldilocks zone," in which we find planet Earth, is evidence of intelligent design of some sort. The additional conjunctions of sentience coming from sentience and life coming from life necessarily foreclose any interpretation of man's existence other than as the work of a divine, personal Creator.

The theorized "big bang" supposedly threw outward, from some infinitesimally small point, all the matter in the universe. We note at the same time that there are observable stars at least twelve billion light years away from us. Assuming that light has always traveled at a speed of 186,000 miles per second, Bible apologists have needed to explain how, in a universe said by God's Word to be no more than a few thousand years old, we observers here on Earth can now be receiving starlight that theoretically should have taken twelve billion years to reach us. The theories of "mature creation" or creation of light "in transit," mentioned earlier (see the "On the Trampoline" section of my Relativity chapter) are possible explanations, but "young earth" creationists have wanted theories that are more observation-based.

As suggested earlier, two explanations for "distant starlight" that may be scientifically feasible are (1) Russell Humphreys' "white hole" cosmology (matter spewed out from a point of beginning near Earth's present location, in a universe that has a center and a boundary that expands and thus dilutes gravity so as to allow faster passage of time in the outer universe), and (2) Barry Setterfield's light-speed decay theory (the light now reaching Earth was previously traveling many times faster than it is now). The Humphreys "white hole" cosmology comes out of the same equations (under Einstein's general relativity theory) as support the "big bang" theory. As previously explained, scientific equations may have more than one solution, and in choosing one solution over another the scientist often simply votes his bias. The Setterfield theory comes out of observation and extrapolation.

Action implies purpose. If a sentient God created us, He must have had a purpose in doing so. The pleasantness of the world in which Adam and Eve found themselves prior to the fall suggests that God's purpose was not torment, but rather that He was motivated by love for us and a desire to use us for His glory (Ephesians

3:20-21; Philippians 2:9-11). As a personal God, however, He was surely desirous of love in return (see Deuteronomy 6:5; also Mark 12:28-30). For that love to be meaningful, as C. S. Lewis observed in *Mere Christianity*, it had to be the product of free will. Otherwise, it would not have been love given to God with all a person's heart, soul, mind, and strength (Mark 12:30). It would have been robotic love.

Here, again, science appears on the scene to help us understand things more fully. The Copenhagen/Feynman interpretation of quantum physics posits alternative realities, as we've seen, and this translates metaphorically in such a way as to confer reality on any human future that might occur and to declare that no outcome is foreordained. We're free to love God or not love Him, although one chooses the latter course at one's peril. Salvation may be freely chosen or it may be freely rejected. And, since God by virtue of His infiniteness already sees as real all possible outcomes, His knowledge cannot be added to by events that happen in what to us is the future, such as our enjoying the blessings of salvation or suffering the curse that results from rejection of Christ's gift. This means that God is immutable, just as Scripture says He is (Malachi 3:6; James 1:17).

Another explanation of how man can have free will, even though God is omniscient, has to do with the fact that man resides in time while God inhabits eternity. This situates God at a vantage point from which He can see man freely living his life as he moves through time, making choices that are genuine choices inasmuch as they're not foreordained. As far as I'm concerned, however, the "view from eternity" explanation, of how God's knowledge of events that to us are future can co-exist with man's autonomy, still leaves open the question of whether an omnipotent God's knowing of events and allowing them to happen is not the same as His foreordaining them. For this reason, I prefer the "alternative realities" explanation, as I've already indicated.

The existence of alternative realities in particle physics, which allows retroactive creation of reality, also provides a clue as to how fulfilled prophecy can be possible in a free-will milieu. See, again, the discussion in the "Prophecy" section of my Radiating

Reality chapter, supra. What it boils down to is that collapse of the probability wave function of a subatomic particle gives the particle a reality that it did not previously have, in effect allowing the particle to have had a concrete existence previously. If the particle is equated to a prophesied event and collapse of the wave function is analogized to a prophecy concerning that event, then the prophecy predates the event and accurately describes the event. Since alternative realities exist with respect to both outcomes and prophecies concerning outcomes, God can instead concretize the prophecy—i.e., mold it to fit the event rather than vice versa. In either case, God operates from a position which is like that of a quantum observer and, being outside of time, is not bound by any constraint as to what must come before what. In either case, the effected scenario must be within God's will, of course.

Similarly validated is the idea of interactive prayer, as noted in the "Speaking to God" section of my Radiating Reality chapter. Pre-collapse multiple quantum realities analogously permit any outcome of prayer to remain open as a possibility until God chooses to respond, one way or the other, to the supplicant's request. God's position in eternity as a vantage point allows Him to give what is a genuine response to prayer rather than an announcement of a predetermined result. Whether the (concretized) answer will be "yes," "no," or "maybe" is not foreknown by God, but in any event the believer's request will receive a hearing on the merits. Prayers to our God do not fall on deaf ears.

Another question touched on earlier in this book was that of how the soul can remember anything after death. Physicist/writer Paul Davies' thoughts on this matter, set forth in his classic *God and the New Physics*, culminate in no firm conclusions, although Davies suggests, in his chapter concerning the self, that one possible solution would be to view the question from the perspective of computer technology and allow for the possibility of a "back-up" system in which information might be stored when the biological brain is no longer available.[1061] Davies continues in the same vein a few pages later, remarking that the newly realized possibilities

---

[1061] See Davies, *God and the New Physics*, 91-92; also the discussion in this regard in the "Locke and Hume" section of my Essence chapter.

of mind transplants and of computers having a human-like awareness of self "hold out the hope that we can make scientific sense of immortality."[1062] The key, he says, might be that the essential ingredient of "mind" is information.[1063] If the mind is basically organized information, then the medium through which that information is expressed could be anything at all; it need not be a particular brain or indeed any brain.[1064] This leaves open the possibility that the "program," which confers memory, could be re-run after death in a system that we don't recognize as part of the physical universe.[1065] Heaven?

Does an information system that allows an individual to survive death and remember having lived also allow that person to exist as a distinct entity that is not limited by his or her temporal experience—i.e., to have a transcendent self? We raised this issue earlier without settling on a solution.[1066] Under a transcendent-self hypothesis, the person in question would be the same person after as before death, and he or she would be able to have postmortem experience. This would mean that the subject could properly be held responsible for unforgiven sin upon standing in judgment before God the Son (John 5:22-29). If, on the other hand, the individual is a Lockean sum total of his or her life experience and nothing more, then there's both good news and bad news. Adverse judgment may not fairly be imposed on the person in this case, since he or she was a different person from moment to moment while on Earth, but his or her ultimate fate will be eternal death inasmuch as there is no transcendent self to which punishment or reward may be affixed.

Linguist Noam Chomsky (b. 1928) has theorized that the human mind is hardwired with a universal grammar, which enables persons of all linguistic backgrounds to understand language structure similarly, and that therefore the mind is far from being a blank

---

[1062] See Davies, *God and the New Physics*, 96-98.

[1063] Ibid., 98.

[1064] Ibid.

[1065] Ibid., 98-99.

[1066] See the "Locke and Hume" and "Hegel and Selah" sections within my Essence chapter, supra.

slate or *tabula rasa* at birth.[1067] This gets us around the front-end empiricism of Locke's *tabula*, but it doesn't help with the question of memory after death, nor does it establish permanent personal surviving uniqueness (i.e., that there is a self that endures postmortally). We must also ask, as we did in another context, where this leaves the physically or mentally impaired person who is either unable to acquire knowledge through the senses or intellect or cannot recall things previously learned.

Here, postmodernist thinking is actually a help to me, to the extent that it holds that no two individuals experience the same thing in the same way. While implicit in this premise is the idea that faulty perception (faulty because it varies) precludes recognition of a faultless God, this should not prevent us from seeing something else: there are recognizable individual human selves out there by reason of our varying funds of information. I speak of distinct bodies of information possessed by each person in the world that reflect different personal experiences, certainly, but more to the point of the present discussion, that show a different way for each person of interpreting experience. An un-universal grammar, as it were. To the extent that we can posit God-implanted perception that is personal to each individual, we advance the idea that there is a permanent surviving self.

The severely mentally disabled person would not necessarily be unable to have uniquely-interpreted experience (and therefore lack a self), since who is to say what actually goes on in the mind of an amnesiac or the comatose! Nor are the questions posed above necessarily mooted out because believers are promised new physical selves in heaven (1 Corinthians 15:42-54; Philippians 3:20-21). The potential for perception that varies from individual to individual would still remain, all other things being equal.

The computer software/hardware analogy may help us answer the question of whether, in the world we go to after death, we experience what God enjoys in the sense of being able to see the entirety of history and humanity all at once, or whether instead our enjoyment of the triune God and the other features of heaven involves our taking pleasure in persons and things sequentially, in a series of

---

[1067] Stokes, *Essential Thinkers*, 183; see also Stephens, *Philosopher's Notebook*, 182-83.

experiences. Is heaven "eternity" (no connection with events that happen in time other than to view them) or is it "everlastingness" (never-ending time)? Paul Davies notes that it is only during the running of a program that the flow of time has any significance.[1068] If the program merely exists but is not run, like a symphony that is written but not performed, then time is not part of the dynamic.[1069] Time and eternity can co-exist on these terms, it would appear, and perhaps this is the framework within which we should view the relationship between an eternal God and a subject who will be with Him everlastingly.

It's not surprising to see so many phenomena in nature (science), on both the micro- and the macro- level, that appear to confirm the truth of the Gospel. God provides everything that we need for our walk with Him (see Matthew 10:9-10; also Mark 6:8-9), and His provision includes guidance by whatever means are needed to open men's eyes to the truth. Thus, He proclaims through the prophet Isaiah, He will "bring the blind by a way that they knew not;…will lead them in paths that they have not known,…will make darkness light before them, and crooked things straight" (Isaiah 42:16). The above statement should not be read as implying that it is wrong to have blind faith such as that described by Christian existentialist Soren Kierkegaard;[1070] it only means that God assists belief.

Why should God have to give us any crutches at all to lean on? some will ask. Without regard to the apparent encouragement of nature mysticism found in Romans 1:20,[1071] Paul notes that believers "walk by faith, not by sight" (2 Corinthians 5:7). This might be read as implying that neither the saved nor the unregenerate should be spending time actively searching for tangible proof of the Gospel. At John 20:29, after encouraging doubting disciple Thomas to put his finger into the nail holes in His hands and the spear hole in His side, the risen Christ says, "Thomas, because thou hast seen me, thou hast believed: blessed are they that *have*

---

[1068] Davies, *God and the New Physics*, 99.

[1069] Ibid.

[1070] See the "Whence Cometh Life" section of my Essence chapter, also the "Where We Stand Now" section of my Proving God chapter.

[1071] See the "Romans 1: 20" discussion in my Region of Awe chapter.

*not seen, and yet have believed"* (emphasis supplied). Consistently with the above statement, the author of Hebrews defines "faith" as "the substance of things hoped for, *the evidence of things not seen*" (11:1; emphasis again supplied).

There is, of course, no biblical prohibition against initially believing the Gospel on the basis of physical proof; in fact, the Bible is filled with accounts of miracles fostering fledgling faith. The point here is that one can get along nicely as a believer without miracles. Call it the work of the Holy Spirit (Romans 8:16; John 14:26; 1 John 5:16). The same is true with respect to Bible-supporting physics principles: you don't need them; just go ahead and believe the Word regardless of whether there are confirming scientific correspondences. But if the correspondences are available to you, exult in them.

At the risk of speaking repetitiously, I call the reader's attention once again to Soren Kierkegaard's statements regarding religious faith. "An objective uncertainty," Kierkegaard says, is "the highest truth there is," provided the "uncertainty" is entertained with passion.[1072] Passion, for Kierkegaard, is "existence at its very highest."[1073] To say with honest conviction that something is true when there is a chance that it might not be entails risk, he explains, but without risk there is no faith.[1074] Conversely, if he is able to apprehend God objectively, then he does not have faith, Kierkegaard says.[1075] It is faith, not empirical certainty, that is "counted unto [the believer] for righteousness" (Romans 4:3; Genesis 15:6). Thus, the unsaved person reading any book purporting to prove the Gospel must still affirmatively make a decision to believe. A suggestion as to how this decision might be implemented is set forth in my Epilogue, infra.

As I've repeatedly urged herein, the idea that God is unchanging is affirmed by the alternative or radiating nature of quantum reality, which when transposed to God and His universe means that God does not have the problem of being continually presented with

---

[1072] Hong and Hong, eds., *Essential Kierkegaard*, 207.

[1073] Ibid., 204.

[1074] Ibid., 207.

[1075] Ibid.

events that enlarge the body of data available to Him and thereby change Him by altering the state of His consciousness. But why should we need to turn to science for confirmation of God's immutability? I ask. The Bible emphatically rejects the notion that God is changeable. The psalmist declares, of the heavens and the earth:

> They shall perish, but thou (O Lord) shalt endure: yea, all of them shall wax old like a garment; as a vesture shalt thou change them, and they shall be changed.
> But thou art the same, and thy years shall have no end.
>
> (Psalm 102:26-27)

That God is not affected by the second law of thermodynamics (ever-increasing entropy of all systems) is made doubly clear by the prophet Isaiah's statement that "the everlasting God, the Lord, the Creator of the ends of the earth, fainteth not, neither is weary" (40:28). Elsewhere, Scripture states that the Lord "change[s] not" (Malachi 3:6), and that with Him there is "no variableness, neither shadow of turning" (James 1:17).

The above are instances in which Scripture makes a plenary statement—i.e., a statement that "says it all." There's no need to look elsewhere in Scripture, much less in nature, for assistance as to the meaning of what has been said. It should be noted that the Bible is full of such statements, and also that faith of the kind encouraged by Soren Kierkegaard is what enables me to take this sort of utterance seriously. In the final analysis, we believe because we want to believe. We want to believe because we've felt a need to believe. We feel a need to believe because someone or something makes us aware of our need. This is where the Bible, written by the Holy Spirit, comes in. The Bible alerts us to the realities of heaven and hell and God's plan of salvation. Natural phenomena, also, can foster God-awareness in us (Romans 1:20), as can the witness of a Christ-like spirit (John 13:15); however, the written Word is where the rubber meets the road for twenty-first century believers who are serious about embracing the traditional Christian faith.[1076]

It would therefore seem elementary that Christian faith, properly defined, is a feeling of confidence about the truth of our Bible, which is said to record both history and revelation from God. Implicit in much of my earlier discussion was the idea that if a thinking person is going to believe that the content of this central document of Christianity is true, the thinker must be able to say with assurance that the Bible's collected writings do not contradict each other.

Possible conflicts in Scripture include the seeming tug of war between divine omniscience and human autonomy, which is another way of referring to a possible clash between God's unchangeability and man's freedom. Are we free to choose salvation over condemnation, heaven over hell, or has the decision already been made for us by virtue of foreknowledge on the part of God that cannot be changed? Likewise, does it do any good to pray to God when He already knows what the resolution of any problem will be? I've suggested a way of resolving these apparent conflicts as well as the question of how fulfilled prophecy and human freedom can co-exist—basically, by our recognizing alternative realities, as God surely must, an approach suggested by modern physics.

I hasten to add, nonetheless, that an analytical, science-based approach is not necessarily the ticket as regards matters such as those mentioned above. As previously noted, the author of Hebrews observes that without faith it is impossible to please God (11:6). If we are to please God, we must take Him at his word. This means, among other things, that where Scripture seems to be inconsistent we should assume that there is a reading of the supposedly conflicting verses under which they are both or all true. We must in every instance believe that what God has said in His Word is true in some sense, and He will then bless us for our belief (Genesis 15:6; Romans 4:3; Galatians 3:6; James 2:23). God's good pleasure

---

[1076] Author David Limbaugh notes that the Gospel initially existed only in the form of an oral tradition, which was eventually reduced to writing and canonized as the New Testament. See David Limbaugh, *The True Jesus: Uncovering the Divinity of Christ in the Gospels* (Washington, D.C.: Regnery Publishing, 2017), 42-46.

is occasioned not only when we show our love for Him by keeping His commandments (John 14:15; 1 John 2:3), but also when we believe what He says in his Word.

One of the other Bible truths that we're asked to believe is that if we trust in Christ's atoning sacrifice, we'll be preserved harmless for eternity (John 3:16; Romans 6:23; Ephesians 2:8). This is so, Scripture explains, because a believer's faith counts as righteousness on his or her part (Genesis 15:6; Romans 4:3-5; Galatians 3:6-7). We're given the witness of Christ's resurrection as objective proof of the availability of eternal life through Him (1 Corinthians 6:14; 2 Corinthians 4:14). The necessary faith is not based on a review of empirical evidence, however; instead, it comes to the believer by way of the Holy Spirit (see John 14:16-17; also Romans 8:16), the person of the Trinity sent by God the Father to the disciples at the request of God the Son (John 14:16, 26) and, by extension, to all of mankind.[1077]

The objective proof of Christ's resurrection is substantial. We can believe, based on the overwhelming testimony of Scripture, that Christ died, rose from the dead, and thereafter walked the earth exhibiting physical proof of His identity as the crucified Lord (Matthew 28:9-10, 16-20; Mark 16:9-14; Luke 24:13-48; John 20:11-29; 21:1-23; Acts 1:3). He was visually recognized, post-resurrection, by persons who had previously known Him (Matthew 28:9; Mark 16:9-10, 12-14; Luke 24:30-31, 33-37; John 20:16, 18, 26-28; 21:7; Acts 1:3). He displayed physical abilities which showed that He was not merely a spirit or a figment of the imagination (Luke 24:38-43; John 20:19-20, 27-29; 21: 13). He displayed wounds that were proof of His crucifixion (Luke 24:38-40; John 20:20, 24-29).[1078]

The man who came forth from the grave was therefore the same man—indeed, God-man— as had lived on Earth a short time earlier. Since the Word unequivocally proclaims that we have eternal life through Jesus (Romans 6:23; 1 John 5:11-13), we need no other

---

[1077] See Matthew 28:18-20; also Acts 5:32; and Morris, *Study Bible*, nn. to John 14: 16, 26; Acts 2:4.

[1078] For further analysis of the evidence of Christ's physical resurrection, see Morris, *Study Bible*, nn. to Matthew 28:9, Luke 24:39, 43, and Acts 1:3.

proof of our immortality than the eternality of the risen Christ (see 1 Timothy 6:14; also Revelation 1:18). Our confidence in this regard rests not in science, either directly or metaphorically applied, but rather in the historical record contained in the Bible. The available Old Testament history, in the words of David Limbaugh, "underscores the reliability of" the gospels of Matthew, Mark, Luke, and John.[1079] In advocating history-mindedness, I do not mean to downplay the importance of faith; however, the empirical evidence in this case is impressive.

The faith part of the salvation equation is irrelevant with respect to children of tender years. There is ample support in Scripture for the conclusion that they're given a free pass into the Kingdom should they die during their formative years (see, for example, Matthew 18:3, 14; 19:13-15; Mark 10:14-15; Luke 18:15-17). Thus, my grandson Eli, whose beloved cousin Amanda passed away a few days shy of her sixth birthday, can rest assured, when he becomes fully able to comprehend, that Amanda still exists; she just has a new address.

It's not always so easy for the rest of us. Children and mental incompetents excepted, we must all enter the Kingdom through the faith door. Some of us have needed stronger factual tailwinds than others to get them across the threshold. The important thing is that the path from death to life has been negotiated, even if the way was narrow and difficult as in the case of C. S. Lewis.[1080] I hope that the above (completion of the course) has been true of your journey also, gentle reader, that you, too, have "passed from death unto life" (1 John 3:14). If you're feeling any doubt about this, you need to pay special attention to my Epilogue, which follows.

---

[1079] David Limbaugh, *The True Jesus*, 31; see also comments at 43-44, 48, 52, 56-57, 60-62.

[1080] See, again, Brown, *A Life Observed*, esp. 111-58

# EPILOGUE

Heaven is my focus these days, as you may be able to tell. What happened with respect to our family's ownership of The Cottage is therefore of secondary importance at this point, at least to me. If you must know, we sold The Cottage so that we could have fuller lives in our principal venues in the Rochester and Albany areas. We—my wife and I and our sons and daughters-in-law—had begun to feel that The Cottage owned us rather than we it. High water on the Great Lakes beginning in the spring of 2017 is making our decision to sell look like prescience on our part in view of what we would have had to do to continue to maintain The Cottage. I miss the sight of grandchildren frolicking on the beach, certainly. I don't miss the sound of jet skis. I also don't miss the work and expense that was normally (even in non-flood years) associated with owning a summer home that we were unable to use for most of the short upstate New York summer.

The crucial consideration, in any event, is not whether you're on the water someplace; it's where everyone is situated in relation to God. I can therefore now demote "The Cottage" to "the cottage." I've come to the conclusion that the only site I should be reverencing is one of another description, my place and the place of loved ones, which includes the rest of the human race, in God's eternal kingdom (John 14:2-3). Wherever the location of the one true God is—actually, He's everywhere (Psalm 139:7; Proverbs 19:3; Acts 17:27)—that's where the believer's spiritual presence should be (see Acts 7:49).

From the fact that you've bought or borrowed this book, I'm inferring that if you're not yet actually a believer, you're at least considering acknowledging Christ as your Lord and Savior. Also, it appears, there's a substantial likelihood that you've become involved in a biblically-grounded fellowship, perhaps only as a church attender but involved nonetheless. Even if both belief and Bible-based fellowship are already the case for you and it's now merely a matter of your attempting to bring someone else into the fold, I'm suggesting that before proceeding farther you do two things to insure that you can stand before God and man knowing that you've been redeemed by the blood of the Lamb (John 1:29; Revelation 13:8).

First, you should assume, on faith and based on the overwhelming evidence presented by the natural world or the history or revelation recorded in the Bible or any combination thereof, that there's a Creator to pray to, i.e., that the universe didn't just happen. Essentially, you're asking your intellect to assist you in believing that there's something beyond your intellect. Easily doable if you have the will to do it.

Secondly, speaking earnestly to the Creator of the cosmos, you should offer a prayer in which you (1) acknowledge that you're a sinner and separated from Him because of your sin, (2) say that you're now voluntarily repenting and turning from your sinful (i.e., selfish) ways, (3) acknowledge that He, Jesus, is the Son of God and therefore was the perfect sacrifice to atone for your sins, and (4) say that you're trusting solely in what He did on the cross to purchase your redemption (i.e., to deliver you from sin and its consequence death).

Having in the fullness of your heart (i.e., not just with your lips) made the above confession, known in evangelical circles as the Sinner's Prayer, you're now able to "walk in newness of life" (Romans 6:4). Paul's second letter to the Corinthians states, "if any man be in Christ, he is a new creature: old things are passed away; behold, all things are become new" (5:17).

# EPILOGUE

The new birth that former nonbelievers experience is "of the spirit" (John 3:6, 8), meaning that it reflects the regenerating work of the Holy Spirit.[1081] It also entails the converted humbling themselves and becoming "as little children" (Matthew 18:3-4); thus, one might wonder whether there is any element of purposefulness in the behavior of new believers. Paul's epistle to the Ephesian church definitely says that there is, since the apostle urges the Ephesian Christians to "walk not as other Gentiles walk, in the vanity of their mind" (Ephesians 4:17), to "put off the old man, which is corrupt according to the deceitful lusts" (Ephesians 4:22), and to "put on the new man, which after God is created in righteousness and true holiness" (Ephesians 4:24). The directives to "walk not," "put off," and "put on" require action on the believer's part, an affirmative response. So you see, gentle reader, we don't just become holy inertness when we tell God that He can take over our lives.

What we do become, upon receiving Jesus as Savior and Lord, is children of God, able to cry "Abba, Father" by reason of the Spirit that has been sent forth into us (John 1:12; Romans 8:14-16; 2 Corinthians 6:17-18; Galatians 4:5-6). Our relationship with God the Father is now a more intimate one than was formerly the case.[1082] Our adoption into His family reflects the great love that He holds for us (1 John 3:1).

Having said The Sinner's Prayer, you and anyone else you convert along the way should begin attending a church where the Bible is taught, if this is not happening already. By "the Bible" I don't mean any of those paraphrases that have become popular in recent years. I mean an actual translation from the original languages (i.e., from the Hebrew, the Aramaic, and the Greek), such as the Authorized (King James) Version. How do you find such a church? You may have found it already if you began going to the church you're now attending upon the invitation of someone who carries a Bible. Churches where everyone comes to services with the written Word of God in hand are almost invariably churches in which the traditional Bible is held in high esteem, as opposed

---

[1081] See, in this regard, Morris, *Study Bible*, n. to John 3:8.

[1082] See Morris, *Study Bible*, n. to Galatians 4:6.

to congregations where the true Bible is disregarded in favor of humanist (man-centered) philosophy or New Age (Emergent Church) mysticism.

Carrying a Bible is more than a symbolic gesture; it has certain practical advantages. We're talking about a book the pages of which can be turned either in the customary fashion or by sliding a finger across an electronic touch screen. Personal and pew Bibles, sad to say, are being replaced, even in some of the more traditional evangelical churches, by giant digital screens that display the words of choruses, sermon outlines, and (occasionally) Scripture, none of which the individual churchgoer has any control over. Contrastingly, your personal Bible—Genesis through Revelation plus study notes—is there in your lap or the palm of your hand, for you to consult without restriction before, during, and after worship or teaching. You don't have to accept "on faith" what the pastor or anyone else is saying about the content of Scripture. You can conduct research on the spot with respect to any independent thoughts you might have.

In conclusion, gentle reader, I urge you to see things horizon to horizon. We've followed a meandering trail through various ideologies, but in all of this there has been only one reality a person needs to know or communicate to anyone else: Christ crucified. Jesus, the Father's overwhelming love personified; He only is needful. Philosophy, mysticism, and even science are, at the end of the day, unable to give you what you need. Only God the Son, revealed and invoked by a process both natural and supernatural, is able to do that. He is able to be our Savior because He is pure love and pure life, seeking not to profit Himself but rather to bless and preserve the creatures for whom He died.

I'm pleased that you've read my book, and hope that it's been a blessing to you. I rejoice that you're concerned about your future for eternity and also about the final destinations of others (I'm assuming a yes as to both of these suppositions). If I don't see you before then, I'm looking forward to meeting you "(i)n the sweet by and by."[1083]

In the meantime, grow in grace (2 Peter 3:18).

---

[1083] Samuel F. Bennett (words), Joseph P. Webster (music), "In the Sweet By and By," in *Majesty Hymns* (Greenville, S.C.: Majesty Music, 1997), no. 564.

# ACKNOWLEDGMENTS

First, some posthumous words of thanks. My mother and father (Mother and Dad throughout this writing) were country folk, but not bumpkins. Although they were both literate and in fact avid readers, neither got past the eighth grade. This stopping point for schooling was fairly normal for rural people in the early twentieth century, since your local one-room schoolhouse took you only as far as the eighth grade. If you wished to continue your education after that, you needed to be enrolled in a central high school that might be miles away from where you lived. Dad had sufficient mechanical skills to make further schooling for him seem unnecessary. In Mother's case, the high school was some five miles from her home over horse-and-buggy roads, which necessitated her rooming with a family in the village where the school was located. She quickly proceeded to get sick on food that Grandma Lane had sent with her in an effort to avoid having to pay for board; thus, her formal education ended a few days into the ninth grade.

Mother and Dad were persistent in their efforts to see that their children were educated, however, and therefore my sister and brother and I, with the encouragement of our parents and financial assistance they could ill afford to provide, eventually amassed a total of some thirteen years of schooling beyond high school. Mother and Dad were also firm in their expectation that each of their children would attend church. Thus, the pew was not a place that was foreign to me. For their establishment of church and classroom as norms for me, which in time have brought me to the writing of this book, I will be eternally grateful to Mother and Dad.

Other than my parents, the person at the Navarino Methodist Church who most helped me along the road to becoming a Christian was a Sunday School teacher, Mrs. Beatrice Sherman. Mrs. Sherman was more than a babysitter, as some Sunday School teachers were; indeed, she took personal responsibility for the spiritual outcome of each of the members of her junior-high classes. Although I was a know-it-all adolescent who at that point was not ready to accept Jesus as Lord and Savior, "Aunt Bea" did all that a mere human being could have done to prepare me for the momentous step of allowing the Holy Spirit to save me. Thank you, Aunt Bea, now gone to your reward, for sowing seeds in a situation where you would only see your efforts bear fruit once we both reached heaven.

Another person in the distant past who also deserves special mention is Miss Barbara Eveland, one of my high school English teachers. It was she who first encouraged me to think that I might have some writing ability. Indeed, one day as she was handing back recently graded papers, Miss Eveland announced to the class that (she felt) I had an exceptional power of self-expression. This praise couldn't have come at a better time, since I had recently learned of my exclusion from National Honor Society because of a low grade in Latin.

In the more recent past, I've been helped greatly by my loving wife, who has served as my secretary during the permissions-gathering process, tirelessly answered my computer questions, and offered helpful comments regarding the content of the present work. Without her forbearance regarding maintenance and other work that I would otherwise have been doing around home, I could never have brought this project to fruition.

Also providing input have been my son Jeremy and his wife, Jen, he as the holder of a degree in physics and therefore an able reviewer of the science contained in this book, she as a former high-school English teacher (Rochester City School District) and

## ACKNOWLEDGMENTS

present continuing-education writing teacher and thus someone qualified to tell me whether I'm using our native tongue properly and effectively. Jeremy took the author photo that graces the back cover of this book. Almost as important as Jeremy's and Jen's collective expertise, which includes the ability to talk to the computer, have been their frequent inquiries regarding my progress on "the book," giving me additional motivation to push ahead.

My Albany-area son, Geof, has likewise contributed. The fine photo seen on the cover of this book was taken by him on the beach in front of The Cottage, where approximately half of the research, thinking, and writing that went into this book took place. Geof has also supported me in this endeavor by chiding me about my physical fitness, so as to help me live long enough to see this work in print. His friend Andrea, a writing instructor and librarian, has also provided valuable input.

Also contributing pictorially was graphic designer Dan Donohue, who did the maps to which the text herein refers. Everything happens in a place, including thoughts regarding things we cannot see. That's why I've included graphics to help the reader visualize lakes and other physical features.

Another person who has assisted in the present effort, as a fount of information concerning publishing and self-publishing and as an often-needed encourager, has been Webster Central School District community education teacher Mary Beth Egeling, herself a self-published author. Thanks, Mary Beth, for making me infinitely more knowledgeable than I otherwise would have been regarding the process by which writings get into print.

Finally, I would like to thank my pastor, Dr. James Gugino, "first among equals" of the many members of the congregation at Lighthouse Bible Baptist Church who have shown an interest in this enterprise. We are one body in Christ (Ephesians 4:4), one organic whole that has many parts each of which needs all the others, particularly when the glory of God is involved (Romans 12:4-8; 1 Corinthians 12:12-31).

# BIBLIOGRAPHY

Appignanesi, Richard; Garratt, Chris; Sardar, Ziauddin; and Curry, Patrick; *Introducing Postmodernism* (New York: Totem Books, 1995).

Aquinas, Saint Thomas, *Summa Theologica*, transl. by Fathers of English Dominican Province, 5 vols. (Westminster, Md.: Christian Classics, 1981 reprint).

Associated Press, "Physicists exult over particle discovery," *Rochester Democrat and Chronicle* (July 5, 2012), 3A.

Augustine, Saint, *Confessions*, transl. by Albert C. Outler (New York: Barnes & Noble Books, 2007).

Bartlett, John; Kaplan, Justin, gen. ed.; *Bartlett's Familiar Quotations: A collection of passages, phrases, and proverbs traced to their sources in ancient and modern literature*, 17th Ed. (New York: Little, Brown & Co., 2002).

Barrow, John D., *Theories of Everything: The Quest for Ultimate Explanation* (Oxford, U.K.: Oxford University Press, 1992).

Basinger, David, *The Case for Freewill Theism: A Philosophical Assessment* (Downers Grove, Ill.: InterVarsity Press, 1996).

Beauvoir, Simone de, *The Second Sex*, transl. by Constance Borde and Sheila Malovany-Chevallier (New York: Alfred A. Knopf, 2010).

Behe, Michael J., *Darwin's Black Box: The Biochemical Challenge to Evolution* (New York: Simon & Schuster, 1996).

Bennett, Samuel F. (words) and Webster, Joseph P. (music), "In the Sweet By and By," in *Majesty Hymns* (Greenville, S.C.: Majesty Music, 1997).

Berger, Arthur Asa, *The Portable Postmodernist* (Lanham, Md.: Rowman & Littlefield Publishers, 2003).

Bernstein, Jeremy, *Quantum Profiles* (Princeton, N.J.: Princeton University Press, 1991).

Boettner, Loraine, *The Reformed Doctrine of Predestination* (Phillipsburg, N.J.: Presbyterian & Reformed Publishing Co., 1932).

Bonhoeffer, Dietrich, *Letters and Papers from Prison*, Enlarged Ed., ed. by Eberhard Bethge, transl. by Reginald Fuller and others (New York: Simon & Schuster, 1971).

Bowden, M., *The Rise of the Evolution Fraud* (Bromley, Kent, U.K.: Sovereign Publications, 1982).

Boyd, Gregory A., *God of the Possible: A Biblical Introduction to the Open View of God* (Grand Rapids, Mich.: Baker, 2000).

Breese, Dave, *Seven Men Who Rule the World from the Grave* (Chicago: Moody Press, 1990).

Broadbent, E. H., *The Pilgrim Church* (Grand Rapids, Mich.: Gospel Folio Press, 1931, 1999).

Brooks, David, *The Road to Character* (New York: Random House, 2015).

Brown, Devin, *A Life Observed: A Spiritual Biography of C. S. Lewis* (Grand Rapids, Mich.: Brazos Press, 2013).

Brown, Walter T., Jr., *In the Beginning: Compelling Evidence for Creation and the Flood* (Phoenix: Center for Scientific Creation, 2008).

Bryan, Denis, *Einstein: A Life* (New York: John Wiley & Sons, 1996).

Burns, Ken, and Novick, Lynn, *Baseball: The Tenth Inning* (Florentine Films, 2010).

Cadwallader, Mark W., *Creation Spelled Out for Us All* (Conroe, Tex.: CTS Publications, 2007).

Calaprice, Alice, ed., *The Ultimate Quotable Einstein* (Princeton, N.J.: Princeton University Press, 2011).

# BIBLIOGRAPHY

Calvin, John, *Institutes of the Christian Religion*, transl. by Henry Beveridge (Peabody, Mass.: Hendrickson Publishers, 2008).

Capra, Fritjof, *The Tao of Physics: An Exploration of the Parallels between Modern Physics and Eastern Mysticism*, 35th Anniversary Ed. (Boston: Shambhala Publications, 2010).

Castaneda, Carlos, *The Teachings of Don Juan: A Yaqui Way of Knowledge*, 30th Anniversary Ed. (London: Univ. of Calif. Press, 1968, 1996, 1998).

Clark, David K., and Geisler, Norman L., *Apologetics in the New Age: A Christian Critique of Pantheism* (Grand Rapids, Mich.: Baker Book House, 1990).

Cloud, David W., "Calvinism on the March Among Evangelicals," 8/22/13 *Calvinism*, cited in *http://www.wayoflife.org/index_files/calvinism_on_the_march.html*.

_____, *Seeing the Non-Existent: Evolution's Myths and Hoaxes* (Port Huron, Mich.: Way of Life Literature, 2011).

Collins, Francis S., *The Language of God: A Scientist Presents Evidence for Belief* (New York: Free Press, 2006).

Cooper, John W., "Pantheism", in *The Encyclopedia Americana* (Danbury, Ct.: Scholastic Library Publishing, 2006).

Crease, Robert, and Mann, Charles, *The Second Creation: Makers of the Revolution in Twentieth-Century Physics* (New York, Macmillan, 1986).

Cristiani, Leon, *St. John of the Cross: Prince of Mystical Theology*, transl. by Leon Cristiani (Garden City, N.Y.: Doubleday, 1962).

Davies, Paul, *About Time: Einstein's Unfinished Revolution* (New York: Simon & Schuster, 1995).

_____, *God and the New Physics* (New York: Simon & Schuster, 1984).

_____, *The Mind of God: The Scientific Basis for a Rational World* (New York: Simon & Schuster, 1992).

Dawkins, Richard, *The God Delusion* (New York: Houghton Mifflin Co., 2008).

Derrida, Jacques, *Of Grammatology*, Corr. Ed., transl. by Gayatri Chakravorty Spivak (Baltimore: Johns Hopkins Univ. Press, 1974, 1976, 1997).

Dicken, E. W. Trueman, *The Crucible of Love: A Study of the Mysticism of St. Teresa of Jesus and St. John of the Cross* (New York: Sheed and Ward, 1963).

Downing, David C., *Into the Region of Awe: Mysticism in C. S. Lewis* (Downers Grove, Ill.: InterVarsity Press, 2005).

Eastman, Mark, and Missler, Chuck, *The Creator Beyond Time and Space* (Costa Mesa, Calif.: The Word for Today, 1996).

Eastman, Mark, *Creation by Design* (Costa Mesa, Calif.: The Word for Today, 1996).

Emerson, Ralph Waldo, "Divinity School Address," in Lawrence Buell, ed., *The American Transcendentalists: Essential Writings*, Modern Library Ed. (New York: Random House, 2006).

_____, "The Oversoul," in Larzer Ziff, ed., *Ralph Waldo Emerson: Nature and Selected Essays* (New York: Penguin Books, 1982).

Erickson, Millard J., *What Does God Know and When Does He Know It?: The Current Controversy over Divine Foreknowledge* (Grand Rapids, Mich.: Zondervan, 2003).

Foster, Richard, *Celebration of Discipline: The Path to Spiritual Growth* (New York: HarperCollins, 1998).

Foucault, Michel, *Discipline and Punish: The Birth of the Prison*, transl. by Alan Sheridan (New York: Vintage Books, 1977, 1995).

Freile, Victoria, "Small steps home for daughter Selah," *Rochester Democrat and Chronicle* (December 12, 2012), 3B, 6B.

Gardner, Martin, *Relativity Simply Explained* (Mineola, N.Y.: Dover Publications, 1962, 1976, 1997).

Garvie, A. E., Pantheism article, *The Encyclopedia of Religion and Ethics* (New York: Charles Scribner's Sons, 1917).

Geisler, Norman, *Chosen But Free: A Balanced View of Divine Election*, 2nd Ed. (Minneapolis: Bethany House Publishers, 2001).

_____, *Christian Apologetics*, 2nd Ed. (Grand Rapids, Mich.: Baker Academic, 2013).

Gilley, Gary E., *Out of Formation: Spiritual Disciplines, of God and Men* (Darlington, U.K.: Evangelical Press, 2014).

_____, "Roots of the Spiritual Formation Movement," (August 2014 *Think on These Things*, at *http://www.svchapel.org/resources/articles/133-spiritual-formation-m.*).

# BIBLIOGRAPHY

_____, *This little church had none: a church in search of the truth* (Darlington, U.K.: Evangelical Press, 2009).

_____, *This Little Church Stayed Home: A faithful Church in deceptive times* (Darlington, U.K.: Evangelical Press, 2006).

_____, *This Little Church Went to Market: The Church in the Age of Entertainment* (Darlington, U.K.: Evangelical Press, 2006).

Gleick, James, *Genius: The Life and Science of Richard Feynman* (New York: Random House, 1992).

Goldman, Steven L., *Science Wars: What Scientists Know and How They Know It* (DVD) (Chantilly, Va.: The Teaching Company, 2006).

Goodman, James, "Big Bang Boon," *Rochester Democrat and Chronicle* (July 3, 2012), 1A, 6A.

Greene, Brian, *The Elegant Universe: Superstrings, Hidden Dimensions, and the Quest for the Ultimate Theory* (New York: W. W. Norton & Co., 1999).

Grenz, Stanley J., *A Primer on Postmodernism* (Grand Rapids, Mich.: Eerdmans, 1996).

Guliuzza, Randy J., "Engineering Principles Point to God's Workmanship," *Acts & Facts* 46, no. 6 (June 2017).

Harkness, Georgia, *Mysticism: Its Meaning and Message* (Nashville: Abingdon Press, 1973).

Hartshorne, Charles, *The Divine Relativity: A Social Conception of God* (New Haven: Yale University Press, 1941).

_____, *Man's Vision of God and the Logic of Theism* (New York: Harper & Brothers, 1941).

_____, Pantheism and Panentheism article, *Encyclopedia of Religion* (New York: Macmillan, 1987).

Hawking, Stephen, *A Brief History of Time*, Updated and Expanded 10th Anniversary Edition (New York: Bantam Books, 1988, 1996).

Hebert, Jake, "'Smoking Gun' Evidence of Inflation?" *Institute for Creation Research* March 21, 2014, cited in *http://www.icr.org/article/8031/?utm_source=d/vr.it&utm_medium=fa*

Heilprin, John, "Physicists say they have found a 'God particle,'" *Rochester Democrat and Chronicle* (March 15, 2013), 1A, 6A.

Hong, Howard V., and Hong, Edna H., eds., *The Essential Kierkegaard* (Princeton, N.J.: Princeton University Press, 2000).

Hume, David, *A Treatise of Human Nature*, 2nd Ed. (Oxford, U.K.: Oxford University Press, 1987).

Humphreys, D. Russell, *Starlight and Time: Solving the Puzzle of Distant Starlight in a Young Universe* (Green Forest, Ark.: Master Books, 1994, 2006).

Hunt, Dave, *Cosmos, Creator, and Human Destiny: Answering Darwin, Dawkins, and the New Atheists* (Bend, Ore.: The Berean Call, 2010).

_____, and McMahon, T. A., *The Seduction of Christianity: Spiritual Discernment in the Last Days* (Eugene, Ore.: Harvest House Publishers, 1985).

_____, *What Love Is This? Calvinism's Misrepresentation of God*, 4th Ed. (Bend, Ore.: The Berean Call, 2013).

Huxley, Aldous, *The Doors of Perception and Heaven and Hell* (New York: HarperCollins Publishers, 1954, 1955, 1966).

James, William, *The Varieties of Religious Experience: A Study in Human Nature* (New York: Barnes & Noble Classics, 2004; orig. publ. 1902).

John of the Cross, Saint, *The Collected Works of John of the Cross*, transl. by Kieran Kavanaugh and Otilio Rodriguez (Garden City, N.Y.: Doubleday & Co., 1964).

Johnson, Luke Timothy, *Mystical Tradition: Judaism, Christianity, and Islam* (DVD) (Chantilly, Va.: The Teaching Company, 2008).

Jung, Carl G., *Psychology of the Unconscious: A Study of the Transformations and Symbolisms of the Libido*, transl. by B. M. Hinkle (New York: Dodd, Mead & Co., 1916, 1963).

Kaku, Michio, *Parallel Worlds: A Journey Through Creation, Higher Dimensions, and the Future of the Cosmos* (New York: Random House, 2005).

Keats, John, *Letter to George and Georgiana Keats [March 19, 1819]*, quoted in *Bartlett's Familiar Quotations*, 440.18.

*King James Study Bible* (Nashville: Thomas Nelson Publishers, 1988).

Kowalski, Gary, *Science and the Search for God* (New York: Lantern Books, 2003).

## BIBLIOGRAPHY

Kuhn, Thomas S., *The Structure of Scientific Revolutions* (Chicago: The University of Chicago Press, 1962, 1970, 1996).

Lerner, Eric J., *The Big Bang Never Happened: A Startling Refutation of the Dominant Theory of the Origin of the Universe* (New York: Random House, 1991).

Lewis, C. S., *Letters to Malcolm: Chiefly on Prayer* (New York: Harcourt, Brace & World, 1963, 1984).

_____, *Mere Christianity* (San Francisco: HarperSanFrancisco, 1952, 1980).

_____, *Surprised by Joy: The Shape of My Early Life* (New York: Harcourt, Inc., 1955).

_____, *That Hideous Strength: A Modern Fairy Tale for Grown-Ups* (New York: Macmillan, 1968 reprint).

_____, "Transposition," in *The Weight of Glory and Other Addresses* (New York: Macmillan, 1949, 1980 Rev. Ed.).

Lightman, Alan, and Brawer, Roberta, *The Lives and Worlds of Modern Cosmologists* (Cambridge, Mass.: Harvard University Press, 1990).

Limbaugh, David, *Jesus on Trial: A Lawyer Affirms the Truth of the Gospel* (Washington, D.C.: Regnery Publishing, 2014).

_____, *The True Jesus: Uncovering the Divinity of Christ in the Gospels* (Washington, D.C.: Regnery Publishing, 2017).

Lisle, Jason, *The Ultimate Proof of Creation: Resolving the Origins Debate* (Green Forest, Ark.: Master Books, 2009).

Locke, John, *An Essay Concerning Human Understanding*, Great Books in Philosophy Ed. (Amherst, N.Y.: Prometheus Books, 1995).

Lyotard, Jean-Francois, *The Postmodern Condition: A Report on Knowledge*, transl. by Geoff Bennington and Brian Massumi (Minneapolis: University of Minnesota Press, 1979, 1984).

Mandelbrot, Benoit B., *Fractals: Form, Chance, and Dimension* (San Francisco: W. H. Freeman & Co., 1977).

Martin, Walter, and Zacharias, Ravi, gen. ed., *The Kingdom of the Cults* (Bloomington, Minn.: Bethany House Publishers, 2003).

McGrath, Alister, and McGrath, Joanna Collicutt, *The Dawkins Delusion? Atheist Fundamentalism and the Denial of the Divine* (Downers Grove, Ill.: InterVarsity Press, 2007).

Michener, James A., *Chesapeake* (New York: Random House, 1978).

Morris, Henry M., *The Henry Morris Study Bible* (Green Forest, Ark.: Master Books, 1995, 2006, 2012).

Mote, Edward (words), Bradbury, William B. (music), "The Solid Rock," in *Majesty Hymns* (Greenville, S.C.: Majesty Music, 1997).

Nichols, Shaun, *Great Philosophical Debates: Free Will and Determinism* (DVD) (Chantilly, Va.: The Teaching Company, 2008).

Nietzsche, Friedrich, *The Will to Power*, transl. by Walter Kaufmann and R. J. Hollingdale, ed. by Walter Kaufmann (New York: Random House, 1967).

Oakland, Roger, *Faith undone: the emerging church—a new reformation or an end time deception?* (Eureka, Mont.: Lighthouse Trails Publishing, 2007).

Oulton, John Ernest Leonard, and Chadwick, Henry, transl. and eds., *Alexandrian Christianity*, vol. 2 (Philadelphia: Westminster Press, 1954).

Paley, William, *Natural Theology* (England: 1802; reprint, Houston: St. Thomas Press, 1972).

Palin, Sarah, *Good Tidings and Great Joy: Protecting the Heart of Christmas* (New York: HarperCollins, 2013).

Patterson, Roger, *Evolution Exposed* (Hebron, Ky.: Answers in Genesis, 2006).

Pirsig, Robert M., *Zen and the Art of Motorcycle Maintenance: An Inquiry into Values* (New York: HarperCollins, 1974, 1999).

Poe, Edgar Allan, "Eureka: A Prose Poem," in *The Complete Works of Edgar Allan Poe*, 17 vols. (New York: Fred de Fau & Co., 1902) (orig. publ. New York: G. P. Putnam, 1848).

Price, Richard, *Augustine*, Great Thinkers Series (Liguori, Mo.: Liguori Publications, 1996).

Ramos, Nestor, "Lost in the Canal," *Rochester Democrat and Chronicle* (October 8, 2012), 1A, 9A.

*Random House Webster's Unabridged Dictionary*, 2$^{nd}$ Ed. (New York: Random House Reference, 1987).

# BIBLIOGRAPHY

Rice, Doyle, "Is anyone out there?--and 2 other big questions," *USA Today* for *Rochester Democrat and Chronicle* (March 18, 2014), 3B.

Rohmann, Chris, *A World of Ideas: A Dictionary of Important Theories, Concepts, Beliefs, and Thinkers* (New York: Random House, 1999).

Russell, Bertrand, *Why I Am Not a Christian and Other Essays on Religion and Related Subjects* (New York: Simon & Schuster, 1957).

Sarfati, Jonathan, *Refuting Compromise: A Biblical and Scientific Refutation of "Progressive Creationism" (Billions of Years), As Popularized by Astronomer Hugh Ross* (Green Forest, Ark.: Master Books, 2004).

_____, *Refuting Evolution: A Handbook for Students, Parents, and Teachers Countering the Latest Arguments for Evolution* (Green Forest, Ark.: Master Books, 1999).

_____, *Refuting Evolution 2* (Green Forest, Ark.: Master Books, 2002).

Sartre, Jean-Paul, *Being and Nothingness: A Phenomeno-Logical Essay on Ontology*, transl. by Hazel E. Barnes, English Eds. 1956, 1984 (New York: Washington Square Press).

Schepisi, Fred, dir., *I. Q.* (DVD) (Hollywood: Paramount Pictures, 1994).

Schilpp, Paul Arthur, ed., *Albert Einstein: Philosopher-Scientist* (New York: Tudor Publishing Co., 1951).

Scofield, C. I., *The Scofield Study Bible, King James Version* (Oxford: Oxford University Press, 2003).

Setterfield, Helen, "History of the Light-Speed Debate Part I," July 2002 *Personal Update NewsJournal*, cited in *http//www.khouse.org/news_article 2013/2044*.

_____, "History of the Light-Speed Debate Part II," July 2002 *Personal Update NewsJournal*, cited in *http//www.khouse.org/news_ article 2013/2044*.

Shelley, Bruce L., *Church History in Plain Language*, updated 3rd Ed. (Nashville: Thomas Nelson Publishers, 1982, 1995, 2008).

Smith, James K. A., *Who's Afraid of Postmodernism? Taking Derrida, Lyotard, and Foucault to Church* (Grand Rapids, Mich.: Baker Academic, 2006).

Smith, Warren B., *A "Wonderful" Deception: The Further New Age Implications of the Emerging Purpose Driven Movement* (Silverton, Ore.: Lighthouse Trails Publishing, 2009).

Spielberg, Steven, *Back to the Future* (DVD) (Universal City, Calif.: Universal Studios, 2002).

Steele, David N., and Thomas, Curtis, *The Five Points of Calvinism* (Phillipsburg, N.J.: Presbyterian and Reformed Publishing Co., 1963).

Stein, Ben, featured interviewer, *Expelled: No Intelligence Allowed* (DVD) (Universal City, Calif.: Premise Media Corp., 2008).

Stephens, Mark, *The Philosopher's Notebook: A Creative Journal for Thinkers and Philosophers* (New York: Sterling Publishing, 2015).

Stokes, Philip, *Philosophy 100 Essential Thinkers* (New York: Enchanted Lion Books, 2002).

Strogatz, Steven, *Chaos*, Pt. 2 (DVD) (Chantilly, Va.: The Teaching Company, 2008).

Teilhard de Chardin, Pierre, *Christianity and Evolution*, transl. by Rene Hague (New York: Harcourt Brace Yovanovich, 1971).

Teresa of Avila, Saint, *Interior Castle*, Image Books Ed., transl. and ed. by E. Allison Peers (Garden City, N.Y.: Doubleday & Co., 1961).

_____, *The Interior Castle*, in *The Collected Works of St. Teresa of Avila*, transl. by Kieran Kavanaugh and Otilio Rodriguez (Washington, D.C.: Institute of Carmelite Studies, 1976-1985).

Thoreau, Henry David, *A Week on the Concord and Merrimack Rivers* (Orleans, Mass.: Parnassus Imprints, 1987).

Underhill, Evelyn, *Mysticism: A Study in the Nature and Development of Spiritual Consciousness* (Mineola, N.Y.: Dover Publications, 2002).

_____, *The Mystics of the Church* (New York: Schocken Books, 1964).

Voltaire (Francois-Marie Arouet), *Candide and Other Stories*, transl. by Roger Pearson, Oxford World's Classics Ed. (New York: Oxford University Press, 2006, 2008).

Watson, Traci, "'Smoking gun' rocks the universe," *USA Today* for *Rochester Democrat and Chronicle* (March 18, 2014), 3B.

# BIBLIOGRAPHY

Webster, Noah, *First (1828) Edition of An American Dictionary of the English Language*, Facsimile Ed. (San Francisco: Foundation for American Christian Education, 1967, 1995).

Weinberg, Steven, *Dreams of a Final Theory: The Search for the Fundamental Laws of Nature* (New York: Pantheon Books, 1992).

Whitcomb, John C., and Morris, Henry, *The Genesis Flood: The Biblical Record and Its Scientific Implications* (Phillipsburg, N.J.: Presbyterian and Reformed Publishing Co., 1961).

Whitehead, Alfred North, *Process and Reality*, Corr. Ed., David Ray Griffin and Donald W. Sherburne, eds. (New York: The Free Press, 1978).

Wilder, Patrick A., *The Battle of Sackett's Harbor: 1813* (Baltimore: The Nautical and Aviation Publishing Co. of America, 1994).

Wolfson, Richard, *Einstein's Relativity and the Quantum Revolution: Modern Physics for Non-Scientists*, Pt. I (VCR) (Chantilly, Va.: The Teaching Company, 2000).

_____, *Einstein's Relativity and the Quantum Revolution: Modern Physics for Non-Scientists*, Pt. II (VCR) (Chantilly, Va.: The Teaching Company, 2000).

Yungen, Ray, *A Time of Departing: How Ancient Mystical Practices are Uniting Christians with the World's Religions*, Expanded 2nd Ed. (Eureka, Mont.: Lighthouse Trails Publishing, 2006).

Zaehner, R. C., *Mysticism Sacred and Profane: An Inquiry into Some Varieties of Praeternatural Experience* (Oxford, U.K.: Oxford University Press, 1957).

# SUBJECT INDEX

Aaron, Hank, 325-26

Abiogenesis, 71, 197

About Time (Davies), 204, 272

Absolute truth non-existent, 318-20

Access to water, 90-99

Actions as prayer, 284-85

Adoption into family of God, 365

Age of Earth
    carbon dating, 159-60
    cosmologies positing
        young or old
        Earth, 157-59, 240-41
    light speed decay, 157-59
    mature creation, 241
    radiometric dating, 159-60

Age of Reason, 218

Age of universe, 165-66, 171, 219-20

Alcohol
    mysticism and, 121, 123-24
    Navarino history, 303-4

Alexandrian Christianity, 284

"All fact, no meaning," 107-8

Allegorization of Scripture, 140-41

"Altered traits, not altered states," 123-24, 141

Alter ego decisions, 272

Alternative realities, 258, 262-63, 266-78, 282-83, 287-88, 291, 298, 319, 352, 357-58

Amiel, 131-32

Amnesiac, 355

Analogy, understanding things by, xlii, 61, 98, 104-6, 116-17, 130, 145

Ancient/modern mysticism, 344-45

Animal souls, 9

Animism, 61-62

Anselm, 8, 111, 210, 216, 218

Answers in Genesis, 183

Anthropic principle, 65, 67, 74-75, 291

Antigravity force, 163, 171

Antimatter, 168

Apes and humans, 73

Approaching Absolute systematically, 134-42

Aquinas, Thomas, 208-11, 216, 218, 350

Aristotle, 9-10, 20, 206, 208

Arminianism, 32, 293-301

Arminius, Jacobus, 293

Artificial intelligence, 72-75

Association Island, 36-37, 87-88, 92

Atheism
awareness of God's provision, 243
New Atheists, 67-69, 240

Atoms, 38, 249

Atoning sacrifice of Christ, 283, 295, 300, 360, 364

Attorney, author as, xxxiii, 52, 217, 343

Augustine, 17-18, 29-31, 203-4, 281

"Awakening" stage in mysticism, 137


*Back to the Future*, 231

Bacon, Francis, 153, 155

Baltimore, 259

*Baseball: The Tenth Inning*, 325

Bass Island, 36, 87-88, 92, 96, 145

Baudrillard, Jean, 322-23

Beauvoir, Simone de, 63

"Before the Bang," 247

Benches at Campbell's Point, 4

Berger, Arthur Asa, 321-27

Berkeley, Bishop George, 75, 242

"Be still, and know that I am God," 279

"Best of all possible worlds," 56

"Bible" defined, 365

"Big Bang," 160-80
atheist reliance on, 69-70
"before the Bang," 247
Bible, beginning consistent with, 248
"bookend," 252
"distant starlight" puzzle, 351
"inflation" remedy, 254-55
pantheism and, 60
paradigm, 155-57
point of beginning, 265, 350-51
quantum effect, 246
secularly-based attacks on theory, 181
singularity, 245

*Big Bang Never Happened, The* (Lerner), 177-78

"Big crunch," 165, 187, 251-52, 265

"Big freeze," 187

## SUBJECT INDEX

"Big story," Bible presenting, 312, 333-35

Biogenesis, 70, 197, 349

Black holes, 156, 250-53

Black River, 87, 91

"Blind" faith, 356-57

"Block time," 228-31

"Blue shift" of light, 172

Boat, 97, 113-14

Bonds, Barry, 325-26

Bonhoeffer, Dietrich, 194-95, 197-98, 290

"Born again," xxxvi. See also Salvation, infra

Born, Max, 269

Branchport, New York, 126

"Breath" prayers, 345

*Brief History of Time, A* (Hawking), 2, 227, 246

Broglie, Louis de, 268

Brown, Walt, 174

"Buck stops here," 276

Buddhism
    Christianity and, 41
    reincarnation under, 39-41
    self, view of, 39-42, 60

Burns, Ken, 325

Bushnells Basin disaster, 150

Calvinism, 31-32, 99, 271, 292-301

Camp Gregory, 125-26

Camus, Albert, 63

Canal, 50-52, 149-53

*Candide* (Voltaire), 56

Candle boats, 127-28

Canoeing, 182, 259

Canseco, Jose, 325

Capra, Fritjof, 265

Carbon dating, 159-60

Casowasco, 126-28

Castaneda, Carlos, 122, 124

Cat experiment, 268-77

Cause and effect
    Aristotle's "uncaused cause," 20
    beginninglessness of God, 207-8
    determinism, infra
    "first cause," 280, 290
    "hidden variables," 289-90
    quantum physics and, 230, 289
    relativity theory and, 232-35
    salvation, scheme of, 232-35
    time, arrow of, 19-23

Causeway
    Association Island, 95
    Otisco Lake, xxix-xxx, 95, 176

"Cave" parable of Plato, 107-8

Cayuga Lake, 125-27

Cazenovia, 302

Centering prayer, 345

Chadwick, Henry, 284

Change
    Anselm and, 7-8
    arrow of time, 19-23
    architecture of cottages, 4
    Earth, 5
    entertainment habits, 3
    God changing, 24-34, 280, 319, 349, 352, 357-58
    Hartshorne and, 13-14
    holistic thinking about, 37
    Parmenides and, 5, 7
    personal change, 4-5
    Plato and, 7, 9-10
    relativity theory and, 5
    salvation and, 22
    universe and, 24
    Whitehead and, 12-13
    Zeno and, 18-19

Charismatic movement, xxxvii, 5-6, 110-11

"Charity," 123

Charles, Ray, 66

Chaos theory, 266

Chaumont, 91-92

Chautaqua
    adult education system, xliii-xliv
    Buildings, style of, 4
    experience similar to, 97

Chemical mysticism, 121-24

Chesapeake Bay, 259-62

*Chesapeake* (Michener), 259, 262

Children
    euthanizing of, 51, 52
    Russell ideas on raising of, 152-53
    salvation, 362

Chomsky, Noam, 354-55

Choptank Indians, 262

*Christian Apologetics* (Geisler), 211-16

Church attendance, 365, 367-68

Churchill, Winston, xlv

Clemens, Roger, 325

Closed loop, 275

Cloud, David W., 151

Colgate Rochester Divinity School, 131, 262

"Collective unconscious," 39

College
    attendance, xxx-xxxi
    girls and sex, xxxii
    Hobart, xxxi, 100-101, 103
    Keuka, xxxi-xxxii
    romance, xxxi-xxxiii

Collins, Francis, 190

Comatose persons, 355

Commercialization of postmodern culture, 306

Communism, 152-53

Completeness of Bible, 331

Computer "consciousness," 72-74

## SUBJECT INDEX

Condemnation, 261

Cones of light or information, 227-28, 236, 254-55

Confidence in Bible, 359

"Conformed" to will of God, 141-42

Consciousness, effect in quantum physics, 269-70, 272-75, 347-48

Conservation of mass-energy, law of, 240

Consideration, lack of, 337

Consumerism, 316-17

Contemplative prayer, 119-20, 278-79, 340, 345

Conversion. Salvation, infra

Copenhagen Interpretation, 266, 269-270, 281, 283-4, 286, 291, 352

Copernicus, Nicolaus, 181

Correspondences, xlii, 61, 98, 104-6, 116-17, 137, 145, 235-36, 252, 264, 277, 281-82, 284-85, 305, 357

Cosmic Background Explorer, 178-79

"Cosmic censorship" hypothesis, 251-52

"Cosmological constant," 171

Cottage
    Central New York
        word for, 91
    condition of, 23
    demoted, 363
    ownership, xxxiii-xxxx, 90, 117
    refinancing, 191
    sand in front of, 84
    site of writing, 369
    sold, 363
    view from, 57, 84, 88-90
    winter use of, 112, 192-94

"Courtship," use of term, xxxii

Cow Flop creek, xxviii-xxix

Creation, 128-30, 152, 164-65, 167, 175, 187-88, 203, 241, 248-49, 265, 288, 331-32, 338, 350, 364

Creative visualization, 346

Cryogenization, 323-24, 336

Crucifixion
    accounts of, 310-11
    Resurrection, 199-201, 218, 256, 360-61

Curvature of space-time, 165, 231, 236, 239-41, 245-6, 250, 256

Daily Bible reading, 341

"Dark ladder," 139

"Dark matter," 178-80, 251-2

"Dark night of the soul," 137-38

Darwin, Charles
    "bulldog" Thomas Huxley, 122
    theory of evolution. Evolution, infra

Davies, Paul, 186, 188, 204-5, 208, 229, 232, 246-47, 265, 272, 277, 353

Davisson, Clinton, 268-69

Dawkins, Richard, 67, 183-84

"Day-age" theory, 130

Death unreal, 323-24, 329

"Decoherence," 272-73, 276

"Deconstruction" strategy, 309

Deductive reasoning, 194-203

"Deeper knowledge" sought, xxxviii-xlv

Deification of self, 135

Deism, 184-85

Democritus, 38

Dennett, Daniel, 67

Depravity of man, 293-94, 299

Derrida, Jacques, 308-11, 330

Descartes, Rene, 58-59, 72, 81-82

"Desert fathers," xxiii, 136, 140, 344

Determinism, 52-56, 59, 79-80, 229-35, 277, 282, 286, 289

Dexter, New York, 87, 91

Dicken, E. W. Trueman, 141

Dimension, flexibility of concept, 226-27, 236-37

Disciplinary society, 316-17

*Discipline and Punish* (Foucault), 315-17

Disney, Walt, 323

"Divine reading," 119-20

Dog sentience, 72, 107-8

*Don Juan, Teachings of* (Castaneda), 122

*Doors of Perception, The* (Huxley), 122-23

Doppler effect, 172-73, 244

Doubting Thomas, 356-57

Downing, David, 118, 143

Dramatic productions, lives as, 322-23, 336

"Drowning" incident, xxvii-xxviii

Drugs supporting mysticism, 121-24

Earth
    Age of Earth, supra
    rotation of, 241

Easton, Maryland, 260

Eclecticism, 321

Ecstasy in mysticism, 139, 141

Eddington, Arthur, 244-45

Education of family, 367-68

"Effluvia," 336

Einstein, Albert, 18-19, 24, 53, 66, 105, 156-59, 161, 163, 165-67, 170-73, 176, 185-87, 204, 219-57, 264, 280, 291-92, 349

"Elect," 295-96

Electronic worship displays, 366

## SUBJECT INDEX

Elements, periodic table of, 176-77

Ellison Park, 182

"Elsewhere" under relativity theory, 228

"Emanation" approach to mysticism, 134-35

Emergent Church, xxii-xxiii, 116-17, 120, 136, 143, 194, 278, 307-8, 340-41, 344-45, 366

Emerson, Ralph Waldo, 77-80, 99-100, 104-5

Empiricism, 10, 37, 42-45, 75-76, 151-52, 357, 361

Energy, mass and, 240

Entertainment
change of habits, 3-4
Mother Nature providing, 4
"seeker-friendly" churches, xxii-xxiii, 344

Entropy, law of, 19-20, 60, 71, 74-75, 128-29, 187, 254

Epicurean philosophy, 84-85

"Epistemological turn" of philosophy, 81-82

"Equivalence" principle, 238, 244

Erickson, Millard, 267

Erie Canal, 50-52, 149-53

Eschatology, 265

Estuaries of Chesapeake, 261-62

Eternality of God (Christ), 129, 279, 281, 286-88, 338, 349-50, 352, 360-61

"Eternal present," 280-81, 295

Eternal security, 296-97, 301

Eternal universe, 248

"Eternity," 279-80, 349, 355-56

Eternity and time joined together, 283

Ether, 223-25

Euthanasia, 50-52

Evangelism, 343, 364

Event horizon, 250-51

"Event" under relativity theory, 227-28

"Everlastingness," 356

Everett, Hugh III, 265-66, 269-71

Evocative stimuli, xxxix-xl

Evolution
Bible, versus, 306
"Darwin's bulldog," 122
historical science, 152
natural selection producing, 183
"old earth" cosmologies, 129-30, 157-59
origin of life and, 66, 70-71, 157
social change brought by theory, 306
Teilhard's embrace of, 14-15

389

Existence of God of Bible, 211-16

Existentialism
    atheist wing, 65, 67
    authenticity under, 321
    postmodernism, roots of, 318-19
    self and, 62-65, 321

*Ex nihilo* creation, 167, 350

Expansion of space-time, 161-65, 171-76, 220, 246-47

Fables supporting faith, 313-15

Faithfulness of God, 301, 338-9

Faith of believer, xxxv, 22, 64-65, 75, 139-40, 145-46, 184, 188-90, 194, 203-4, 292, 294-95, 301, 313, 342, 356-57, 359-60, 363

False Ducks Islands, 113

"False vacuum" theory, 166-67

"Falsification" theory, 195-96, 200-201, 218

Feynman, Richard, 235, 266, 276, 287, 291, 352

Finger Lakes
    Cayuga Lake, 125-127
    geologic history, xxx-xxxi
    Keuka Lake, 126
    Otisco Lake, xxvii-xxx
    Owasco Lake, 126-27
    Skaneateles Lake, xxxi

Fire
    Heraclitus principle, 11-12
    "soul" of, 105

"First cause," 280, 290, 292

Fishing, 258-61, 303

Fish's Pond
    dried up, 305
    good times there, 304
    location, 303-4
    postmodernism tie-in, 305-6, 327-28

Five Points of Calvinism, 292-301

"Flat Earthers," Christians as, 2, 181

"Flatness" problem, 165-66

Fleck, Ludwig, 155

Flood, world-wide, xxx-xxxi, 160, 334

Flux
    God in state of, 280
    Heraclitus principle, 11

Foreknowledge of God, xxxix, 15-19, 24-34, 282-83, 295, 298-99, 353, 359

Forgiveness, 261

Foster, Richard, xxiii, 98, 136

"For ever," 279

Foucault, Michel, 308-9, 315-17, 335

Fractals, 262-64, 267

Free will of man, xxxix, 16, 24-25, 52-56, 60, 79, 143, 232-35, 263, 288-93, 352-53, 359

French influences on postmodernism, 309

Friedmann, Alexander, 161, 165, 170-71, 175-76, 246

## SUBJECT INDEX

Fundamentalism revival, 306

Future under relativity theory, 227-29

Galaxies
    expansion of space-time, 161-65, 171-76, 220, 246-247
    spin of, 251-52

Galileo thought experiments, 205-6, 238

Galloo Island, 90, 94, 97

Gallou Shoal, 113

Gamow, George, 160, 176-77

"Gap" theory, 130

"Gaps" God, 195, 197

Gardner, Martin, 170, 229-30, 232

Geisler, Norman, 211-216, 218, 299

Gender-neutral references, xxiv

General relativity theory, 156-57, 161, 165, 170-72, 204, 226, 238-41, 244-48, 250, 254, 256, 265, 351

Geocentrism, 181, 255

Geodesics in space-time structure, 241-45

Germer, Leslie, 268-69

Gilley, Gary, 317, 345

"God abhors a naked singularity," 251

*God and the New Physics* (Davies), 272, 353

*God Delusion, The* (Dawkins), 183

"Godhead," 128-30

"God of the gaps," 195, 197, 290

"Golden Rule," 335

"Goldilocks zone," 74-75, 351

Goldman, Steven, 327

Gospels, 75, 198-200, 361

Grand recits, 312, 335

Grand Unified Theory, 167-69

Gravity, 156-57, 165-68, 170-72, 180, 231, 236-38, 242-45, 250-52, 265, 351

Great Lakes, 11

Greene, Brian, 161, 171-72, 186-87, 265

Grieg, Edvard, 342

Gull Island, 36, 87-88, 92, 96, 145

Guth, Alan, 160-66, 169


Harris, Sam, 67

Harris Shoal, 113

Hartshorne, Charles, 13-14

Harvey, Paul, 344

Hawking, Stephen, 82, 162, 184, 227-28, 234, 245-46, 250-51, 254-55, 265

Healings, 197-201, 218

Heart of man, 297

Herd instinct, 153-55

Heaven
    accountability after death, 47
    admission to, 80, 232-35, 368
    Calvinist election to, 292-301
    description of, xxxvi-xxxvii, xxxix
    Emerson and, 79-80
    focus these days, 363
    "many worlds" theory frustrating attainment of, 271
    optional, 258
    Plato's, 8
    "sweet by and by," 366

Hegel, Georg W. F., 49

Heidegger, Martin, 64, 213

Heisenberg, Werner, 167, 230, 249, 265, 289

Hell
    accountability after death, 47
    alternative reality, 263
    Calvinism and, 292-301
    consignment to, 80, 232-35, 261, 292-301
    description of, xxxvii
    "many worlds" theory sending persons to, 271
    possibility of, 258

Henderson Bay, 37, 87, 93-96

Henderson Harbor, 93-94

Heraclitus, 11

Hermeneutics, 143, 299-300, 321-22

Heroes vanishing, 319, 326-27, 329, 336

Hesse, Mary, 105-6, 285

Hesychasm, 278

"Hidden variables," 289-90

Hinduism, 39-41

Historical science, 151-52

Hitchens, Christopher, 67

Hitler, Adolph, 185

Hoaxes against post-modernists, 327-28

Hobart College, xxxi, 100-101, 103

Holism, 37, 67, 71, 147-48

"Holy Place," 103

Holy Spirit, 21-22, 106, 128-30, 146, 180, 189, 203, 216-17, 232, 279, 293-96, 343, 357-58, 360, 365, 368

"Horizon" problem, 161-62, 170, 255

Horizon seen from Cottage, 90-91, 145

"Host," 243, 338

Hovey Island, 95

Hoyle, Fred, 177

Hubble, Edwin, 171-73, 244, 246

Human Genome Project, 190

Hume, David, 21, 44-45, 195

## SUBJECT INDEX

Humphreys, Russell, 156-57, 163, 173-74, 240-41, 255, 351

Hunt, Dave, 299-300

Hunting trip, 101

Huxley, Aldous, 122-24

Huxley, Thomas, 122

Huygens, Christian, 242

Hybels, Bill, 344

"Hybrid reality," 266-78, 282, 286

Hyperreality, 322-23, 328

"I AM," 207, 253

Idealism, 76

"Identity," 303, 331, 341

Idolatry, 278, 317

"Idols of the marketplace," 155

"Idols of the mind," 153

"Illumination" stage in mysticism, 137, 141

"Image," 322-23, 336

"Imitation" game, 72-74

Immanence of God, 134-35, 184-85, 211, 339

Immutability of God, 24-34, 319, 349, 352, 357-9

Imputed righteousness of God, 188-89

Incarnation, 283

Inconsistent Scripture, 359-60

"Incredulity toward metanarratives," 308-9, 311-15, 317-20, 336, 344

Individuality, 355

Inductive reasoning, 192-203, 218, 314

Ineffability of mystical experience, 116

Inertia, 238, 244

Infinity of God, 247-48, 263, 338

"Inflation" hypothesis, 161-66, 169-70, 178-80, 254-55

Information
    cones of light or information, 227-28, 234
    existence supported by, 274-76

"Innate" ideas, 10

Intellect, reliance on, 117-18, 136, 142-44, 364

Intelligence
    artificial, 72-75
    design demonstrating, 65-66, 182-3, 186, 209, 364

Intensity of prayer, 284-5

Interference of waves, 268

*Interior Castle* (Teresa of Avila), 138-39

Interpretation
    community context, within, 310
    "deconstruction," 309

everything as, 309-11
hermeneutics, 143, 299-300, 321-22
linguistic idealism, 309
universality of supporting knowledge, 311, 313

"In transit" creation of light, 351

Intuition, 99, 117-18, 145, 148, 203, 264, 278

Irondequoit Creek, 182

"Irresistible grace," 295, 300

Islam, 37

Islands
chain spanning Ontario, 113
The, 36-37, 86-97, 114, 145

Isotropy, 162, 170, 176, 178-79

"It from bit" theory, 274-76

James, William, 116-18, 121-22, 131-32, 133-34

Jaspers, Karl, 63

Jeans, James, 277

Jehovah's Witnesses, 47

Jet skis, 36-37, 337, 363

Joel, Karl, 132

John of the Cross, 139-41

Johnson, Luke Timothy, 346

"Joy in the morning," 260

Judaism, 37

Judgment, 261, 354

Jung, Carl, 39, 132-33

Kaku, Michio, 161, 165-66, 168, 231, 245, 248, 250, 265, 268, 270-71, 275, 291

Keats, John, 274

Keuka College, xxxi-xxxii

Keuka Lake, 126

Kennedy, John F., 101

Kierkegaard, Soren, 62, 64-65, 194, 203, 216-17, 356

King James Bible, 365

Kingston, Ontario, 92, 113

Kuhn, Thomas, 153-55

"Label idolatry," 317

Ladder of faith, 139-40

La Jolla, California, 83

Lambda, 165

"Lamb of God," 283

Language, inadequacy of, 110-12

*Language of God, The* (Collins), 190

Laplace, Pierre, 52-53, 289

"Latent capacity for God," 137

Laws of physics, 202-3, 241, 243-44, 247-49, 250-51, 280

# SUBJECT INDEX

Lawyer, author as, xxxiii, 52, 217, 344

Lazarus, 56, 200-201

"Learned" persons, xxi-xxii

Leary, Timothy, 122

*Lectio divina*, 119-20, 345

Leeches, xxviii-xxix

Leibniz, Gottfried Wilhelm von, 55-56, 242

Lerner, Eric, 177-78, 251

*Letters and Papers from Prison* (Bonhoeffer), 194-95

Lewis, C. S., xl-xli, xliv, 16, 61-62, 86, 98, 106-8, 114-16, 118, 128, 143-45, 147, 182-83, 264, 280-81, 296, 305, 342, 346, 352, 361

Life
    abiogenesis, 91, 197
    biogenesis, 70, 197
    Christ as, 366
    "God of the gaps" and, 197
    materialist view, 65-71
    ontology supporting, 349
    origin of, 65-71, 196-97
    personal in nature, 349
    sentience and, 72, 348
    shared by Father and Son, 283
    shared by subject and God, 347-49
    shortness of, 101

Light
    black holes, 156, 250-53
    cones of light or information, 227-28, 234
    creation week, 164-65
    curved by matter, 244-45
    distant starlight, 156-57, 219-20, 351
    Doppler effect, 244
    New Jerusalem, 220
    Olbers' paradox, 219-20
    particle property, 250
    "red shift," 172-75, 244
    speed of, 157-59, 161-62, 219-20, 226, 236, 254-55, 351
    spheres of light or information, 227-28, 234
    waves, traveling in, 244, 250

Limbaugh, David, 361

Lime Barrel Shoal light buoy, 88, 90, 92, 97

Limited atonement, 295, 300

"Linguistic turn" of philosophy, 82

Lisle, Jason, 183-84

Little Gallou Island, 113

"Live and let live," 338

Location of God, 363-64

Locke, John, 37, 42-44, 46, 354-55

Locutions, 141

Logic, God of, 187-90, 203

Long Point State Park, 92

Love
    affair replacing "love of my life," 324, 329
    freedom in loving

God, 54, 296, 352
God as, 17, 56, 351, 365
God's demonstrated, 365-66
John of the Cross,
love stages per, 140
man's love for God,
16, 54, 359-60
romantic, 324, 329

Luminiferous ether, 223-25

Lyotard, Jean-Francois, 308-9, 311-15, 319-21

McGuire, Mark, 325

Mach, Ernst, 241-43, 255

Maid of the Mist, 57

Main Duck Island, 113

Major league baseball unreality, 324-26, 329

*Malcolm, Letters to* (Lewis), 143-44

Mantle, Mickey, 326

Mantras, 345

"Many worlds" theories, 265-66, 269-72, 276, 284

Map, use of, 91, 112-15, 135, 144-45

Maris, Roger, 325

"Market-driven" evangelism, xxii-xxiii

Marks of mystical experience, 116-17

Marriage, spiritual, 139-40

Martin, Walter, 47

Maryland, summer trip to, 259-62

Maslow, Abraham, 39

Mass
energy,
interchangeable with, 240
flexibility of concept, 226
space-time curvature
affected by, 240

Materialism
disagreement with dualism and idealism, 96
life's origin, 65-71

Mathematics
bullying by means of, 176
infinity and, 246-47
logic underlying
universe, 188-90
unworkable concepts,
producing, 176,

"Mature creation," 241, 351

Maxwell, James Clerk, 222-23

May, Rollo, 39

Medical truth claims, 317

Memory transcending death,
47-49, 353-54

*Mere Christianity* (Lewis),
280-81, 352

Mescaline, 122

Metanarratives
postmodernist incredulity as to,
308-9, 311-15,
317-19, 344
progress, 326-27, 329
search for, 118, 344

# SUBJECT INDEX

Metaphor, understanding things as, xlii, 61, 98, 104-6, 116-17, 130, 145, 235-36, 252, 264, 278, 281-82, 284-8, 347-8

"Metaphysics," 10

Methodist church, xxi, xxxvii
John Wesley, founder, 342
Navarino Methodist, infra

Meysenburg, Malwida von, 133

Michelson, Albert, 223-25

Michener, James A., 259, 262

Microwave background
Biblical explanation, 164-5
"ripples" in, 179-80
theories re, 164-5, 170
uniformity, 162, 170, 176, 178-9

Miller-Urey experiment, 66-67

"Mindless" mysticism, 98

Mind/matter duality, 72, 76

Minkowski, Hermann, 228-30, 236, 254-55

Miracles, 197-201, 218, 310-11, 314, 334-35, 357

Misner, Charles, 185

Missler, Chuck, 169

Modernism
arts reflecting, 305-6
Bible challenged by, 306
failure of, 307
legitimation of claims, 313
postmodernism predecessor, 305
"reason" (science) ascendant over revelation, 305

"Monads," 55-56

Monastic prayer practice, 278

Monism, 76

Monopole problem, 169

Moon, observations re, 1

Moral law argument for God, 24-25

Morley, Edward, 223-25

Morris, Henry, 241

"Most Holy Place," 103

"Mother Nature," xxiv, 4

Motion of observer, 349

Multiple prayers, 280-81

Multiple prophecies, 287-88

Multiple universes, 65-66, 197; see also "many worlds theories" supra

Mysticism, xxii-xxiii
actions as prayer, 284-85
adopting God's thinking, 106-7
ancient/modern, 345-48
chemical, 121-24
Emergent Church, 278, 346, 366
evangelical faith increasingly, 218
"extrovertive," 103
"foothills" of, 264
facts and, 107-8
Holy Spirit, and, 109, 147
Kierkegaard, 194
marks of, 116
"mindless," 98
nature mysticism, xl-xlv, 86-134, 217-18, 305
postmodern resort to, 307-8, 319

prayer of mystics, 278
radiating nature of reality, 283
Scripture, and, 108-10, 115, 117-21, 204
soul searching re, 142-48, 257
systematic approach, 134-42

*Mysticism Sacred and Profane* (Zaehner), 131

Narrative knowledge, 314

Natural God, 187, 254, 277

Natural selection, 66, 183

*Natural Theology* (Paley), 209-10

Nature
interpretation of, 160
jet skis and, 36-37
mystic experience of, xl-xlv, 86-134, 217-18, 305, 331-32, 364
object in itself, 117
recreational vehicles and, 36-37
term "(n)ature," use of, xxiv, xxv
Word (Bible) explained by, 331-32

Nature mysticism, xl-xlv, 86-134, 217-18, 331-32

Navarino
church. Navarino Methodist, infra
family living patterns, 102-3
Fish's Pond, supra
mysticism and, 104-5
writing of book, and, 104

Navarino Methodist
attendance at, 367-68
Chautauqua attendees from, xliii
Sunday School teacher, 368
temperance efforts, 304-5
Youth Fellowship summer camp, 125-26

Navigation, 113-15

Nazism, 64, 318

Necessity of God's existence, 211-16

"New Age" Christianity, xxiii, 116-17, 120-21, 204, 340-41, 346-47, 366

New Atheists, 67-69, 240

New England transcendentalism, xxv

"New heaven and a new earth," 220

New Jerusalem, 200

"New man," putting on, 365

Newton, Isaac, 167-68, 237-38, 264, 289, 291

Niagara Falls, 57

Nietzsche, Friedrich, 317-18

Night sky, 219-20

Nihilism, 317-18, 329-330, 344

Nitrous oxide and mysticism, 121

Noetic quality and mystical experience, 116

"Normal" prayer, 347

"Nothing is real," 329

## SUBJECT INDEX

"Nothing outside the text," 308-11, 318, 330-32, 336

Nucleosynthesis, 176-77

"Objective uncertainty," 357

Olbers, Heinrich, 219-20

"Old man," putting off, 365

Old Testament, 361

Omega, 165-66

Omnipotent God, 230, 280, 287-88, 338, 352

Omnipresent God, 363-64

Omniscient God, 15-19, 24-34, 230, 263, 339, 352, 359

On-line shopping, 322

"Ontological" argument for God, 7-8, 210-11

"Openness" theology, xliii, 15-19, 23-28, 33, 62, 267

Operational science, 151-52, 241

Oppenheimer, J. Robert, 250

"Orientation," 302, 330, 340

Origen, 140-42, 284

Origins science, 151-53, 241

Otisco Lake, xxvii-xxx, 90, 95, 100-101, 176, 259, 302

Oulton, John, 284

*Out of Formation* (Gilley), 345

"Oversoul," xxv, 77-79, 81, 99

Owasco Lake, 126-27

Page, Don, 185

Paley, William, 209-10

Pan(en)theism, xxiv, xlii-xliii, 12, 57-62, 76, 99, 103, 117, 134-35, 220, 254, 346

Paradigm paralysis, 149-90, 264

Paradoxes, 24, 332

*Parallel Worlds* (Kaku), 268

Paraphrases, 365

Parmenides, 5, 7, 18

Particle property of quanta, 268

Pascal, Blaise, 45

Passion of believer, 357

Passive euthanasia, 50-52

Passive nonelection, 299-300

Passivity of mystic, 116

Pastiche, postmodernist, 321, 328

Past under relativity theory, 227-29

Pasteur, Louis, 197

Penrose, Roger, 250-52

Pentecostalism, xxxvii, 5-6, 110-11, 123-24

"People of the foothills," 143-44

Perception
  drugs altering, 122
  reality conferred by, 75
  unique in each person, 355

Permanence, 83-84

Perseverance of the saints, 296, 301

Personal Bible, 365-66

Personal evangelism outreach, 343

Perspective
  deaths imposing, 101-2
  Emerson view, 99-100
  science and, 1
  Thoreau view, 100
  Westcott Beach upper
    campsites, 35-37

Peyote and mysticism, 121

Philosophers
  Anselm, 8, 111, 210, 216, 218
  Aquinas, Thomas,
    208-11, 216, 218
  Aristotle, 9-10
  Democritus, 38
  Descartes, Rene, 58-59
  Hartshorne, Charles, 13-14
  Hegel, Georg W. F., 49
  Heraclitus, 11
  Hume, David, 21, 44-45
  James, William, 116-17
  Joel, Karl, 132
  Kierkegaard, Soren, 62,
    64-65, 194
  Locke, John, 37, 42-44, 354-55
  Olbers, Heinrich, 219-20
  Parmenides, 5, 7, 18-19
  Pascal, Blaise, 45
  Plato, 6-7, 8-10, 107-8, 196
  Popper, Karl, 195-96
  Protagoras, 8
  Pythagoras, 18

Russell, Bertrand, 38, 46-47,
  152-53, 195
Spinoza, Baruch, xxiv, 58-60
Teilhard de Chardin, 14-15
Whitehead, Alfred North,
  12-13, 58
Wittgenstein, Ludwig, 82
Zeno, 18-19

Philosophy, 257, 304-5, 366

Physical proof, 357, 360

Physicists
  Born, Max, 269
  Capra, Fritjof, 265
  Davies, Paul, 186, 188, 204-5,
    208, 229, 232, 246-47, 265,
    272, 277, 353
  Einstein, Albert, 18-19, 24,
    53, 66, 105, 156-59, 163,
    165-67, 170-173, 176,
    185-87, 280, 291-92, 350
  Feynman, Richard, 235, 266,
    276, 287, 291, 352
  Greene, Brian, 161, 171-72,
    186-87, 265
  Guth, Alan, 160-66, 169
  Hawking, Stephen, 82, 162,
    227-28, 234, 245-46, 250-
    51, 254-55, 265
  Heisenberg, Werner, 167, 230,
    249, 265, 289
  Humphreys, Russell, 156-57,
    163, 173-74, 240-41, 255
  Kaku, Michio, 161,
    165-66, 168, 231, 245, 248,
    250, 265, 268, 270-71,
    275, 291
  Lerner, Eric, 177-78, 251
  Mach, Ernst, 241-43
  Michelson, Albert, 223-25
  Morley, Edward, 223-25
  Schrodinger, Erwin, 268-73
  Sokal, Alan, 327-28

## SUBJECT INDEX

Wheeler, John, 240
Wigner, Eugene, 273-74, 276
Wolfson, Richard, 265, 270
Young, Thomas, 268
Zeh, Dieter, 272

Pillar Point, 86-87, 89, 91-92

Pirsig, Robert, xxxix-xl, xliii, 86

Pittsford, village of, 149-50

Plant soul, 9

Plasma physics, 177

Plato, 6-7, 8-10, 107-8, 196, 349

Playacting, 322, 328

Plenary statement, 358

Poe, Edgar Allan, 298

Point Peninsula, 91-92, 94

Point Salubrious, 91

"Poisoned tree of power relations," 315, 335

Polarity of water, xxviii

Political correctness, xxiv

Popper, Karl, 195-96

Popular science, 155, 157, 160, 165-67, 169, 176, 248, 254

*Portable Postmodernist, The* (Berger), 321

Postmodernism, xxii, 111-12, 118, 127-28, 140-41, 218, 302-41, 343-44, 355

Post-resurrection appearances of Christ, 360

Power augmentation rather than truth, 320

"Power is knowledge," 308-9, 315-19, 335, 344

Prayer, 18, 29, 119-20, 125, 143, 185, 278-85, 297, 337-41, 345-47, 353, 359, 364

Predestination, 29-33, 232-35, 262-63, 292-302

Preparedness, 343

Presley, Elvis, 265-66

Probability wave function collapse, 272-76, 281, 286, 353

"Process" thinking, 12-13, 17

Property rights, xxix-xxx

Prophecies, 201-2, 206, 218, 243, 256, 285-88, 352-53, 359

Protagoras, 8

Proving God, 191-218

Providence of God, xxxv

Psyche Shoal, 113

Psychedelic substance, 122

*Psychology of the Unconscious* (Jung), 132-33

Psychotropic substance, 122

"Purgation" stage in mysticism, 137, 141

Purpose driven movement, 120

Pythagoras, 18

Quakerism, xxiii, 278

Quantum physics
- cause and effect in, 230, 249
- collapse of probability wave function, 272-76, 281-82, 286, 291
- consciousness, effect of, 269-70, 273-76
- fluctuations, 167-68, 249
- Geiger counter observation, 269-70
- God as observer, 288
- human observation, 269-70, 273-76
- "hybrid reality," 265-77, 281
- indeterminism, 276, 288-92
- information-based existence, 274-76
- measurement, 269-71
- merger with gravity theory, 167-68, 170-72
- mind determinative in, 269-70, 273-76
- observation, role of, 269-71, 273-77, 287-88
- origin of universe, 245-48, 255-56
- paired electrons communicating, 105
- particles, quanta as, 268
- probability rather than certainty, 289
- quanta, 266
- "red shift" quantization, 174
- regressing layers of observation, 276
- retroactive creation of reality, 275, 286-88
- scale, 249
- "superposition" of quantum states, 268-73
- uncertainty principle, 167-68, 230, 265
- waves, quanta as, 268

Radiating reality, 258-301, 319, 357-58

Radiometric dating, 159-60

Rapture in mysticism, 139, 141

Rationalism, 76

Reading of author, xliv

"Reality" television programming, 322

"Reason"
- Age of Reason, 218
- assisting faith, 204, 364
- Christianity supplanted by, 218
- deductive reasoning, 194-203
- God of, 187-90
- inductive reasoning, 192-203, 218
- Kierkegaard position re, 194, 216-17
- modernism cleaving to, 305
- self uncovered by, 98
- subservience to revelation, 84-85

Recreational vehicles, 36-37, 95

"Red shift" of light, 172-75, 244

Reductionism, 37-38

*Region of Awe, Into The* (Downing), 118, 143

Reincarnation, 39-41, 252

Relativistic invariants, 236, 254

## SUBJECT INDEX

Relativity theory, 156-59, 161, 170-72, 219-57, 264-65, 350-51

"Religious experience," 127

Repentance, 115, 234-35, 364

Resolve to not believe in God, 216

Resurrection, 199-201, 218, 256, 360-61

Retroactive creation of reality, 275, 286, 348, 352-53

Revelation
  God shown by, 364
  modernism's subservience to "reason," 306
  preeminent over "reason," 84-85

"Rightly dividing the word of truth," 343

Rock
  Christ as, 84
  permanence of, 83-84

Rodriguez, Alex, 325

Roemer, Olaf, 158

Roman Catholic mysticism, 135-36, 307

Roman Empire, 198-99

Romanesco, 267

Romans 1: 20 scripture, 128-30, 332

Romantic love, 324, 329, 336-37

Rubin, Vera, 252

Russell, Bertrand, 38, 46-47, 152-53, 181, 195

Sackets Harbor, 86-88, 91

"Sacred reading," 119-120

Salvation
  alternative reality, 263
  author's, xxxv-xxxviii, 5
  Bible and, 358, 364
  Calvinistic election to, 292-301
  cause, effect, and, 20, 232-35
  children of tender years, 361
  faith and, xxxv, 22, 64-65, 75, 139-40, 145-46, 184, 188-90, 194, 203-4, 291, 294-95, 301, 313, 342, 359-60
  Heraclitus and, 12
  Holy Spirit enabling, 146, 360
  justice principle, xxxviii, 145-46
  Lewis, C. S., 342
  "many worlds" theory frustrating intent as to, 271
  "once for all" redemption, 12
  option, as, 258, 292
  prophecy and, 286-88
  questionable, 38
  reading habits and, 341-42
  spiritual matter, 254-55
  time element, 22
  works and, 21, 145-46, 297

Sandy Point ferry, 259-60

Sartre, Jean-Paul, 62

Schooling of family, 367-68

School taxes, 88

Schrodinger, Erwin, 268-73

Science
  direct support of Bible, 2, 264
  equations of, 351
  hoaxes against postmodernists, 327-28

"ignorance" regarding, 1
"K. Wilhelm Abian"
    hoax, 327, 330
legitimation of, 313-14
metaphor, things as, xlii, 61, 98,
    104-6, 116-17, 130, 145,
    235-36, 252, 264, 281, 285
modernism challenge to
    Bible, 306
normal and abnormal
    science, 153-60
operational science, 151-52, 241
origins science, 151-53, 241
perspective, matter of, 1
postmodernism and, 311-12, 320
religion, and, generally, 180-190
right interpretation of nature, 160
Sokal hoax, 327-29
truth of observations, 313-14, 317

*Science Wars* (Goldman), 327

*Scofield Study Bible, The*, 279

"Seeker-friendly" churches, xxii-xxiii

Selah, 50-52

Self, 35-39
    boundary, 147, 182
    Buddhist view, 39-42
    Christian focus on, ,38, 124
    culpability, 52-56
    "deification," 135
    Eastern thinking, 39-42
    euthanasia and, 50-52
    existentialist view, 62-65
    Hindu view, 39-41
    holistic view, 37
    Hume's view, 44-45
    Locke's view, 42-44, 46, 354
    mystic cocoon of, 148
    "Oversoul," 77-79, 99
    pan(en)theism, xxiv, xlii-xliii,
        12, 57-62

performance-enhancing drugs
    and, 326
postmodern negation of, 321-22
"reason" uncovering, 98
reductionist view of, 37-38
Russell's view, 38
solipsism, 75-76, 275-77
"spiritual formation" and, 99
time travel negating, 231
transcendence, 354
transpersonal psychology, 39

Senses
    reliance on, 117-18

September tranquility, 3

Servanthood, 335

Setterfield, Barry, 158, 174-75, 351

"(S)harper than any
    two-edged sword," 284, 346

Shelley, Bruce, 218

Shopping mall, performing in, 322

"Signs and wonders," xxxvi

Silence, discipline of, 278-79,
    340, 345

Simians and humans, 73

Simulation, God as, 323

Simultaneous prayers, 280-81

Singularities, 245-50, 255
    "Big Bang," supra
    "black holes," supra

Sinner's Prayer, 364-65

Six Town Point Island, 36, 87-88,
    92, 96, 145

## SUBJECT INDEX

Skaneateles, village of, xxxi, 302

Smith Hollow pond. Fish's Pond, supra

Smith, Huston, 123-24, 141

Smith, James K. A., 306-17

Snowshoe Bay, 94

Socialism, 152-53

Sokal, Alan, 327-28

Solar eclipse, use of, 245

"Sola scriptura," 308

Solid food of the Word, 207

Solipsism, 75-76, 275-77

Sosa, Sammy, 325

Space
 beginning of, 248
 flexibility of concept, 226-27, 236-37
 inseparable from time, 227, 236-37, 247, 265, 280, 350

Space-time interval, 236, 254

Special theory of relativity, 158-59, 204, 226, 236-37, 240, 264-65, 349

Spheres of light or information, 227-28, 236, 254-55

Spinoza, Baruch, xxiv, 58-60, 117, 124

Spirit guides, 345

Spiritual directors, 340, 345

"Spiritual disciplines," xxiii, 98, 345

"Spiritual formation," xxiii, 99

Spiritual marriage, 139-40

"Spiritual reading," 119-20

Sports performance, unreality of, 324-26, 329, 336

Standard Model of Cosmology, 161-62, 168, 170

*Starlight and Time* (Humphreys), 156-57

Stars
 black holes, becoming, 250
 necessity of, 241-43, 255

Static universe, 171, 219-20

Stengel, Casey, 266

Steroid use by baseball players, 324-26

"Still small voice" of God, 279, 340

Stoic philosophy, 84-85

Stokes, Philip, 210

Stony Island, 90, 94, 113

Stony Point, 36, 90, 93-97, 113

Storrs Point, 91, 94

String theory, 171-72

Strogatz, Steven, 267

*Structure of Scientific Revolutions, The* (Kuhn), 153-55

Sufficiency of Scripture, 108-10

*Sui generis*, God as, 208

Summer camp, 125-128

Summer trip to Maryland, 259-62

"Sum over histories/paths," 266, 287

Sun, observations re, 1

Sunday School teacher, 368

Susquehanna River, 259

"Superposition" of quantum states, 268-78

Synod of Dort, 293

Tabula rasa, 37, 354-55

Taxes, 88

Teaching Company, 264-65

Teilhard de Chardin
    Christology, 15
    evolution, and, 14-15

Telos, 209, 316

Tense change, xliv-xlv, 83

Teresa of Avila, 138-42

*That Hideous Strength* (Lewis), 147

Theism, 211

"Theory of everything," 186-87, 311-12

Thermodynamics first law, 168, 240

Thermodynamics second law, 19-20, 60, 71, 74-75, 128-29, 187, 254, 358

Thoreau, Henry David, 80-81, 86, 100, 180-82

Thoreau, John, 181-82

"Thought collectives," 155

Thought experiments, 24, 203-16, 268-69, 272-73

Tifft, William, 174

Tilghman Island, 260

Time
    arrows of, 19-23
    atoning sacrifice of Christ, 283
    beginning of, 248, 250, 350
    "block time," 228-31
    cause, effect, and, 19-23
    end of, 265
    "eternity," versus, 279-80, 355-56
    flexibility of concept, 226-27, 236-37, 349
    future under relativity theory, 227-29
    God and, 16-19, 264, 281, 353
    gravity affecting, 240-41, 349
    inseparable from space, 227, 236, 247, 265, 280, 350
    joinder of with eternity, 283
    Lewis view of, 251
    motion of observer affecting, 349
    past under relativity theory, 227-29
    present under relativity theory, 229
    prophecy, time sequence affecting, 287-88
    relativity theory generally, under, 226

## SUBJECT INDEX

retroactive creation of reality, 286
salvation and, 21-22, 288
travel in, 231
"world line" going back in, 266

*Time of Departing, A* (Yungen), 345

Total depravity of man, 293-4, 299

Trances, 139-41

Transcendentalism, 77-82, 99, 116

Transiency of mystical experience, 116

Transpersonal psychology, 39

Transpositions, 106-8

Trinity, 21-22, 128-29, 145-47, 253, 255

Trivialization of postmodern culture, 306

Troitskii, V. S., 158

Truthfulness of God, 338-9

"TULIP," 293

Turing, Alan, 72-74

"Turn on, tune in, drop out," 122

"Turtles all the way down," 2, 310

"Twins" paradox, 24, 205, 226

Two-slit experiment, 268-69

Uncertainty principle, 167-68, 230, 249, 265, 289

Unconditional election, 294, 297, 300

Underhill, Evelyn, 117, 134-35, 144

Understanding as an object, xxxviii-xliv

"Union" stage in mysticism, 138, 141-42, 147

Unitarianism, 99

Universal grammar, 354-55

Universe
  age, 165-66, 171, 219-20, 246
  change, 24
  curvature of space-time, 165, 231, 236, 239-41, 245-46
  Earth age, see Age of Earth, supra
  ether, 223-25
  expansion, 161-65, 246-47
  initial singularity, 245-48
  mathematics underlying, 188-90
  microwave background radiation, supra
  multiple universes, 65-66, 197
  omega value as to age, 165-66
  origin of, 245-48
  pan(en)theism, xxiv, xlii-xliii, 12, 57-62
  perception of, conferring reality, 75
  static universe, 171, 219-20

"Unmediated contact" with God, 346

Unrepented sin, 234-35

"Vain repetitions," 340, 347

*Varieties of Religious Experience, The* (James), 131-32, 133-34

Veil in tabernacle, temple, 103

Vietnam service, xxxviii

Visions, 141

Visualization, 346

Voltaire, 56

War of 1812, 87

Warren, Rick, 344

"Watchmaker" argument for God, 209-10

Wave function. See Waves, infra

Waves
   collapse of
      probability wave function, 272-76, 281-82, 286, 353
      decoherence, 272-73, 276
   interference of, 268
   light as, 244
   quanta as, 268

*Week on the Concord and Merrimack Rivers, A* (Thoreau), 80, 100, 180-82

Weight
   flexibility of concept, 226
   Mass, supra

Wesley, John, 341

*What Does God Know and When Does He Know It?* (Erickson), 267

Wheeler, John, 240, 274-76

Whewell, William, 160

Whitehead, Alfred North, 12-13, 58

"White hole," 156-57, 351

*Who's Afraid of Postmodernism?* (Smith), 306-17

Wigner, Eugene, 273-74, 276

"Wigner's friend," 274, 276

Williams, Ted, 323

*Will to Power, The* (Nietzsche), 317

Wind farm proposed, 94

Wittgenstein, Ludwig, 82

Wolfson, Richard, 265, 270

Women
   early Church, xxiv
   college, xxxii
   courtship, xxxii

"Word," both Bible and Christ as, 121

Works salvation, xxxv, 21, 145-46, 297

"World line," 229-30, 235, 263, 266

Yorkshire Island, 113

Young Earth. See Age of Earth, supra

Young, Thomas, 268

Yungen, Ray, 345

Zaehner, R. C., 122-24, 131

## SUBJECT INDEX

*Zen and the Art of Motorcycle Maintenance* (Pirsig), xxxix-xl, xliii

Zeh, Dieter, 272

Zwicky, Fritz, 251-52

# ILLUSTRATION CREDITS

**TEXT ILLUSTRATIONS**

Figs. 2, 3, 5, 6, 7, 8, 10—Dan Donoghue

**PHOTOGRAPHS**

Front cover—Geoffrey Case
Fig. 1—Arlo Case
Fig. 9—Susan Case
Fig. 11—Lester Fish
Back cover (author photo) — Jeremy Case